T0231127

SEISMIC DESIGN GUIDELINES FOR PORT STRUCTURES

INTERNATIONAL NAVIGATION ASSOCIATION

ASSOCIATION INTERNATIONALE DE NAVIGATION

SEISMIC DESIGN GUIDELINES FOR PORT STRUCTURES

Working Group No. 34 of the Maritime Navigation Commission
International Navigation Association

A.A. BALKEMA PUBLISHERS LEIDEN / LONDON / NEW YORK / PHILADELPHIA / SINGAPORE

Library of Congress Cataloging-in-Publication Data

Applied for

First print 2001
Second print 2006

Cover design: Studio Jan de Boer, Amsterdam, The Netherlands.
Typesetting: Macmillan India Ltd., Bangalore, India.
Printed by: Antony Rowe Ltd, Chippenham, Wiltshire, UK.

© 2001 Taylor & Francis Group plc, London, UK

Technical Commentary 5 is exempt from the copyright because the original material was produced by the US Navy for the state of California, USA. No one can exercise the copyright.

The figures and table in Technical Commentary 6 have been reprinted from:
Handbook on liquefaction remediation of reclaimed land, Port and Harbour Research Institute (ed.), ISBN 90 5410 653 0, 1997, 25 cm, 324 pp.

ISBN 90 265 1818 8 (hardback)

Contents

Preface

Although the damaging effects of earthquakes have been known for centuries, it is only since the mid-twentieth century that seismic provisions for port structures have been adopted in design practice. In 1997, the International Navigation Association (PIANC; formerly the Permanent International Association for Navigation Congresses) formed a working group, PIANC/MarCom/WG34, to focus international attention on the devastating effects of earthquakes on port facilities. This book, entitled 'Seismic Design Guidelines for Port Structures,' is the culmination of the efforts of this working group.

This book is the first of its kind in presenting international guidelines for seismic design. The provisions reflect the diverse nature of port facilities. Although constructed in the marine environment, the port facilities are associated with extensive waterfront development, and provide multiple land-sea transport connections. The port must accommodate small to very large vessels, as well as special facilities for handling potentially hazardous materials and critical emergency facilities that must be operational immediately after a devastating earthquake.

The primary goal of the working group was the development of a consistent set of seismic design guidelines that would have broad international support. The diverse characteristics of port structures led the working group to adopt an evolutionary design strategy based on seismic response and performance requirements. Proven simplified methods and state-of-the-art analysis procedures have been carefully selected and integrated in the guidelines in order to provide a flexible and consistent methodology for the seismic design of port facilities.

This book consists of a main text and eight technical commentaries. The main text introduces the reader to basic earthquake engineering concepts and a strategy for performance-based seismic design. The technical commentaries illustrate specific aspects of seismic analysis and design, and provide examples of various applications of the guidelines.

The working group members, technical editors, and graphics experts are listed below. The working group would like to express their sincere gratitude to a group of New Zealand wharf designers/earthquake engineering experts, including Mr. Dick Carter, Prof. Bob Park, Dr. Rob Park, Mr. Grant Pearce and Mr. Stuart Palmer, who are not a part of the working group, for their contributions on New Zealand design code and practice. Many colleagues in Greece, Japan and USA provided constructive critical reviews of the book, including Mr. Hisao Oouchi, Prof. Kyriazis Pitilakis, Dr. Craig Taylor, Dr. Tadahiko Yagyu, Dr. Shuji Yamamoto, and Dr. Hiroshi Yokota. Sincere thanks are due to them for suggestions that improved the quality of the book. The working group would also like to express their sincere appreciation to the Coastal Development Institute of Technology, Japan, for funding the publication, to the PIANC regional sections and organizations for sponsoring the working group meetings and activities, and many of the member's organizations for partially funding activities associated with the development of these guidelines.

The working group hopes that the recommended seismic guidelines for port facilities will make a significant contribution towards mitigating seismic disasters in port areas around the world.

December 15, 2000

Susumu Iai
Chairman
PIANC/MarCom/WG34

Members of PIANC/MarCom/Working Group 34

Chairman:
Susumu Iai, Port and Harbour Research Institute, Japan

Secretaries:
Takahiro Sugano, Port and Harbour Research Institute, Japan
Koji Ichii, Port and Harbour Research Institute, Japan

Primary Authors:
Alberto Bernal, ByA Estudio de Ingenieria, Spain
Rafael Blazquez, Universidad Politecnica de Madrid, Spain
Hans F. Burcharth, Aalborg University, Denmark
Stephen E. Dickenson, Oregon State University, USA
John Ferritto, Consulting Engineer, USA
W.D. Liam Finn, University of British Columbia, Canada/Kagawa University, Japan
Susumu Iai, Port and Harbour Research Institute, Japan
Koji Ichii, Port and Harbour Research Institute, Japan
Nason J. McCullough, Oregon State University, USA
Piet W.H. Meeuwissen, Delta Marine Consultants bv, Netherlands (from May 1998)
Constantine D. Memos, National Technical University of Athens, Greece
M.J.N. Priestley, University of California, San Diego, USA
Francesco Silvestri, Universita della Calabria, Italy
Armando L. Simonelli, Consiglio Nazionale delle Ricerche, Italy
R. Scott Steedman, Whitby Bird & Partners Ltd., UK
Takahiro Sugano, Port and Harbour Research Institute, Japan

Contributing Working Group Members:
Steve J. Bowring, Delta Marine Consultants bv, Netherlands (through May 1998)
Valery M. Buslov, Han-Padron Associates, USA
Brad P. Erickson, TranSystems Co., USA

List of Tables and Figures: Main Text

Tables

Figures

MAIN TEXT

MAIN TEXT

CHAPTER 1

Introduction

The occurrence of a large earthquake near a major city may be a rare event, but its societal and economic impact can be so devastating that it is a matter of national interest. The earthquake disasters in Los Angeles, USA, in 1994 (61 fatalities and 40 billion US dollars in losses); Kobe, Japan, in 1995 (over 6,400 fatalities and 100 billion US dollars in losses); Kocaeli, Turkey, in 1999 (over 15,000 fatalities and 20 billion US dollars in losses); Athens, Greece, in 1999 (143 fatalities and 2 billion US dollars in losses); and Taiwan in 1999 (over 2,300 fatalities and 9 billion US dollars in losses) are recent examples. Although seismicity varies regionally as reflected in Fig. 1.1, earthquake disasters have repeatedly occurred not only in the seismically active regions in the world but also in areas within low seismicity regions, such as in Zones 1 or 2 in the figure. Mitigating the outcome of earthquake disasters is a matter of worldwide interest.

In order to mitigate hazards and losses due to earthquakes, seismic design methodologies have been developed and implemented in design practice in many regions since the early twentieth century, often in the form of codes and standards. Most of these methodologies are based on a force-balance approach, in which structures are designed to resist a prescribed level of seismic force specified as a fraction of gravity. These methodologies have contributed to the acceptable seismic performance of port structures, particularly when the earthquake motions are more or less within the prescribed design level. Earthquake disasters, however, have continued to occur. These disasters are caused either by strong earthquake motions, often in the near field of seismic source areas, or by moderate earthquake motions in the regions where the damage due to ground failures has not been anticipated or considered in the seismic design.

The seismic design guidelines for port structures presented in this book address the limitations inherent in conventional design, and establish the framework for a new design approach. In particular, the guidelines are intended to be:
- performance-based, allowing a certain degree of damage depending on the specific functions and response characteristics of a port structure and probability of earthquake occurrence in the region;
- user-friendly, offering design engineers a choice of analysis methods, which range from simple to sophisticated, for evaluating the seismic performance of structures; and

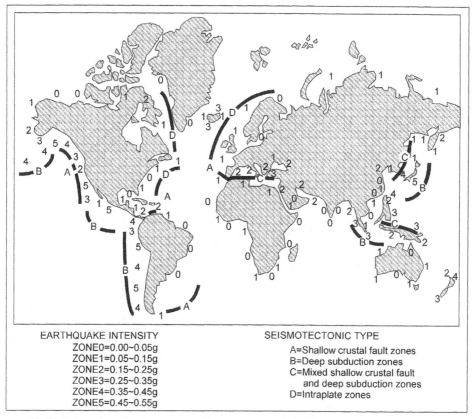

EARTHQUAKE INTENSITY	SEISMOTECTONIC TYPE
ZONE0=0.00~0.05g	A=Shallow crustal fault zones
ZONE1=0.05~0.15g	B=Deep subduction zones
ZONE2=0.15~0.25g	C=Mixed shallow crustal fault
ZONE3=0.25~0.35g	and deep subduction zones
ZONE4=0.35~0.45g	D=Intraplate zones
ZONE5=0.45~0.55g	

NOTE: Values of acceleration corresponding to a return period of 475 years.
Some areas of low average seismic hazard have historically experienced major destructive earthquakes.

Fig. 1.1. Worldwide zoned average earthquake hazard (modified from Bea, 1997; GSHAP, 1999).

– general enough to be useful throughout the world, where the required functions of port structures, economic and social environment, and seismic activities may differ from region to region.

The expected users of the guidelines are design engineers, port authorities, and specialists in earthquake engineering. The applicability of the guidelines will reflect regional standards of practice. If a region has no seismic codes or standards for designing port structures, the guidelines may be used as a basis to develop a new seismic design methodology, or codes applicable to that particular region. If a region has already developed seismic codes, standards, or established design practice, then the guidelines may be used to supplement these design and analysis procedures (see Technical Commentary 1 for existing codes and guidelines). It is not the intent of the authors to claim that these guidelines should be used instead of the existing codes or standards or established design practice in the region of interest. It is anticipated, however, that the guidelines will, with continual modification and upgrading, be recognized as a new and useful basis for mitigating seismic disasters in port areas. It is hoped that the guidelines may eventually be accepted worldwide as recommended seismic design provisions.

Earthquake engineering demands background knowledge in several disciplines. Although this background knowledge is not a pre-requisite to understanding the guidelines, readers may find it useful to have reference textbooks readily available. Pertinent examples include Kramer (1996) on geotechnical earthquake engineering and Tsinker (1997) on design practice for port structures.

This Main Text provides an overview of the seismic design guidelines. More details in the particular aspects of the seismic design guidelines can be found in the following Technical Commentaries (TC):

TC1: Existing Codes and Guidelines
TC2: Case Histories
TC3: Earthquake Motion
TC4: Geotechnical Characterisation
TC5: Structural Design Aspects of Pile-Deck Systems
TC6: Remediation of Liquefiable Soils
TC7: Analysis Methods
TC8: Examples of Seismic Performance Evaluations

CHAPTER 2

Earthquakes and Port Structures

This chapter addresses issues fundamental to understanding seismic effects on port structures. As illustrated in Fig. 2.1, seismic waves are generated along a crustal fault and they propagate through upper crustal rock, travelling to the surface of the bedrock at a site of interest. The ground motions then propagate through the local soil deposits, reaching the ground surface and impacting structures. Depending on the intensity of shaking and soil conditions, liquefaction of near-surface soils and associated ground failures may occur and could significantly affect the port structures. If an offshore fault motion involves vertical tectonic displacement of the seabed, tsunamis may be generated. The engineering aspects of these phenomena are collectively important in the evaluation of the seismic effects on port structures.

2.1   EARTHQUAKE MOTION

(1) Bedrock Motion

The bedrock motions used for seismic analysis and design at a particular site are characterized through seismic hazard analysis. If a specific earthquake scenario is assumed in the seismic hazard analysis, the bedrock motion is defined deterministically based on the earthquake source parameters and wave propagation effects along the source-to-site path. Most often, however, the bedrock motion is defined probabilistically through the seismic hazard analysis, taking into account uncertainties in frequency of occurrence and location of earthquakes.

One of the key parameters in engineering design practice is the intensity of bedrock motion defined in terms of peak ground acceleration (PGA), or in some cases peak ground velocity (PGV). This parameter is used either by itself or to scale relevant ground motion characteristics, including response spectra and time histories. In the probabilistic seismic hazard analysis, the level of bedrock motion is defined as a function of a return period, or a probability of exceedance over a prescribed exposure time. An example is shown in Fig. 2.2, in which PGAs at bedrock are obtained based on geologic, tectonic and historical seismic activity

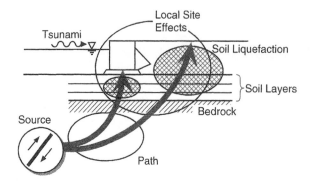

Fig. 2.1. Schematic figure of propagation of seismic waves.

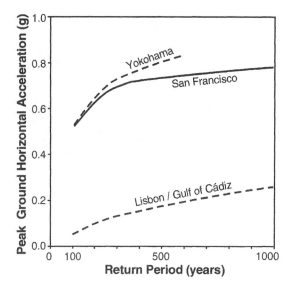

Fig. 2.2. Peak horizontal bedrock accelerations over various return periods.

data available for the region in consideration. It should be noted that the results of the seismic hazard analysis, such as shown in Fig. 2.2, can depend on the degree of knowledge about the regional tectonic environment, quality of the earthquake data base, and on the probabilistic methods employed. For these reasons, seismic hazard studies are commonly updated based on the most recent research. The bedrock motion for a prescribed return period is often specified in codes and standards for a region. See Technical Commentary 3 for more details on bedrock motion and seismic hazard analysis.

(2) Local Site Effects

The soil deposits at a particular site may significantly modify the bedrock ground motion, changing the amplitude, frequency content, and duration. This is due to the dynamic response characteristics of the soils, and it has been termed 'local site effects.' Local site effects depend on the material properties of the subsoil an stratigraphy, as well as the intensity and frequency characteristics of the bedrock motion.

As strong ground motion propagates upwards, towards the ground surface, the reduction in the strength and stiffness of soil deposits tends to amplify the ground motions. Depending on their depth and properties, soft soil layers can strongly amplify particular frequencies of motion. For very soft soils, the shaking motion may be attenuated and large strains may develop where the imposed cyclic shear stresses approach the strength of the deposits. Care must be taken to ensure the site response analysis is appropriate to the soil strain levels.

In engineering practice, local site effects are evaluated either by using prescribed site amplification factors based on statistical analysis of existing data or a site-specific response analysis (Kramer, 1996; ISSMGE-TC4, 1999). The site amplification factors are often specified in codes and standards, and used to scale the bedrock PGA or PGV to obtain the corresponding values at the ground surface, or used to scale bedrock response spectra to define the ground surface response spectra. Site response analysis of horizontally layered subsoil is generally accomplished using a one-dimensional model to obtain time histories of ground surface motion. Non-linear behaviour of soil is often idealized through the equivalent linear model, in which strain dependent material parameters such as shear modulus and damping ratio are defined to idealize non-linearity and energy dissipation imposed by design ground motions. See Technical Commentaries 3 and 4 for more details on site response analysis and geotechnical characterization.

2.2   LIQUEFACTION

As saturated soil deposits are sheared rapidly back and forth by the shaking motion, the water pressure in the pores of the soil starts to rise. In loose saturated cohesionless soils, the pore water pressure can rise rapidly and may reach such a level that the particles briefly float apart and the strength and stiffness of the soil is temporarily lost altogether. This is a condition called soil liquefaction, and it is shown diagrammatically in Fig. 2.3. The strength of soil comes about as the result of friction and interlocking between the soil particles. At any depth in the ground, before the earthquake, the weight of the soil and other loads above is carried in part by forces between the soil particles and in part by the pore water. When loose soil is shaken, it tries to densify or compact. The presence of the water, which has to drain away to allow the compaction, prevents this from happening immediately. As a consequence, more and more of the weight above is transferred to the pore water and the forces between the soil particles reduce. Ultimately, the pore water pressures may reach such a level that

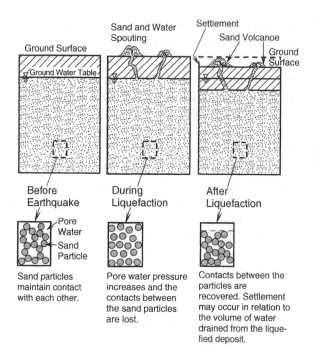

Fig. 2.3. Mechanism of liquefaction.

they cause water spouts to break through the overlying layers and the whole weight of the overlying material is transferred to the pore water. In this condition, the liquefied soil behaves as a heavy fluid, and large ground movements can occur. This liquefaction condition will continue until the high pore water pressures can drain again, and the contact between the soil particles is restored. Some layers in the ground will densify as a result of this process, and ground settlements will be observed. Other layers will remain in a very loose condition, and will be prone to liquefy again in future earthquakes.

The potential for earthquake-induced liquefaction is related to the resistance of the soil to the generation of excess pore pressures, and to the magnitude and duration of the cyclic shear stresses/strains that the soil is subjected to. The cyclic resistance of cohesionless soil is primarily a function of its density, permeability, and the effective confining stresses acting on the material. Given the influence of these factors on the results of in-situ tests such as the Standard Penetration Test (SPT), Cone Penetration Test (CPT), and shear wave velocity (V_S), these geotechnical investigations have been used to characterize the cyclic resistance of the soil. Laboratory testing methods involving the undrained cyclic loading of high quality specimens are also used in practice to ascertain the liquefaction resistance of the soil. See Technical Commentary 4 for more details.

Fig. 2.4. Evidence of sand boils due to liquefaction at Hakodate Port, Japan, during the Hokkaido-Nansei-oki earthquake of 1993.

In order to adequately assess the liquefaction susceptibility of a soil, both the cyclic resistance of the material and the seismic actions on the soil by design-level earthquake motions must be determined. The seismic actions are represented by the intensity and duration of the earthquake motions. This information is obtained from the site response analysis previously discussed. In the most sophisticated analysis of soil liquefaction, its effects are taken into account through effective stress formulation of the problem, incorporating pore pressure generation and dissipation models into a finite element or finite differ-ence computational scheme.

Dynamic pore pressures due to the earthquake are important since they lead to the development of hydraulic gradients within the mass of soil. As a result, the pore water begins to migrate and moves quickly to free-draining surfaces. This aspect of liquefaction is responsible for the evidence of liquefaction in level ground, such as sand boils (Fig. 2.4), water springs and vent fractures. Other phe-nomena associated to liquefaction are lateral spreading and flow failure of the soil, which are manifestations of the existence of a liquefied sand layer at some depth below a crest or embankment (e.g. Kramer, 1996).

2.3   TSUNAMIS

Tsunamis are long period sea waves that are generated by vertical seafloor movements. They are often associated with seismic fault ruptures, but occasionally with submarine landslides. Although wave amplitudes may be small in the open ocean, the wave height increases as the tsunamis approach shallower depths, occasionally reaching tens of meters at the coastline. The wave height of tsunamis is also amplified toward the end of V-shaped bays. The predominant wave period of tsunamis ranges from five to ten minutes, when produced by earthquakes at close proximity. Tsunamis can easily propagate long distances, such as across the Pacific Ocean. In this case, the predominant wave period typically ranges from forty minutes to two hours. Arrival time ranges from within five minutes for locally generated tsunamis to one day for distant tsunamis travelling across the Pacific Ocean. Destructive forces by tsunamis can be tremendous. An example of damage to breakwaters is shown in Fig. 2.5, in which tsunamis due to the Hokkaido-Nansei-oki earthquake of 1993 arrived at Okushiri Island, Japan, within five minutes and claimed 200 lives.

Engineering approaches to mitigate tsunami-related disasters are currently based on either issuing tsunami warnings or implementing engineering measures at bay mouths and coastlines. For warning of distant tsunamis, the Pacific Tsunami Warning Centre (PTWC) in Hawaii, USA, functions as the Operation

Fig. 2.5.   Damage to a breakwater at Okushiri Port, Japan, during the Hokkaido-Nansei-oki earthquake of 1993.

Centre for the Tsunami Warning System in the Pacific (TWSP). The time from earthquake occurrence to the issuance of a warning ranges from 34 to 92 minutes, 55 minutes on average (Blackford, 1998).

For warning of tsunamis due to earthquakes in close proximity, regional tsunami warning centres issue the warnings to local governments and citizens. For example, a highly advanced tsunami warning system has been established in Japan, where Japan Meteorological Agency (JMA) acts as the regional tsunami warning centre. Immediately after an earthquake, JMA computes the hypocentre and magnitude based on data from the seismic motion monitoring network deployed throughout Japan, and issues a tsunami warning within three minutes after an earthquake occurrence. The warning is issued over 66 regions along the entire Japanese coastline. Tsunami height and arrival time are the contents of the warning. The warning is issued based on a digital catalogue compilation of the results of computer simulations of hypothetical tsunamis assuming 100,000 fault plane models around Japan. The height of the tsunami at the shore (H_t) is obtained from a height at an offshore site (H_o) with water depth of h_o through Green's law: $H_t = (h_o/h_t)^{1/4}H_o$, with $h_t = 1$ m (Sekita et al., 1999). The warning is typically issued for earthquakes in the sea with a magnitude larger than 6.5 and a hypocentre shallower than 60 km, and is aired to the citizens immediately through the national TV network (NHK).

Difficulty lies in warning of tsunamis due to submarine landslides and 'tsunami earthquakes,' the latter of which are known to be associated with a slow seismic fault movement. Both are distinctive in that large displacement of the seabed is induced without strong earthquake shaking. The Meiji-Sanriku, Japan, earthquake of 1886 (magnitude 7.2) is a typical example, which claimed 22,000 lives without significant ground motion. The Nicaragua earthquake of 1992 is the first case of a tsunami earthquake that was instrumentally monitored and determined as a tsunami earthquake.

Implementing engineering measures for mitigating tsunami disasters has also been undertaken. For example, a composite breakwater over a length of 2 km at a bay mouth with a water depth of 63 m has been under construction in Kamaishi, Japan, since 1978. This breakwater is designed to reduce the height of the scenario tsunami based on case histories to 2.9 m at the end of the bay. In addition to the breakwater, a 4-m high tidal wall has also been constructed along the coastline at the bay to significantly reduce the potential for a tsunami disaster.

2.4   PORT STRUCTURES

From an engineering point of view, port structures are soil-structure systems that consist of various combinations of structural and foundation types. Typical port structures are shown in Fig. 2.6. This figure indicates that some port structures are

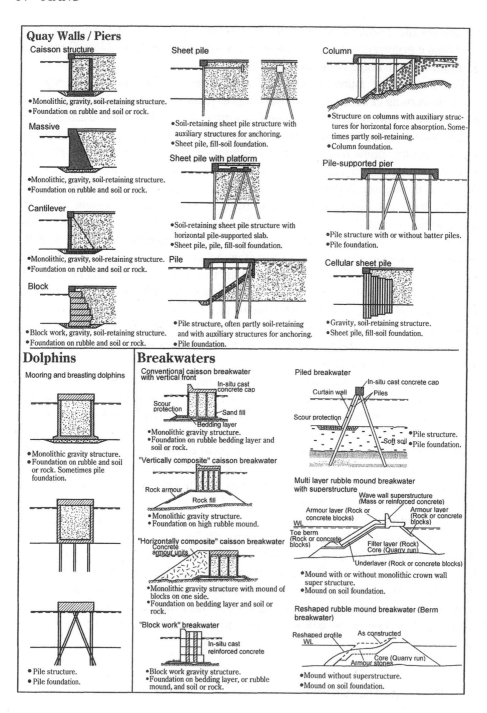

Fig. 2.6. Typical port structures.

'mixed' and can not be fully characterized by a single structural or foundation type, or stability mechanism. It is also difficult to produce an alternative characterization based on the vulnerability of the structure to damage by strong ground motions. This is due to the relative importance of the soil-fill conditions on the seismic performance of the structure. In fact, one of the major issues in this book is a discussion on how to evaluate the seismic performance of a port structure based on the specific structural and geotechnical conditions. Included in the definition of port structures in this book are quay walls with or without cranes and breakwaters. In particular, the quay walls discussed include those of gravity and sheet pile types, pile-supported wharves, and other mixed structures such as cellular quay walls. The breakwaters being discussed are vertical face and rubble mound types.

2.5   EXAMPLES OF SEISMIC DAMAGE

Typical examples of seismic damage to port structures are as follows.

(1) Damage to gravity quay walls

Damage to a gravity quay wall is shown in Figs. 2.7 and 2.8. This quay wall was located in Kobe Port, Japan, and shaken by the Great Hanshin earthquake of 1995.

Fig. 2.7. Damage to a caisson quay wall at Kobe Port, Japan, during the Great Hanshin earthquake of 1995.

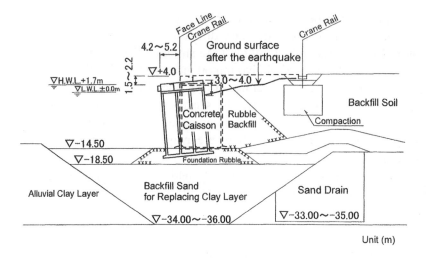

Fig. 2.8. Cross section of a caisson quay wall at Kobe Port.

The damage involves large seaward displacement, settlement and tilt. The damage was caused mainly by deformation in the loosely deposited foundation soil beneath the caisson wall.

(2) Damage to sheet pile quay walls

Representative damage to a sheet pile quay wall is shown in Figs. 2.9 and 2.10. This quay wall was located in Akita Port, Japan, and shaken by the 1983 Nihonkai-Chubu earthquake. The damage was mainly due to large pressures applied to the sheet pile wall by the liquefied backfill sand. The sheet pile bending moment became excessively large, resulting in the opening of a crack in the wall. Additional examples of damage to sheet pile quay walls are shown in Figs. 2.11 and 2.12. This quay wall was located in the same port and subjected to ground motions of similar intensity. The anchor was embedded in liquefiable soil, resulting in deformations in the entire wharf as shown in these figures. These two examples indicate that the mode of damage can differ significantly according to the geotechnical and structural conditions of a quay wall.

(3) Damage to pile-supported wharves

Damage to a pile-supported wharf is shown in Fig. 2.13. This quay was located in Kobe Port, Japan, and was shaken by the Great Hanshin earthquake of 1995. The damage was caused by the large seaward movement of a dike over a liquefiable sand layer. The bending moment in the piles became excessively large, resulting

Fig. 2.9. Damage to a sheet pile quay wall at Ohama Wharf, Akita Port, Japan, during the Nihonkai-chubu earthquake of 1983.

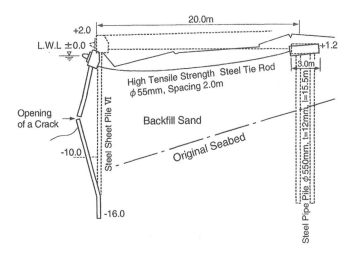

Fig. 2.10. Cross section of a sheet pile quay wall at Ohama Wharf, Akita Port.

in formation of plastic hinges, both at the pile-deck connection and within the embedded portion of the pile. The piles, extracted after the earthquake for investigation, clearly exhibited signs of damage in the embedded portion as shown in Fig. 2.14.

Fig. 2.11. Damage to a sheet pile quay wall at Shimohama Wharf, Akita Port, Japan, during the Nihonkai-chubu earthquake of 1983.

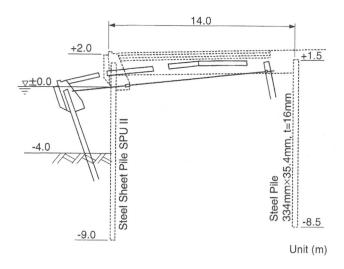

Fig. 2.12. Cross section of a sheet pile quay wall at Shimohama Wharf, Akita Port.

(4) Damage to cranes

Damage to a crane on rails is shown in Fig. 2.15. This crane was located in Kobe Port, Japan, and shaken by the Great Hanshin earthquake of 1995. The damage was caused not only by strong shaking, but also by the movement of the caisson

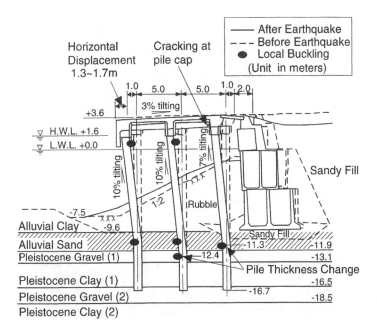

Fig. 2.13. Cross section of a pile-supported wharf at Takahama Wharf, Kobe Port, Japan, and damage during the Great Hanshin earthquake of 1995.

Fig. 2.14. Damage to piles at Takahama Wharf, Kobe Port (after extraction for inspection).

Fig. 2.15. Damage to a crane at Kobe Port, Japan, during the Great Hanshin earthquake
of 1995.
(a) Overview.
(b) Close-up.

quay wall toward the sea, resulting in formation of plastic hinges at the legs of the crane.

(5) Damage to breakwaters

Damage to a composite breakwater with a vertical face is shown in Fig. 2.16. This breakwater was located in Kobe Port, Japan, and shaken by the Great Hanshin earthquake of 1995. The damage was mostly in the form of excessive settlement caused by significant deformation of the loose soil deposit below the breakwater.

Damage to a rubble mound breakwater is shown in Fig. 2.17. The site was a southern extension of the detached breakwater of Patras Port, Greece. This involved large settlements of the order of 3 to 4 m and was triggered by a series of moderate earthquakes in 1984, and amplified considerably by the local geotechnical conditions. The damage was mostly in the form of flattening of the cross section and of intrusion of the lower mound material into the soft clay.

(6) Common features of damage to port structures

The evidence of damage to port structures previously addressed suggests that:
— most damage to port structures is often associated with significant deformation of a soft or liquefiable soil deposit; hence, if the potential for liquefaction

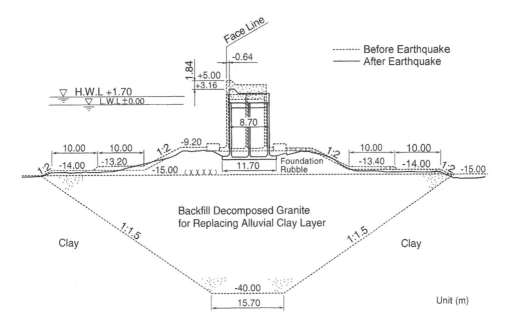

Fig. 2.16. Damage to a composite breakwater at Kobe Port, Japan, during the Great Hanshin earthquake of 1995.

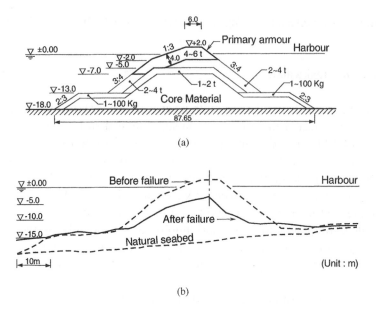

Fig. 2.17. Damage to a rubble mound breakwater at Patras Port, Greece, after a series of
earthquakes in 1984.
(a) Typical cross section.
(b) Cross section before and after the failure.

exists, implementing appropriate remediation measures against liquefaction
may be an effective approach to attaining significantly improved seismic performance of port structures;

– most failures of port structures, from a practical perspective, result from excessive deformations, not catastrophic collapses; hence, design methods based on
displacements and ultimate stress states are desirable over conventional force-
based design methods for defining the comprehensive seismic performance of
port structures; and

– most damage to port structures is the result of soil-structure interaction; hence,
design and analysis procedures should include both geotechnical and structural
conditions of port structures.

Technical Commentary 2 provides more details on case histories of seismic damage to port structures.

CHAPTER 3

Design Philosophy

An evolving design philosophy for port structures in many seismically active regions reflect the observations that:
- deformations in ground and foundation soils and the corresponding structural deformation and stress states are key design parameters;
- conventional limit equilibrium-based methods are not well suited to evaluating these parameters; and
- some residual deformation may be acceptable.

The performance-based methodology presented in this book incorporates this design philosophy into practice-oriented guidelines, and provides engineers with new tools for designing ports.

3.1 PERFORMANCE-BASED METHODOLOGY

Performance-based design is an emerging methodology, which was born from the lessons learned from earthquakes in the 1990s (SEAOC, 1995; Iai and Ichii, 1998; Steedman, 1998). The goal is to overcome the limitations present in conventional seismic design. Conventional building code seismic design is based on providing capacity to resist a design seismic force, but it does not provide information on the performance of a structure when the limit of the force-balance is exceeded. If we demand that limit equilibrium not be exceeded in conventional design for the relatively high intensity ground motions associated with a very rare seismic event, the construction/retrofitting cost will most likely be too high. If force-balance design is based on a more frequent seismic event, then it is difficult to estimate the seismic performance of the structure when subjected to ground motions that are greater than those used in design.

In performance-based design, appropriate levels of design earthquake motions must be defined and corresponding acceptable levels of structural damage must be clearly identified. Two levels of earthquake motions are typically used as design reference motions, defined as follows:

Level 1 (L1): the level of earthquake motions that are likely to occur during the life-span of the structure;

Level 2 (L2): the level of earthquake motions associated with infrequent rare events, that typically involve very strong ground shaking.

The acceptable level of damage is specified according to the specific needs of the users/owners of the facilities and may be defined on the basis of the acceptable level of structural and operational damage given in Table 3.1. The structural damage category in this table is directly related to the amount of work needed to restore the full functional capacity of the structure and is often referred to as direct loss due to earthquakes. The operational damage category is related to the length of time and cost associated with the restoration of full or partial serviceability. Economic losses associated with the loss of serviceability are often referred to as indirect losses. In addition to the fundamental functions of servicing sea transport, the functions of port structures may include protection of human life and property, functioning as an emergency base for transportation, and as protection from spilling hazardous materials. If applicable, the effects on these issues should be considered in defining the acceptable level of damage in addition to those shown in Table 3.1.

Once the design earthquake levels and acceptable damage levels have been properly defined, the required performance of a structure may be specified by the appropriate performance grade S, A, B, or C defined in Table 3.2. In performance-based design, a structure is designed to meet these performance grades.

The principal steps taken in performance-based design are shown in the flow-chart in Fig. 3.1:

1) Select a performance grade of S, A, B, or C: This step is typically done by referring to Tables 3.1 and 3.2 and selecting the damage level consistent with

Table 3.1. Acceptable level of damage in performance-based design.*

Level of damage	Structural	Operational
Degree I : Serviceable	Minor or no damage	Little or no loss of serviceability
Degree II: Repairable	Controlled damage**	Short-term loss of serviceability***
Degree III: Near collapse	Extensive damage in near collapse	Long-term or complete loss of serviceability
Degree IV: Collapse****	Complete loss of structure	Complete loss of serviceability

* Considerations: Protection of human life and property, functions as an emergency base for transportation, and protection from spilling hazardous materials, if applicable, should be considered in defining the damage criteria in addition to those shown in this table.
** With limited inelastic response and/or residual deformation.
*** Structure out of service for short to moderate time for repairs.
**** Without significant effects on surroundings.

Table 3.2. Performance grades S, A, B, and C.

Performance grade	Design earthquake	
	Level 1(L1)	Level 2(L2)
Grade S	Degree I: Serviceable	Degree I: Serviceable
Grade A	Degree I: Serviceable	Degree II: Repairable
Grade B	Degree I: Serviceable	Degree III: Near collapse
Grade C	Degree II: Repairable	Degree IV: Collapse

the needs of the users/owners. Another procedure for choosing a performance grade is to base the grade on the importance of the structure. Degrees of importance are defined in most seismic codes and standards. This procedure is presented in Table 3.3. If applicable, a performance grade other than those of S, A, B, or C may be introduced to meet specific needs of the users/owners.

2) Define damage criteria: Specify the level of acceptable damage in engineering parameters such as displacements, limit stress states, or ductility factors. The details are addressed in Chapter 4, Damage Criteria.

3) Evaluate seismic performance of a structure: Evaluation is typically done by comparing the response parameters from a seismic analysis of the structure with the damage criteria. If the results of the analysis do not meet the damage criteria, the proposed design or existing structure should be modified. Soil improvement including remediation measures against liquefaction may be necessary at this stage. Details of liquefaction remediation from the publication of the Port and Harbour Research Institute, Japan (1997) can be found in Technical Commentary 6.

More details on design earthquake motion and performance evaluation are addressed in the following sections.

3.2   REFERENCE LEVELS OF EARTHQUAKE MOTIONS

Level 1 earthquake motion (L1) is typically defined as motion with a probability of exceedance of 50% during the life-span of a structure. Level 2 earthquake motion (L2) is typically defined as motion with a probability of exceedance of 10% during the life-span. In defining these motions, near field motion from a rare event on an active seismic fault should also be considered, if the fault is located nearby. If the life-span of a port structure is 50 years, the return periods for L1 and L2 are 75 and 475 years, respectively.

In regions of low seismicity, L1 may be relatively small and of minor engineering significance. In this case, only L2 is used along with an appropriately specified damage criteria. Here, it is assumed that satisfactory performance for L2 will implicitly ensure required performance under the anticipated L1 motion. It

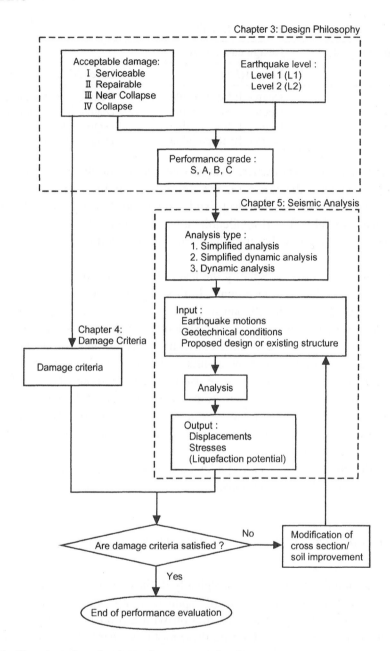

Fig. 3.1. Flowchart for seismic performance evaluation.

may be noted that this single level approach is somewhat similar to conventional design practice; it differs only in that a structure is designed in accordance with a designated acceptable level of damage.

The dual level approach using both L1 and L2 attempts to: 1) ensure a specified level of safety and serviceability for L1, and 2) prescribe the level and modes of seismic damage for L2. This dual level approach is particularly useful in regions of moderate and high seismicity where meeting the specified damage criteria for L2 may not be sufficient to ensure the desired degree of safety and serviceability during L1, or meeting the performance standard for L1 is not sufficient to ensure the specified performance standard for L2. It should be noted here that stronger L2 excitations will not necessarily solely dictate the final design, which may be highly influenced or even dominated by a high performance standard for L1.

In the remainder of this main text, the dual level approach will be the focus of attention. The information needed for the single level approach can be easily inferred by adopting the issues associated with L2 only.

3.3   PERFORMANCE EVALUATION

As a guide for evaluating performance criteria at a specific port, the relationship between degree of damage and the design earthquake motion is illustrated in Fig. 3.2. The curves in this figure form the basis for the performance evaluation procedure. This figure is based on the specification of performance grades in Table 3.2. The curves in Fig. 3.2 indicate the upper limits for the acceptable level of damage over a continuously varying level of earthquake motions, including the

Table 3.3. Performance grade based on the importance category of port structures.

Performance grade	Definition based on seismic effects on structures	Suggested importance category of port structures in Japanese code
Grade S	① Critical structures with potential for extensive loss of human life and property upon seismic damage ② Key structures that are required to be serviceable for recovery from earthquake disaster ③ Critical structures that handle hazardous materials ④ Critical structures that, if disrupted, devastate economic and social activities in the earthquake damage area	Special Class
Grade A	Primary structures having less serious effects for ① through ④ than Grade S structures, or ⑤ Structures that, if damaged, are difficult to restore	Special Class or Class A
Grade B	Ordinary structures other than those of Grades S, A and C	Class A or B
Grade C	Small easily restorable structures	Class B or C

designated L1 and L2 motions. Each curve in this figure is defined by two control points corresponding to the upper limits of the level of damage for L1 and L2 motions defined in Table 3.2. For example, the curve defining the upper limit for Grade B should go through a point defining the upper limit for damage degree I for L1 motion, and another defining the upper limit for damage degree III for L2 motion. The shape of the curves may be approximated by line segments through the controlling points or may be refined by referring to typical results of non-linear seismic analysis of port structures.

The vertical coordinates of Fig. 3.2 are converted into engineering parameters such as displacements, stresses or ductility factors specified by the damage criteria discussed in Chapter 4. This conversion allows direct comparison between required performance and seismic response of a structure. The seismic response of a structure is evaluated through seismic analysis over L1 and L2 motions and plotted on this figure as 'seismic response curve.' As minimum requirement, analysis should be performed for L1 and L2 earthquake motions. For example, if the structure being evaluated or designed has the seismic response curve 'a' in Fig. 3.3, the curve is located below the upper bound curve defining Grade A. Thus, this design assures Grade A performance. If an alternative structural configuration yields the seismic performance curve 'b' in Fig. 3.3 and a portion of the curve exceeds the upper limit for Grade A, then this design assures only Grade B performance.

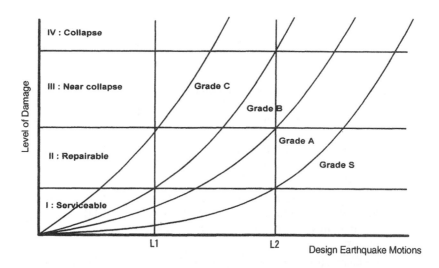

Fig. 3.2. Schematic figure of performance grades S, A, B and C.

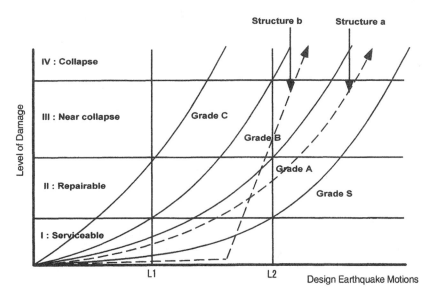

Fig. 3.3. Examples of seismic performance evaluation.

CHAPTER 4

Damage Criteria

In performance-based design, the acceptable level of damage, i.e. the damage criteria, should be specified in engineering terms such as displacements, limit stress state, and ductility/strain limit based on the function and seismic response of the structure. The damage criteria may be established based on Table 3.1 in Chapter 3 by a group of design engineers with assistance and advice from the users and owners of the structure/facility. The guidelines for establishing damage criteria for typical quay walls are described in Sections 4.1 through 4.4. The proposed damage criteria discussed in this chapter have been established for the following specific conditions:

1) The operational damage criteria in Table 3.1 in the previous chapter dictate the Degree I, whereas the structural damage criteria in the same table allow for the Degrees II through IV.

2) The quay walls considered are those where the seismic damage poses no threats to human life, no hazardous material is handled, without cranes on rails, and the sea space in front of the quay wall is unlimited.

Additional guidelines are offered in Section 4.5 for quay walls with cranes on rails. Damage criteria for breakwaters are described in Section 4.6.

4.1 GRAVITY QUAY WALLS

(1) Seismic response of gravity quay walls

A gravity quay wall is made of a caisson or other gravity retaining structure placed on the seabed. Stability against earth pressures from the backfill soil behind the wall is maintained by the mass of the wall and the friction at the bottom of the wall. For gravity quay walls on firm foundations, typical failure modes during earthquakes are seaward displacements and tilting as shown in Fig. 4.1(a). For a loose backfill or natural loose sandy foundation, the failure modes involve overall deformation of the foundation beneath the wall, resulting in large seaward displacement, tilting and settlements (see Fig. 4.1(b)). A wall with a relatively small width to height ratio, typically less than about 0.75, will exhibit a predominant tilting failure mode rather than horizontal displacements.

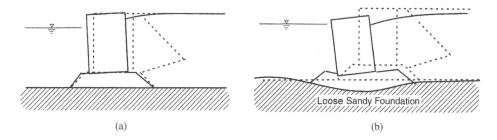

Fig. 4.1. Deformation/failure modes of gravity quay wall.
    (a) On firm foundation.
    (b) On loose sandy foundation.

On the conceptual level of design, several options may be available for increasing the stability of the gravity quay wall. An obvious measure to reduce the earth pressures against the wall is to use backfill material with a large angle of internal friction. Rockfill and similar materials should be preferable to loose sand for such use. The risk of overall deformation as well as of liquefaction would be minimized if such practice could be followed. Other possibilities for increasing the stability of the quay wall include drawing landwards the centre of gravity of the wall and thereby increasing the stabilizing moment, and imposing a higher friction coefficient between the base slab and the foundation layer. In concept, the former may be achieved by allowing for appropriate offsets of the precast blocks along the vertical face of the structure or by filling the rear cells of caissons with heavy materials such as metal slag. The latter has indeed been achieved by using an asphalt or rubber mat beneath the caisson wall. The latter may alternatively be achieved by providing a foundation slab of concrete with a special anti-skid shape or by designing for inclined, rather than horizontal contact surfaces at foundation and higher levels of the block wall.

Horizontal displacement and uniform vertical settlement of a gravity quay wall may not significantly reduce the residual state of stability, and may be generally acceptable from a structural point of view. However, tilting of the wall can significantly reduce the residual stability and result in an unacceptable structural stability situation. Case histories show an overturning/collapse of concrete block type walls when tilting occurs. This type of wall needs careful consideration in specifying damage criteria regarding the overturning/collapse mode.

(2) Parameters for specifying damage criteria for gravity quay walls

Seismic performance of a gravity quay wall is specified based on such serviceability considerations as safe berthing, safe operation of wheeled vehicles and cargo handling, flooding, and level of structural damage in the form of displacements and tilting (including relative displacements between the blocks of a concrete block type quay wall). Parameters that may be used for specifying damage criteria include displacements, settlements, tilting, and differential displacements along the face line of a wall, and the deformations at apron including settlement, differential settlement at and behind apron, and tilting (Fig. 4.2). Damage criteria should be established by choosing and specifying appropriate parameters from those mentioned above. More specifics are given in the next subsection.

(3) Damage criteria for gravity quay walls

Provided the conditions mentioned at the beginning of this chapter are applicable, the damage criteria may be established by referring to Table 4.1. The criteria shown in Table 4.1 are minimum requirements. Thus, in evaluating seismic performance by referring to damage criteria based on a number of different parameters, the highest damage degree should be the final result of the evaluation.

4.2   SHEET PILE QUAY WALLS

(1) Seismic response of sheet pile quay walls

A sheet pile quay wall is composed of interlocking sheet piles, tie-rods, and anchors. The wall is supported at the upper part by anchors and the lower part by embedment in competent soil. Typical failure modes during earthquakes depend on structural and geotechnical conditions as shown in Fig. 4.3.

Structural damage to a sheet pile quay wall is basically governed by structural stress/strain states rather than displacements. It is important to determine the preferred sequence and degrees of ultimate states that may occur in the composite sheet pile quay wall system.

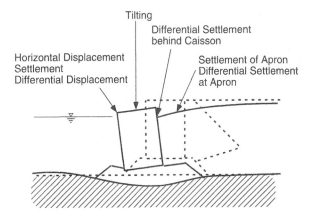

Fig. 4.2. Parameters for specifying damage criteria for gravity quay wall.

Table 4.1. Proposed damage criteria for gravity quay walls.

Level of damage		Degree I	Degree II	Degree III	Degree IV
Gravity wall	Normalized residual horizontal displacement (d/H)*	Less than 1.5%**	1.5~5%	5~10%	Larger than 10%
	Residual tilting towards the sea	Less than 3°	3~5°	5~8°	Larger than 8°
Apron	Differential settlement on apron	Less than 0.03~0.1 m	N/A***	N/A	N/A
	Differential settlement between apron and non-apron areas	Less than 0.3~0.7 m	N/A	N/A	N/A
	Residual tilting towards the sea	Less than 2~3°	N/A	N/A	N/A

* d: residual horizontal displacement at the top of the wall; H: height of gravity wall.
** Alternative criterion is proposed with respect to differential horizontal displacemt less than 30 cm.
*** Abbreviation for not applicable.

(2) Parameters for specifying damage criteria for sheet pile quay walls

Seismic performance of a sheet pile quay wall is specified based on serviceability, similar to that for a gravity quay wall, and in terms of structural damage regarding stress states as well as displacements. Parameters for specifying damage criteria are as follows (refer to Fig. 4.4).
Displacements:
– sheet pile wall and apron: refer to the parameters for a gravity quay wall;
– anchor: differential settlement, ground surface cracking at anchor, pull-out displacement of battered pile anchor.
Stresses:
– sheet pile (above and below mudline);
– tie-rod (including joints);
– anchor.
Damage criteria should be established by choosing and specifying appropriate parameters from those mentioned above.
    The preferred sequence to reach ultimate states with increasing level of seismic load should be appropriately specified for a sheet pile quay wall. If a damaged anchor is more difficult to restore than a sheet pile wall, the appropriate sequence may be given as follows (refer to Fig. 4.5).
1) Displacement of anchor (within the damage Degree I).
2) Yield at sheet pile wall (above mudline).
3) Yield at sheet pile wall (below mudline).

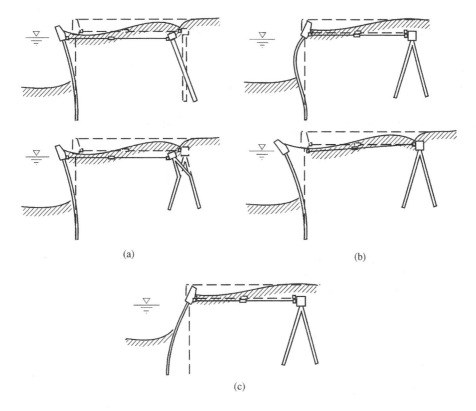

Fig. 4.3. Deformation/failure modes of sheet pile quay wall.
     (a) Deformation/failure at anchor.
     (b) Failure at sheet pile wall/tie-rod.
     (c) Failure at embedment.

4) Yield at anchor.
5) Yield at tie-rod.
If a damaged sheet pile wall is more difficult to restore than an anchor, the yield at anchor should precede the yield at sheet pile wall.

(3) Damage criteria for sheet pile quay walls

Provided the conditions mentioned at the beginning of this chapter are applicable, the damage criteria for a sheet pile quay wall may be established by referring to Table 4.2. The most restrictive conditions among displacements and stresses should define the damage criteria. Structural damage to the embedded portion of a sheet pile is generally difficult to restore, and thus necessitates higher seismic resistance. Brittle fracture of a sheet pile wall, rupture of a tie-rod, and the collapse of anchor should be avoided.

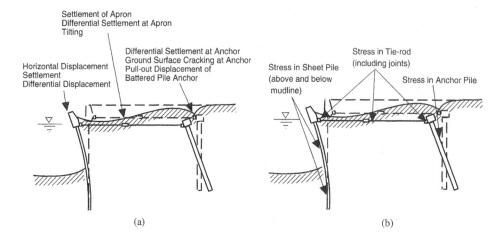

Fig. 4.4. Parameters for specifying damage criteria for sheet pile quay wall.
(a) With respect to displacements.
(b) With respect to stresses.

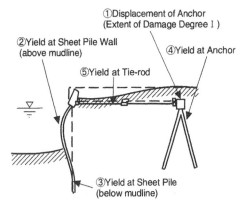

Fig. 4.5. Preferred sequence for yield of
sheet pile quay wall.

4.3   PILE-SUPPORTED WHARVES

(1) Seismic response of pile-supported wharves

A pile-supported wharf is composed of a deck supported by a substructure consisting of piles and a dike/slope. The unsupported pile length above the dike/slope surface is variable. When rockfill suitable for construction of the dike is uneconomical, a gravity or sheet pile retaining structure is also constructed to replace a portion of the dike. The seismic response of pile-supported wharves is influenced to a great degree by complex soil-structure interaction during ground shaking. Typical failure modes during earthquakes depend on the magnitude of the inertia force relative to the ground displacement (Fig. 4.6). In particular, the failure

Table 4.2. Proposed damage criteria for sheet pile quay walls. (when an anchor is more difficult to restore than a wall).

Level of damage			Degree I	Degree II	Degree III	Degree IV
Residual displacements	Sheet pile wall	Normalized residual horizontal displacement (*d/H*)*	Less than 1.5%**	N/A	N/A	N/A
		Residual tilting towards the sea	Less than 3°	N/A	N/A	N/A
	Apron	Differential settlement on apron	Less than 0.03~0.1 m	N/A	N/A	N/A
		Differential settlement between apron and non-apron areas	Less than 0.3~0.7 m	N/A	N/A	N/A
		Residual tilting towards the sea	Less than 2~3°	N/A	N/A	N/A
Peak response astresses/ strains	Sheet pile wall	Above mudline	Elastic	Plastic (less than the ductility factor/strain limit above mudline)	Plastic (less than the ductility factor/ strain limit above mudline)	Plastic (beyond the ductility factor/strain limit above mudline)
		Below mudline	Elastic	Elastic	Plastic (less than the ductility factor/ tstrain limit below mudline)	Plastic (beyond the ductility factor/strain limit below mudline)
	Tie-rod		Elastic	Elastic	Plastic (less than the ductility factor/ strain limit for tie-rod)	Plastic (beyond the ductility factor/strain limit for tie-rod)
	Anchor		Elastic	Elastic	Plastic (less than the ductility factor/ strain limit for anchor)	Plastic (beyond the ductility factor/strain limit for anchor)

*d: residual horizontal displacement at the top of the wall; H: height of sheet pile wall from mudline.
**Alternative criterion is proposed with respect to differential horizontal displacement less than 30 cm.

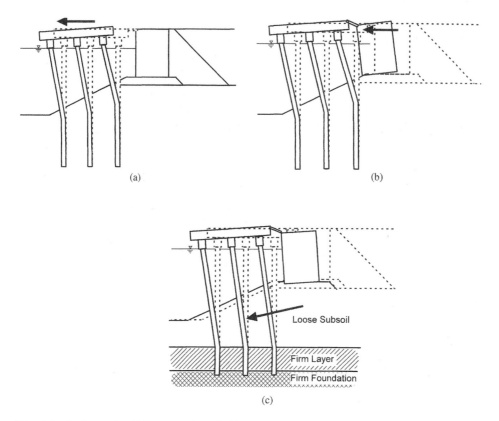

Fig. 4.6. Deformation/failure modes of pile-supported wharf.
    (a) Deformation due to inertia force at deck.
    (b) Deformation due to horizontal force from retaining wall.
    (c) Deformation due to lateral displacement of loose subsoil.

associated with the ground displacement shown in Fig. 4.6(c) suggests that appropriate design consideration is required of the geotechnical aspects.

Among various structural components in the pile-supported wharves, attention should be paid to the application of batter piles. Batter piles remain the most efficient structural component for resisting lateral loads due to mooring, berthing and crane operation. However, the batter pile-deck system results in a much more rigid frame than one with vertical piles. As a result, large stress concentrations and shear failures of concrete batter piles have been observed during earthquakes (e.g. 1989 Loma Prieta earthquake, and 1994 Northridge earthquake in California, USA). This mode of pile failure demonstrates that the structural design of batter piles used in regions of high seismicity must account for displacement demand and appropriate ductility.

When using the concrete piles in the areas of high seismicity, it is becoming more common to design the pile-supported wharves with vertical piles only. This structural system resists the earthquake-induced lateral loads/displacements by bending of the piles and associated moment resistance. Moderate structural damage to vertical piles in bending may be unavoidable during strong earthquake motions. However, bending failure at the pile head is considered preferable to shear failure at the batter pile-deck connection because the former is substantially easier to repair (Buslov et al., 1996).

Innovative structural design and detailing of pile-deck connections has allowed for the continued use of batter piles for pile-supported wharves. The design concepts incorporate a repairable structural 'fuse' at the pile cap. The fuse serves as a weak link or load-limiting device that is robust enough to withstand lateral loads during routine port operation, but yields during strong earthquake motions, thereby precluding damage to the wharf deck itself. The load-limiting device may be designed as a component that relies on sliding friction (Zmuda et al., 1995), or as a structural 'fuse-link' that yields in shear and can be easily replaced (Johnson et al., 1998).

In the areas of high seismicity such as in Japan, it is common practice to use steel piles for pile-supported wharves.

Structural damage to a pile-supported wharf is governed by stress/strain states rather than displacements. It is important to determine the preferred sequence and degrees of ultimate states to occur in the composite pile-supported wharf system.

(2) Parameters for specifying damage criteria for pile-supported wharves

The structure of a pile-supported wharf consists of a deck/pile system and a dike/slope/retaining wall. The seismic performance of the dike/slope/retaining wall may be specified based on the same criteria as for a gravity or sheet pile wall. The effects of ground displacements, including the movement of the retaining wall and the dike/slope below the deck, should be carefully evaluated for the performance of the deck/pile system.

The seismic performance of a deck/pile system is specified based on serviceability and structural damage. Parameters for specifying damage criteria for deck and piles are as follows (refer to Fig. 4.7).

Displacements:
– deck and piles: settlement, tilting, differential displacements;
– apron (deck and loading/unloading area behind the deck): differential settlement between deck and retaining wall, tilting, fall/fracture of bridge.

Stresses:
– piles (pile top and below dike/slope surface or mudline);
– deck (deck body, pile cap);
– bridge.

Damage criteria should be established by choosing and specifying appropriate parameters from these parameters.

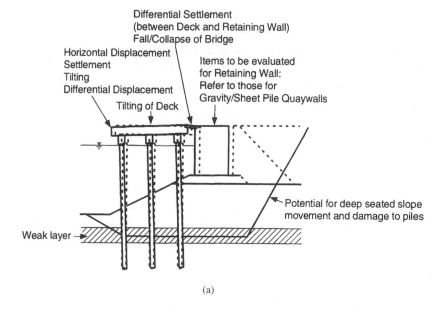

(a)

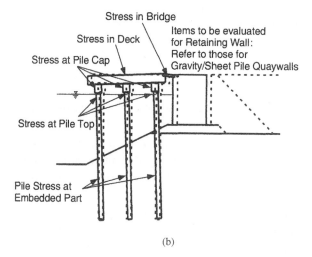

(b)

Fig. 4.7. Parameters for specifying damage criteria for pile-supported wharf.
    (a) With respect to displacements.
    (b) With respect to stresses.

With respect to ease of restoration after an earthquake, the preferred sequence to reach ultimate states with increasing levels of seismic load may be specified for a pile-supported wharf as follows (refer to Fig. 4.8).
1) Pile cap (portion of pile embedded for a pile-deck connection).
2) Pile top (just below the pile cap – concrete beam connection).

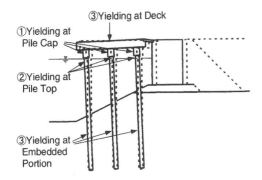

①Yielding at Pile Cap
②Yielding at Pile Top
③Yielding at Deck
③Yielding at Embedded Portion

Fig. 4.8. Preferred sequence for yielding of pile-supported wharf.

3) Deck and pile below dike/slope surface or mudline (within allowable ductility factor).

In the bridging area between the wharf deck and the retaining wall, structural details such as a fail-safe device to prevent fall-down or an easily repairable structure are also important, and, if applicable, a structure to absorb the displacements from the retaining wall can be introduced. This could lead to further development of energy absorption devices for pile-supported wharves.

(3) Damage criteria for pile-supported wharves

The criteria for the retaining wall of a pile-supported wharf may be established by referring to those for a gravity or sheet pile quay wall. Damage criteria for the dike/slope under the deck are less established because of the complex soil-structure interaction between the piles and the dike/slope.

Provided the conditions mentioned at the beginning of this chapter are applicable, criteria for the piles and deck of a pile-supported wharf may be established by referring to Table 4.3. The most restrictive condition among displacements and stresses should define the damage criteria.

Structural damage to the embedded portion of a pile is generally difficult to restore and has the potential to trigger collapse of a pile-supported wharf. Thus, a more restrictive ductility factor should be used for the design. Examples of ductility factors/strain limits for use in practice are found in Ferritto (1997a), Ferritto et al. (1999) and Yokota et al. (1999). More details can be found in Technical Commentaries 5 and 7. No case histories have been reported of brittle fracture of steel piles during earthquakes. However, the case histories of brittle fracture of thick steel columns during the 1995 Kobe earthquake indicate that it is still necessary to study this aspect in steel piles.

Table 4.3. Proposed damage criteria for pile-supported wharves.*

Level of damage		Degree I	Degree II	Degree III	Degree IV
Residual displacements	Differential settlement between deck and land behind	Less than 0.1~0.3 m	N/A	N/A	N/A
	Residual tilting towards the sea	Less than 2~3°	N/A	N/A	N/A
Peak response	Piles**	Essentially elastic response with minor or no residual deformation	Controlled limited inelastic ductile response and residual deformation intending to keep the structure repairable	Ductile response near collapse (double plastic hinges may occur at only one or limited number of piles)	Beyond the state of Degree III

* Table refers only to piles & deck.
** Bending failure should precede shear failure in structural components.

4.4   CELLULAR QUAY WALLS

(1) Seismic response of cellular quay walls

A cellular quay wall is made of a steel plate or steel sheet pile cell with sand or other fill. Resistance against inertia forces and earth pressures is provided by the fill friction at the bottom surface of the cell for a non-embedded cell (i.e. steel plate cofferdam type), or by the resistance of the foundation subsoil at the cell embedment. Typical failure modes during earthquakes depend on cell embedment and geotechnical conditions as shown in Fig. 4.9.

Structural damage to a cellular quay wall is governed by displacements as well as stress states. It is important to determine the preferred sequence of occurrence and degrees of ultimate states in the composite cellular quay wall system. The 1995 Kobe earthquake caused damage in a failure mode, shown in Fig. 4.10, that included deformation along the horizontal cross section of a cell.

(2) Parameters for specifying damage criteria for cellular quay walls

Seismic performance of a cellular quay wall is specified based on serviceability, similar to that for a gravity quay wall, and on structural damage regarding stresses as well as displacements. Parameters for specifying damage criteria are similar to those for a gravity or a sheet pile quay wall regarding displacements as shown in Fig. 4.11. Parameters regarding stresses include stress states in cell and cell joints.

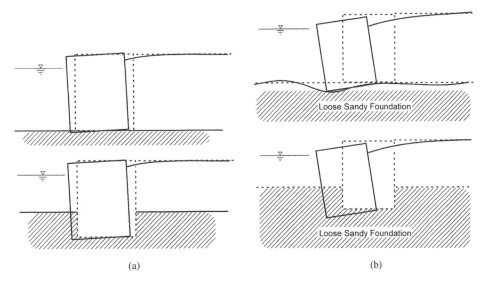

(a)   (b)

Fig. 4.9. Deformation/failure modes of cellular quay wall.
    (a) On firm foundation.
    (b) On loose sandy foundation.

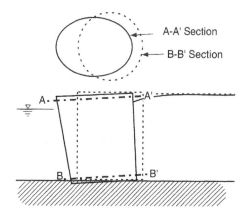

Fig. 4.10. Deformation/failure modes of cellular quay wall involving cross sectional deformation.

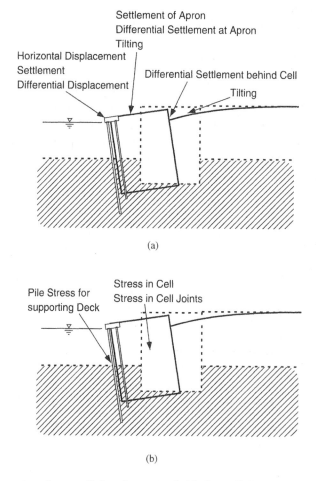

(a)

(b)

Fig. 4.11. Parameters for specifying damage criteria for cellular quay wall.
    (a) With respect to displacements.
    (b) With respect to stresses.

(3) Damage criteria for cellular quay walls

Provided the conditions mentioned at the beginning of this chapter are applicable, the damage criteria for a cellular quay wall may be established by referring to Table 4.4. The most restrictive conditions among displacements and stresses should define the damage criteria.

4.5   QUAY WALLS WITH CRANES

In order to be serviceable after an earthquake, quay walls equipped with facilities such as cranes on rails (hereinafter referred to as cranes) or ferry facilities need to maintain the serviceability of these facilities in addition to general purpose quay wall serviceability. Safety and protection of human life should be guaranteed. Consequently, the criteria for acceptable damage to quay walls equipped with cranes should be specified based on considerations additional to those discussed in Sections 4.1 through 4.4. For example, serviceability of quay walls with conveyor belts is strongly affected by differential displacements and settlements. For ferry wharves, consideration of quay wall displacement and implementation of fail-safe devices for passenger bridges is necessary in order to ensure passenger safety and the safety of the passenger transport operation. Additional criteria for quay walls with cranes, such as container cranes, are discussed in this section. These criteria supplement those presented earlier in Sections 4.1 through 4.4.

(1) Seismic response of cranes

A crane consists of an upper structure for handling cargo, and a supporting structure for holding in place and transporting the upper structure, as shown in Fig. 4.12. The crane is generally made of a steel frame. The supporting structure is either of the rigid frame type or hinged leg type, with one hinge at the location A as shown in Fig. 4.12. The supporting structure rests on rails through the wheels.

A crane at rest is fixed to rails or to a quay wall with clamps or anchors, whose strength provides the upper limit for the crane resistance against external forces. However, a crane in operation is not supported by clamps or anchors, and the lateral resistance of the crane against external forces is from friction and from the wheel flanges.

Typical failure modes during earthquakes are derailment of wheels, detachment or pull-out of vehicle, rupture of clamps and anchors, buckling, and overturning. As shown in Fig. 4.13(a), widening of a span between the legs due to the deformation of the quay wall results in derailment or buckling of the legs.

Table 4.4. Proposed damage criteria for cellular quay walls.

Level of damage			Degree I	Degree II	Degree III	Degree IV
Residual displacements	Cellular wall	Normalized residual horizontal displacement (d/H)*	Less than 1.5%**	1.5~5%	5~10%	Larger than 10%
		Residual tilting towards the sea	Less than 3°	3~5°	5~8°	Larger than 8°
	Apron	Differential settlement on apron	Less than 0.03~0.1 m	N/A	N/A	N/A
		Differential settlement between apron and non-apron areas	Less than 0.03~0.7 m	N/A	N/A	N/A
		Residual tilting towards the sea	Less than 2~3°	N/A	N/A	N/A
Peak response stresses/strains	Cell		Elastic	Elastic	Plastic (less than the strain limit for cell)	Plastic (beyond the strain limit for cell)
	Cell joint		Elastic	Plastic (less than the strain limit for cell joint)	Plastic (beyond the strain limit for cell joint)	Plastic (beyond the strain limit for cell joint)

*d: residual horizontal displacement at the top of the wall; H: height of deck from mudline.
**Alternative criterion is proposed with respect to differential horizontal displacement less than 30 cm.

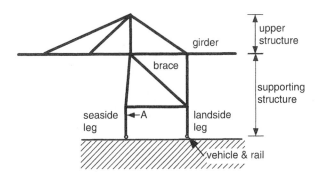

Fig. 4.12. Schematic figure
of gantry crane.

Conversely, as shown in Fig. 4.13(b), narrowing of a leg span can also occur due to the rocking response of the crane. This is due to alternating action of the horizontal component of resisting forces from the quay wall during rocking type response involving uplifting of one of the legs. Derailment and detachment of the wheel can also occur due to rocking. As shown in Fig. 4.13(c), when differential settlement occurs on a quay wall below the crane, tilting or over-turning of the crane may occur. If the crane has one-hinge type legs, the derailment can result in tilting and overturning of the crane as shown in Fig. 4.13(d). Though a clamp or anchor will provide more resistance to motion under the action of external forces, the internal stresses induced in the crane framework will become larger in comparison to the case with no clamp, thus allowing for rocking responses.

Crane rails are often directly supported either by a portion of a retaining wall or by the deck of a pile-supported wharf. When the width of the gravity wall is small, or the quay wall is a sheet pile or cellular type, a separate foundation that often consists of piles, is provided to support the rails. In order to achieve desirable seismic performance of quay walls with cranes, special consideration is required for the rail foundation such as providing a dedicated and cross-tied upper structure to support the rails.

(2) Parameters for specifying damage criteria for quay walls with cranes

Seismic performance of quay walls with cranes is specified based on the serviceability requirements for the crane and the possible structural damage. Serviceability of the crane is specified regarding the function of the upper structure (i.e. cargo handling) and that of the supporting structure (i.e. transportation and support of the upper structure). Structural damage regarding the crane is specified not only in terms of displacements, derailment, tilting, and stress within the crane framework but also as displacements and stresses within the rails and the foundation. With regard to serviceability of the crane, maintaining the power supply should also be considered.

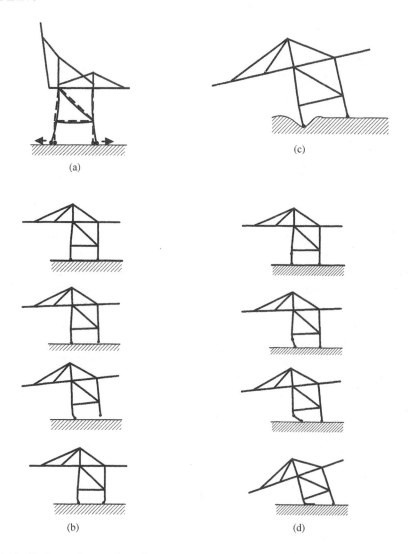

Fig. 4.13. Deformation modes of gantry crane.
    (a) Widening of span between the legs.
    (b) Narrowing span between the legs due to rocking motion.
    (c) Tilting of crane due to differential settlement of foundation.
    (d) Overturning of one-hinged leg crane due to rocking/sliding.

Parameters for specifying damage criteria are illustrated in Fig. 4.14 and are as follows. For rails and the foundation, the parameters include rail span, rail misalignment (including discontinuity), differential settlement (differential levels between the rails; differential levels, inclination and vertical discontinuity along a rail), and displacements and stresses of the rail foundation. For a crane, the parameters include derailment of wheels, detachment of vehicle, rupture of clamps and anchors, displacements of the crane (derailment, tilting, and

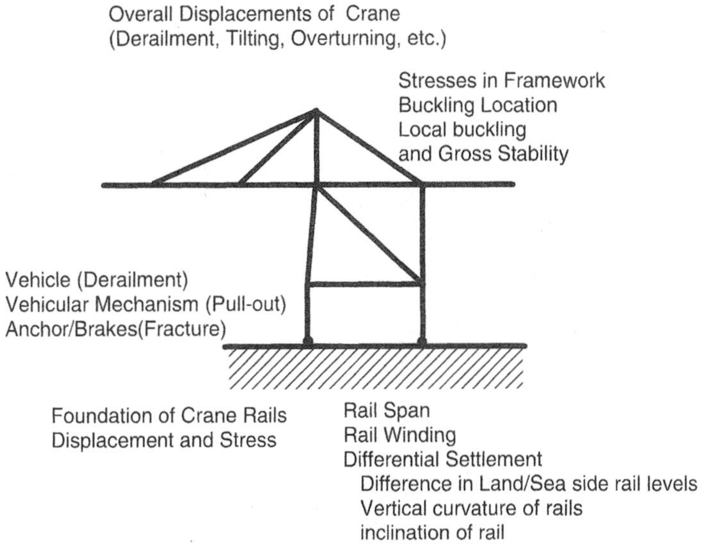

Overall Displacements of Crane
(Derailment, Tilting, Overturning, etc.)

Stresses in Framework
Buckling Location
Local buckling
and Gross Stability

Vehicle (Derailment)
Vehicular Mechanism (Pull-out)
Anchor/Brakes(Fracture)

Foundation of Crane Rails   Rail Span
Displacement and Stress   Rail Winding
                                 Differential Settlement
                                   Difference in Land/Sea side rail levels
                                   Vertical curvature of rails
                                   inclination of rail

Fig. 4.14. Parameters for specifying damage criteria for crane.

overturning) and stresses in the frame (stress levels, location of buckling, possibility of collapse).

Damage criteria should be established by choosing and specifying appropriate parameters from those mentioned. More specifics are given in the next subsection.

(3) Damage criteria for quay walls with cranes

As mentioned earlier, damage criteria are established considering both serviceability and structural damage not only to the quay wall, but also to the crane and rail foundation. The damage criteria for the quay wall required to limit structural damage to the crane are discussed below. Criteria for the quay wall with respect to the rail foundation should be established separately by referring to the damage criteria for both the crane and the quay wall.

Provided the conditions mentioned at the beginning of this chapter are applicable, the damage criteria for a crane may be established by referring to Table 4.5. The most restrictive conditions among displacements and stresses should dictate the damage criteria.

The damage criteria for quay walls with cranes may be established by referring to the following:
1) Damage Degree I of the quay wall should dictate displacements that keep the crane supporting structure within the elastic limit. For some cases, this limit may correspond to a widening of the leg span as large as one meter for a rigid frame crane with a rail span of 30 m. There is no clearly defined limit to

Table 4.5. Proposed damage criteria for cranes.

Level of damage		Degree I	Degree II	Degree III	Degree IV
Displacements		Without derailment	With derailment	Without overturning	Overturning
Peak response stresses/strains	Upper structure	Elastic	Elastic	Plastic (less than the ductility factor/strain limit for upper structure)	Plastic (beyond the ductility factor/strain limit for upper structure)
	Main framework of supporting structure	Elastic	Plastic (less than the ductility factor/strain limit for main framework)	Without collapse	Collapse
	Toe	Elastic	Damage to toe (including pull-out of vehicle, fracture of anchor/brakes)	Damage to toe (including pull-out of vehicle, fracture of anchor/brakes)	Damage to toe (including pull-out of vehicle, fracture of anchor/brakes)

damage Degree I for a one-hinge leg crane because this type of crane can easily absorb the change in the leg span.

2) Damage Degree II of the quay wall should dictate displacements that may allow the yielding of the crane supporting structure, but keep the stress state less than the ductility factor allowable for the main framework of the supporting structure.

3) Damage Degree III of the quay wall should maintain the level difference between the rails and the differential settlements and tilting of the apron within the overturning limit for the crane. A hypothetical possibility is to introduce a measure to increase the surface friction of the apron, and thereby increase the overturning limit of a derailed one-hinge leg crane, or a crane with a plastic hinge at a leg. The required friction coefficient is given by $\mu_b \geq d_l/H_c$, where μ_b: coefficient of friction; d_l: expansion of leg span (m), H_c: height of the hinge from the apron (m) (refer to Fig. 4.15).

4) Damage Degree IV of the quay wall should be the state beyond the upper limit of the damage Degree III.

A set of tolerances for ordinary maintenance of cranes is shown in Table 4.6 (Japan Cargo Handling Mechanization Association, 1996). This set of tolerances

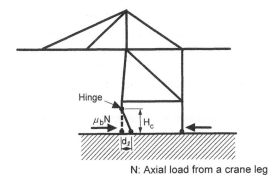

N: Axial load from a crane leg

Fig. 4.15. Parameters for evaluation overturning of crane.

Table 4.6. Tolerance for ordinary maintenance of container cranes.

Parameters	Tolerance
Rail Span L_{span} ($L_{span} < 25$ m)	± 10 mm
$\qquad$ (25 m $\leq L_{span} \leq$ 40 m)	± 15 mm
Level difference between sea and land side rails	$L_{span}/1000$
Curving in vertical direction	5 mm per 10 m
Curving in horizontal direction	5 mm per 10 m
Inclination	1/500
Rail joint $\quad$ Differential displacements (vertical and horizontal)	1 mm
$\qquad$ Gap	5 mm*

*Relative to original layout.

could be used as one of the references to establish more restrictive damage criteria than discussed earlier.

4.6   BREAKWATERS

A breakwater is usually made of a rubble mound, a massive structure such as a caisson, or a combination of both placed on a seabed. Stability against a horizontal external load is maintained by shear resistance of rubble, friction at the bottom of the caisson, and with associated resistance to overturning and bearing capacity failure. Typical failure modes expected during earthquakes are shown in Fig. 4.16.

Breakwaters are generally designed to limit wave penetration and wave overtopping during specific design storms, and at the same time designed to resist the related wave actions. It is unlikely that a major earthquake will occur simultaneously with the design sea state because the two events are typically not related. Consequently, design storm wave action and an earthquake can be treated as two independent load situations. Only wave actions from a moderate sea state should be considered together with the design earthquakes. Decision on this sea state has to be made based on the site-specific long-term statistics of the storm.

Selection of the appropriate design criteria depends on the functions of the breakwater and the type of earthquake-induced failure modes. However, for all breakwaters, the main criterion is the allowable settlement of the crest level because it determines the amount of overtopping and wave transmission. For breakwaters carrying roads and installations, additional criteria for allowable differential settlement, tilting and displacement of superstructures and caissons are needed.

Shaking of the breakwater may cause breakage of concrete armour units. Criteria have been proposed with regard to maximum breakage in terms of number of broken units that may occur while the breakwater remains serviceable (e.g. Zwamborn and Phelp, 1995). The same criteria may be adopted for the earthquake-related damage.

Since the damage criteria for breakwaters have not been fully developed at present, the performance grade is being described in order to indicate the relative degree of allowable damage. The level of damage to various kinds of breakwaters, according to the primary and secondary functions of breakwaters, is as follows:
- reduce wave penetration in basins (Grade C);
- recreational (access for people) (Grade C; but can be Grade A or B depending on the level of acceptable human life safety);
- provision of berthing on the port side of the breakwater, and related access roads (Grade B) (see Fig. 4.17);
- provision of cargo handling facilities on the breakwater, including conveyor belts (Grade B), and pipelines for oil and liquid gas (Grade A or S, depending on the level of threat of explosion).

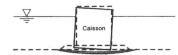

- Crest lowering due to settlement of subsoil
- Differential settlement and tilt of caissons
- Possible damage to caisson shear keys

- Slip failures and subsequent crest lowering due to liquefaction of subsoil
- Differential settlement and tilt of caissons
- Damage to caisson shear keys if they exist

(a)

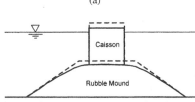

- Crest lowering due to shake down of rubble foundation
- Differential settlement of caissons

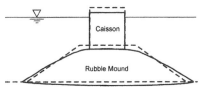

- Crest lowering and lateral spreading due to settlement/liquefaction of subsoil
- Differential settlement of caissons

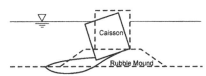

- Slip failures and subsequent crest lowering due to liquefaction of subsoil
- Differential tilt and settlement of caissons
- Damage to caisson shear keys if they exist

(b)

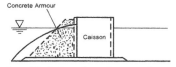

- Breakage of slender types of concrete armour units
- Subsequent increased overtopping and risk of displacement of caisson

(c)

- Crest lowering due to shake down of rubble material
- Differential settlement of superstructure elements

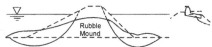

- Crest lowering and lateral spreading due to settlement/liquefaction of subsoil
- Differential settlement of superstructure elements

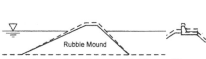

- Failures due to liquefaction of subsoil
- Subsequent lowering of crest
- Possible tilt and displacement of superstructure elements

(d)

Fig. 4.16. Deformation/failure modes of breakwater.
   (a) Caisson resting on sea bed.
   (b) 'Vertically composite' caisson breakwater.
   (c) 'Horizontally composite' caisson breakwater.
   (d) Rubble mound breakwater.

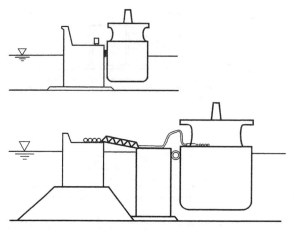

Fig. 4.17. Examples of berthing
behind breakwater.

CHAPTER 5

Seismic Analysis

Seismic analysis of port structures is accomplished in three steps that include assessment of the regional seismicity, the geotechnical hazards, and soil-structure interaction analysis. The first step is to define the earthquake motions at the bedrock (see Fig. 2.1 for the relative location of the bedrock with respect to the seismic source, ground surface and port structures). This is typically accomplished by seismic hazard analysis based on geologic, tectonic and historical seismicity data available for the region of interest. The second step involves the following two interrelated aspects of dynamic soil response: (1) an evaluation of local site effects for obtaining the earthquake motions at or near the ground surface; and (2) an assessment of the liquefaction resistance of the near surface sandy soils and the associated potential for ground failures. Once the ground motion and geotechnical parameters have been established, then seismic analysis of the port structure(s) can proceed.

As in all engineering disciplines, reasonable judgement is required in specifying appropriate methods of analysis and design, as well as in the interpretation of the results of the analysis procedures. This is particularly important in seismic design, given the multidisciplinary input that is required for these evaluations, and the influence of this input on the final design recommendations. This chapter provides general recommendations for the type and level of analysis required for various port structures.

5.1 TYPES OF ANALYSIS

The objective of analysis in performance-based design is to evaluate the seismic response of the port structure with respect to allowable limits (e.g. displacement, stress, ductility/strain). Higher capability in analysis is generally required for a higher performance grade facility. The selected analysis methods should reflect the analytical capability required in the seismic performance evaluation.

A variety of analysis methods are available for evaluating the local site effects, liquefaction potential and the seismic response of port structures. These analysis methods are broadly categorized based on a level of sophistication and capability as follows:

1) Simplified analysis: Appropriate for evaluating approximate threshold limit for displacements and/or elastic response limit and an order-of-magnitude estimate for permanent displacements due to seismic loading.

2) Simplified dynamic analysis: Possible to evaluate extent of displacement/stress/ductility/strain based on assumed failure modes.

3) Dynamic analysis: Possible to evaluate both failure modes and the extent of the displacement/stress/ductility/strain.

Table 5.1 shows the type of analysis that may be most appropriate for each performance grade. The principle applied here is that the structures of higher performance grade should be evaluated using more sophisticated methods. As shown in the index in Table 5.1, less sophisticated methods may be allowed for preliminary design, screening purposes, or response analysis for low levels of excitation.

In the present state of practice, it is desirable to confirm the applicability of methods for analysis of port structures by using suitable case histories or model test results before accomplishing the seismic performance evaluation.

5.2   SITE RESPONSE/LIQUEFACTION ANALYSIS

Site response analysis and liquefaction potential assessments are typically accomplished using the methods outlined in Table 5.2. Useful references that highlight the applicability of the various modeling procedures, have been prepared by several investigators (e.g., Finn, 1988; Idriss, 1991).

(1) Site response analysis

In simplified analysis, local site effects are evaluated based on the thickness of the deposits and the average stiffness to a specified depth (generally 30 m), or over the entire deposit above the bedrock. This information is then used to establish the site classification, leading to the use of specified site amplification factors or site dependent response spectra. This type of procedure is common in codes and standards.

In simplified dynamic analysis, local site effects are evaluated numerically with models such as common equivalent linear, total stress formulations. Soil layers are idealized as horizontal layers of infinite lateral extent (i.e. one-dimensional (1D)). These methods are used to generate time histories of acceleration, shear stress, and shear strain at specified locations in the soil profile.

In both of these categories of analyses, the computed ground surface earthquake motion parameters are used as input for subsequent simplified structural analysis as discussed in Section 5.3.

Table 5.1. Types of analysis related to performance grades.

Type of analysis	Performance grade			
	Grade C	Grade B	Grade A	Grade S
Simplified analysis: Appropriate for evaluating approximate threshold level and/or elastic limit and order-of-magnitude displacements	■ (black)	□ (gray)	□ (gray)	□ (gray)
Simplified dynamic analysis: Of broader scope and more reliable. Possible to evaluate extent of displacement/stress/ductility/strain based on assumed failure modes		■ (black)	□ (gray)	□ (gray)
Dynamic analysis: Most sophisticated. Possible to evaluate both failure modes and extent of displacement/stress/ductility/strain			■ (black)	■ (black)

Index:

■ Standard/final design

□ Preliminary design or low level of excitations

Table 5.2. Methods for site response analysis and liquefaction potential assessment.**

Type of analysis	Method	Simplified analysis	Simplified dynamic analysis	Dynamic analysis
Site response analysis	Method	Site category	1D total stress (equivalent linear) analysis	1D effective stress (non-linear) analysis, or 1D total stress (equivalent linear) analysis*
	Input parameters	Peak bedrock acceleration CPT q_c/SPT N-values Stratigraphy	Time history of bedrock earthquake motion V_s, $G/G_0-\gamma$, $D-\gamma$ curves	For effective stress analysis: Time history of bedrock earthquake motion Undrained cyclic properties For total stress analysis: the same as those for simplified dynamic analysis
	Output of analysis	Peak ground surface motion (PGA, PGV) Design response spectra	Time history of earthquake motion at ground surface and within the subsoil Computed response spectra at ground surface	Time history of earthquake motion at ground surface and within the subsoil

Liquefaction potential assessment	Field correlation (SPT/CPT/V_s)	Laboratory cyclic tests and/or Field correlation (SPT/CPT/V_s) + 1D total stress analysis	Laboratory cyclic tests and/or Field correlation (SPT/CPT/V_s) + 1D effective stress analysis or 1D total stress analysis*
Method	Field correlation (SPT/CPT/V_s)	Laboratory cyclic tests and/or Field correlation (SPT/CPT/V_s) + 1D total stress analysis	Laboratory cyclic tests and/or Field correlation (SPT/CPT/V_s) + 1D effective stress analysis or 1D total stress analysis*
Input parameters	Peak ground surface acceleration (PGA) CPT q_c/SPT N-values/V_s Stratigraphy	Time history of earthquake motion at ground surface, or time histories of shear stresses in the subsoil Liquefaction resistance, (τ/σ'_{v0}) or γ_{cyc}, based on laboratory cyclic tests and/or SPT/CPT/V_s	For effective stress analysis: Time history of bedrock earthquake motion Undrained cyclic properties based on laboratory cyclic tests and/or SPT/CPT/V_s For total stress analysis: the same as those for simplified dynamic anaysis
Output of analysis	Liquefaction potential (F_L)	Liquefaction potential (F_L) Excess pore water pressure ratio (u/σ'_{v0})	Excess pore water pressure ratio (u/σ'_{v0}) Depth and time at the onset of liquefaction

*If the bottom boundary of the domain in a soil-structure interaction analysis differs from the bedrock (i.e. if the bedrock level is too deep for soil-structure interaction analysis), local site effects below the bottom boundary of the soil-structure analysis domain may be evaluated based on 1D effective stress (non-linear) or equivalent-linear (total stress) analysis.

**CPT: cone penetration test, SPT: standard penetration test, PGA, PGV: peak ground acceleration and velocity, V_s: shear wave velocity, G/G_0: secant shear modulus (G) over shear modulus at small strain level (G_0), D: equivalent damping factor, γ: shear strain amplitude, F_L: factor of safety against liquefaction, u/σ'_{v0}: excess pore water pressure (u) over initial effective vertical stress (σ'_{v0}), τ/σ'_{v0}: shear stress ratio, γ_{cyc}: cyclic shear strain amplitude.

Dynamic analyses of soil-structure interaction (SSI) attempt to account for the combined, or coupled, response of the structure and the foundation soils. Unlike the simplified procedures, wherein the response of the structure is evaluated using the free-field soil response as input, SSI analyses incorporate the behaviour of both soil and structure in a single model. The finite element methods (FEM) or finite difference methods (FDM) are commonly used for advanced SSI analyses. In this type of analysis, local site effects are often not evaluated independently but are evaluated as a part of the soil-structure interaction analysis of port structures. If the bottom boundary of the analysis domain for soil-structure interaction analysis differs from the bedrock (i.e. when the bedrock is too deep for soil-structure interaction analysis), the local site effects below the bottom boundary of the soil-structure interaction analysis domain may be evaluated based on 1D non-linear (effective stress) or 1D equivalent linear (total stress) analysis.

In both of the simplified dynamic analysis and dynamic analysis, adequate choice of input earthquake motion at the bedrock should be important. For details, refer to Technical Commentaries 3 and 7.

(2) Liquefaction potential assessment

In simplified analysis, the liquefaction potential of sandy soils are evaluated based on standard penetration tests (SPT) or cone penetration tests (CPT) through empirical criteria. In simplified dynamic analysis, liquefaction potential is evaluated based on a comparison of computed shear stresses during the design earthquake and the results of cyclic laboratory tests, and/or based on SPT/CPT data. The liquefaction potential evaluated through these categories of analysis are used later as input for subsequent simplified deformation analysis of structures at liquefiable sites as discussed in Section 5.3.

In dynamic analysis, liquefaction potential is often not evaluated independently, but is evaluated as part of the soil-structure interaction analysis of port structures. Details are discussed in Section 5.3.

5.3   ANALYSIS OF PORT STRUCTURES

The method of analysis for a port structure depends on structural type. The appropriate method may be chosen by referring to Table 5.3. A list of readily available methods is shown in Table 5.4. As indicated in Table 5.4, methods for analysis of port structures may be broadly classified into those applicable to retaining/earth structures, including quay walls, dikes/slopes and breakwaters, or those applicable to open pile/frame structures, including pile/deck system of pile-supported wharves and cranes.

Table 5.3. Analysis methods for port structures.

Type of analysis	Simplified analysis	Simplified dynamic analysis	Dynamic analysis	
			Structural modelling	Geotechnical modelling
Gravity quay wall	Empirical/pseudo-static methods with/without soil liquefaction	Newmark type analysis Simplified chart based on parametric studies (see Table 5.4)	FEM/FDM* Linear or Non-linear analysis	FEM/FDM* Linear (Equivalent linear) or Non-linear analysis
Sheet pile quay wall				
Pile-supported wharf	Response spectrum method	Pushover and response spectrum methods	2D/3D**	2D/3D**
Cellular quay	Pseudo-static analysis	Newmark type analysis		
Crane	Response spectrum method	Pushover and response spectrum methods		
Breakwater	Pseudo-static analysis	Newmark type analysis		

*FEM/FDM: Finite element method/finite difference method.
**2D/3D: Two/three-dimensional analysis.

Table 5.4. List of analysis methods and references.

Type of analysis		Simplified analysis		Simplified dynamic analysis		Dynamic analysis
Gravity quay wall Sheet pile quay wall Cellular quay wall Dike/slope/ retaining wall of pile-supported wharf Breakwater	**Analysis Method**	Pseudo-static method for evaluating threshold level	Empirical/pseudo-static methods for evaluating order-of-magnitude displacement	Newmark type analysis for evaluating sliding displacement	Simplified chart based on parametric studies for evaluating displacement	FEM/FDM 2D Linear geotechnical modeling (FLUSH) Non-linear geotechnical modeling (FLIP)
	Relevant Technical Commentaries	T7.1.1, T7.1.4, T8.1.2, T8.2.2, T8.2.3	T7.1.2	T7.2.1(1), T7.2.2(1), T7.2.4	T7.2.1(2)(3), T7.2.2(2), T8.1.3, T8.1.4, T8.2.4	T7.2.1(2)(3), T7.3.1, T7.3.2(1)(2)(4), T7.3.3, T8.1.5, T8.3.2
	References	Ebeling & Morrison (1992) Tsinker (1997) Ministry of Transport, Japan (1999)	Uwabe (1983) (without liquefaction) Gazetas et al. (1990) (for sheet pile wall) Iai (1998a) (with liquefaction)	Newmark (1965) (for dikes) Franklin & Chang (1977) (for dikes) Richards & Elms (1979) (for gravity wall) Whitman & Liao (1984) (for gravity wall) Towhata & Islam (1987) (for sheet pile wall) Steedman (1998) (for gravity and sheet pile walls)	Dickenson & Yang (1998) (for gravity wall, liquefaction remediation measures) Iai et al. (1999) (for gravity wall, with/without liquefaction) McCullough & Dickenson (1998) (for sheet pile wall, liquefaction remediation measures)	Lysmer et al. (1975) (equivalent linear/ total stress) Iai (1998b) (non-linear/effective stress)

	Analysis Method	Response spectrum method for evaluating elastic limit response	Pushover and response spectrum methods (SDOF/MDOF)	FEM/FDM 2D Linear geotechnical modeling (FLUSH) Non-linear geotechnical modeling (FLIP)
Pile and deck of pile-supported wharf Crane	Relevant Technical Commentaries	T7.1.3, T8.4.2	T7.2.3, T8.4.3	T7.3.1, T7.3.2(3), T8.5.2
	References	Ferritto (1997a) Werner (1998) Yokota et al. (1999) Ferritto et al. (1999)	Ferritto (1997a) Werner (1998) Yokota et al. (1999) Ferritto et al. (1999)	Lysmer et al. (1975) (equivalent linear/total stress) Iai (1998b) (non-linear/effective stress) Ferritto et al. (1999) (non-linear)

(1) Methods for analysis of retaining/earth structures

a) Simplified analysis
Simplified analysis of retaining/earth structures is based on the conventional force-balance approach, sometimes combined with statistical analyses of case history data. The methods in this category are often those adopted in conventional seismic design codes and standards. In simplified analysis, retaining/earth structures can be idealized as rigid blocks of soil and structural masses. The rigid block analysis is typically applied for gravity, sheet pile and cellular quay walls, dike/slope/retaining walls for pile-supported wharves and breakwaters.

Effects of earthquake motions in simplified analysis are represented by a peak ground acceleration or an equivalent seismic coefficient for use in conventional pseudo-static design procedures. These parameters are obtained from the simplified analysis of local site effects discussed in the previous section. A capacity to resist the seismic force is evaluated based on structural and geotechnical conditions, often in terms of a threshold acceleration or a threshold seismic coefficient, beyond which the rigid blocks of soil and structural masses begin to move. When soil liquefaction is an issue, the geometric extent of liquefaction must also be considered in the analysis.

Results of the simplified analysis are appropriate for evaluating the approximate threshold level of damage, which ensures at least the repairable state of structural performance for L1. Whether or not the approximate threshold level ensures the serviceable state of structural performance for L1 depends on the details in evaluating the design parameters for the pseudo-static method. Order-of-magnitude displacement is also obtained by the pseudo-static method combined with statistical analysis of case history data. This, however, is a crude approximation and should be used only for the preliminary design stage or low levels of excitation.

There is a significant difference between the design approach recommended in these guidelines and the conventional design concept. In the conventional design concept, especially when referring to the simplified analysis, an equivalent seismic coefficient is used as an input parameter representing adequately the ensemble of ground motions, and a factor of safety is applied to determine the dimensions of the structure. In the proposed approach, the design is based on the seismic performance of the structure evaluated appropriately through response analysis for a variety of input earthquake motions. The ensemble of the seismic responses, rather than the ensemble of input motions, is used as a basis for accomplishing the design in the proposed guidelines. For each response analysis, input parameters most appropriate are those well defined in terms of applied mechanics, such as a peak ground acceleration for the simplified analysis and/or an equivalent parameter clearly defined in terms of peak ground acceleration. Consequently, no factor of safety should be applied to input data used in seismic analysis for evaluating the threshold level of the structure. This important distinction between conventional design and performance-based design should also be borne in mind when interpreting the design guidelines included in the various seismic codes.

b) Simplified dynamic analysis
Simplified dynamic analysis is similar to simplified analysis, idealizing a structure by a sliding rigid block. In simplified dynamic analysis, displacement of the sliding block is computed by integrating the acceleration time history that exceeds the threshold limit for sliding over the duration until the block ceases sliding.

Effects of earthquake motions are generally represented by a set of time histories of earthquake motion at the base of a structure. The time histories of earthquake motion are obtained from the simplified dynamic analysis of local site effects discussed in the previous section. In the sliding block analysis, structural and geotechnical conditions are represented by a threshold acceleration for sliding. A set of empirical equations obtained from a statistical summary of sliding block analyses is available (see references in Table 5.4). In these equations, peak ground acceleration and velocity are used to represent the effect of earthquake motion.

In more sophisticated analysis, structural and geotechnical conditions are idealized through a series of parametric studies based on non-linear FEM/FDM analyses of soil-structure systems, and the results are compiled as simplified charts for use in evaluating approximate displacements.

c) Dynamic analysis
As mentioned earlier, dynamic analysis is based on soil-structure interaction, generally using FEM or FDM methods. In this category of analysis, effects of earthquake motions are represented by a set of time histories of earthquake motion at the base of the analysis domain chosen for the soil-structure system. A structure is idealized either as linear or as non-linear, depending on the level of earthquake motion relative to the elastic limit of the structure. Soil is idealized either by equivalent linear or by an effective stress model, depending on the expected strain level in the soil deposit during the design earthquake.

Fairly comprehensive results are obtained from soil-structure interaction analysis, including failure modes of the soil-structure system and the extent of the displacement/stress/strain states. Since this category of analysis is often sensitive to a number of factors, it is especially desirable to confirm the applicability by using a suitable case history or a suitable model test result.

(2) Analysis methods for open pile/frame structures

a) Simplified analysis
Simplified analysis of open pile/frame structures is typically done by idealizing the pile/deck system of pile-supported wharves or the frame of cranes by a single degree of freedom (SDOF) or multi-degree of freedom (MDOF) system. In this

analysis, earthquake motions are generally represented by the response spectrum. Structural and geotechnical conditions are represented by a resonant frequency and damping factor of the deck-pile system and/or the cranes. A ductility factor may also be introduced. The movement of the dike/slope is generally assumed to be negligible. Results of the SDOF/MDOF analysis are useful to evaluate approximate limit state response of a pile-deck system or a crane, which ensures at least the repairable state of structural performance for L1.

b) Simplified dynamic analysis

In simplified dynamic analysis of open pile/frame structures, the SDOF or MDOF analysis of pile-deck structure or cranes is combined with pushover analysis for evaluating the ductility factor/strain limit. The movement of the dike/slope is often assumed to be negligible but sometimes estimated by a sliding block type analysis. Movement of a pile-supported deck could thereby be estimated by summing-up the dike/slope movement and structural deformation. Soil-structure interaction effects are not taken into account, and thus there is a limitation in this analysis. Interaction between the pile-supported wharves and cranes can be taken into account by MDOF analysis. Displacement, ductility factor/strain, and location of yielding or buckling in the structure are generally obtained as a result of the analysis of this category. Failure modes with respect to sliding of retaining walls/dikes/slopes are not evaluated but assumed and there is, thus, another limitation in this type of analysis.

c) Dynamic analysis

Dynamic analysis is based on soil-structure interaction, generally using FEM/FDM methods. Similar comments to those related to the dynamic analysis of earth/retaining structures apply also to the open pile structures and cranes.

See Technical Commentaries 5 and 7 for further details on analysis methods.

5.4   INPUTS AND OUTPUTS OF ANALYSIS

Table 5.5 shows major input parameters required for the various types of analysis. It should be noted that the reliability of the results depends not only on the type of analysis, but also on the reliability of input parameters. It is ideal to use input data based on thorough geotechnical investigations for more sophisticated analysis. However, it may not be practical to demand that only the highest level of geotechnical investigations, including the in-situ soil freezing technique for sand sampling, be used in the design. Some of the parameters for the most sophisticated methods could be evaluated based on the empirical correlation using SPT results, provided that this empirical correlation is calibrated through geotechnical investigations of high level at the site of interest. See Technical Commentary 4 for further details.

Tables 5.6 and 5.7 show major outputs from the analyses. Clearly, more parameters of seismic performance can be evaluated by more sophisticated analysis. These outputs are the final indices that define the level of damage. The performance evaluation can be completed by comparing these outputs to the damage criteria discussed in Chapter 4. Examples of seismic performance evaluation are given in Technical Commentary 8.

The displacements/ductility/strain limits in the proposed damage criteria are intended to serve as general guides which are deemed as acceptable. However, exceeding these values does not mean that performance is unacceptable. Engineering judgement must be applied on a case by case basis in which the amount of deformation/stress strain states and its spatial distribution are examined in terms of performance requirements and consequences for the port facilities.

Ultimately, the economics of initial investment, risk of damage and the consequences of failure of the operation must be considered. If the seismic responses evaluated are less than those cited in the criteria, good performance can be expected. If the seismic responses are greater than those cited in the criteria, it is up to the engineer to demonstrate that the performance objectives can be achieved.

Table 5.5. Major input parameters for analysis.

Type of analysis		Simplified analysis	Simplified dynamic analysis		Dynamic analysis
Gravity quay wall	Method	Pseudo-static/empirical methods	Newmark type method	Simplified chart based on parametric studies	FEM/FDM
	Design parameters	k_e: equivalent seismic coefficient k_t: threshold seismic coefficient (Geometrical extent of liquefiable soils relative to the position and dimensions of a wall for a liquefiable site)	Empirical equations: a_{max}: peak acceleration v_{max}: peak velocity Time history analysis: time histories of earthquake motions a_t: threshold acceleration	a_{max}: peak acceleration at the bedrock Cross section of wall Index properties of soil including SPT N-values	Time histories of earthquake motions at the bottom boundary of analysis domain Cross section of wall For equivalent linear geotechnical analysis: $G/G_0-\gamma$ & $D-\gamma$ curves For non-linear geotechnical analysis:
	Input parameters	Results of site response analysis, including a_{max}, and liquefaction potential assessment Cross section of wall Geotechnical parameters, including c, ϕ: cohesion and internal friction angle of soils; μ_b, δ: friction angles at bottom and back face of wall; ground water level			Undrained cyclic properties and G, K: shear and bulk modulus, in addition to the geotechnical parameters for pseudo-static and simplified analyses

		Pseudo-static/empirical methods	Newmark type method	Simplified chart based on parametric studies	FEM/FDM
Sheet pile quay wall	Method	Pseudo-static/empirical methods	Newmark type method	Simplified chart based on parametric studies	FEM/FDM
	Design parameters	The same as those for gravity quay wall.	The same as those for gravity quay wall	The same as those for gravity quay wall	In addition to those for gravity quay wall, dimensions and material parameters for sheet pile, tie-rod and anchor For linear structural analysis: elastic material parameters, including E, I, A, yield stress For non-linear structural analysis: yield stress, M–ϕ curve
	Input parameters	In addition to those for gravity quay wall structural parameters for sheet pile, tie-rod and anchor, including E, I, A, and allowable stress	In addition to those for pseudo-static method, material parameters defining limit states of structures, including yield and ultimate stress states	In addition to those for pseudo-static method, material parameters defining limit states of structures, including yield and ultimate stress states	
Pile-supported wharf	Method	Response spectrum method	Pushover and response spectrum methods		FEM/FDM
	Design parameters	Pile-deck structure: design response spectra resistance at elastic limit or allowable stress level $$\beta = \sqrt[4]{\frac{k_{h\text{-sub}} D_p}{4EI}}$$ or p–y curve Dike/retaining wall: the same as those for gravity quay wall	In addition to those for pseudo-static method, material parameters defining limit states of structures, including yield and ultimate stress states of pile-deck structure		In addition to those for gravity quay wall, dimensions and material parameters for pile-deck structure: For linear structural analysis: elastic material parameters, including E, I, A, yield stress For non-linear structural analysis: yield stress, M–ϕ curve

Table 5.5. Continued.

Type of analysis		Simplified analysis	Simplified dynamic analysis	Dynamic analysis
	Input parameters	Results of site response analysis and liquefaction potential assessment Pile-deck structure: dimensions & material parameters for piles and deck, including E, I, A, D_p, allowable stress $k_{h\text{-sub}}$: coefficient of lateral subgrade reaction, or p–y curve Dike/retaining wall: relevant parameters for gravity/sheet pile quay walls	The same as above	
Cellular quay wall	Method	Pseudo-static method	Newmark type analysis	FEM/ FDM
	Design parameters	For overall stability: the same as those for gravity/sheet pile quay wall For structural performance: elastic limit/allowable stress	In addition to those for pseudo-static method, material parameters defining limit states of structures, including yield and ultimate stress states of cell structure	In addition to those for gravity quay wall, dimensions and material parameters for cell For linear structural analysis: elastic material parameters, including E, I, A, yield stress For non-linear structural analysis: yield stress, M–ϕ curve

	Input parameters	Results of site response analysis and liquefaction potential assessment. For overall stability: the same as those for gravity/sheet pile quay wall. For structural performance: dimensions and material parameters for cell	The same as above	The same as above
Crane	Method	Response spectrum method	Pushover and response spectrum methods	FEM or framework analysis
	Design parameters	Design response spectra. Resistance at elastic limit or allowable stress level	In addition to those for pseudo-static method, material parameters defining limit states of structures, including yield and ultimate stress states	Dimensions and material parameters of crane
	Input parameters	Results of site response analysis and liquefaction potential assessment. Dimensions and material parameters	The same as above	For structural interaction analysis of quay wall and crane system: The same as those for relevant type of quay wall analysis. For crane response only: Time history of input earthquake motions to the crane
Break-water	Method	Pseudo-static method	Newmark type method	FEM/FDM
	Design parameters	The same as those for gravity quay wall	The same as those for gravity quay wall	The same as those for gravity quay wall
	Input parameters	The same as those for gravity quay wall	The same as those for gravity quay wall	

Table 5.6. Analysis output.

Analysis performance	Analysis type	Simplified analysis	Simplified dynamic analysis	Dynamic analysis
	Performance grade	Grade C L1: Repairable L2: Allow collapse but without adverse effects on surroundings	Grade B L1: Serviceable L2: Near collapse	Grade A/Grade S L1: Serviceable L2: Repairable (Grade A)/ Serviceable (Grade S)
Gravity quay wall		Threshold limit Order of magnitude displacement	Wall displacement	Response/failure modes Peak and residual displacements
Sheet pile quay wall		Threshold limit Order of magnitude displacement	Wall displacement Stress/ductility	Response/failure modes Peak and residual displacements, stress/ductility
Pile-supported wharf		Pile-supported deck: threshold limit Dike/retaining wall: threshold limit order of magnitude displacement	Pile-supported deck: displacement, stress/ductility Dike/retaining wall: displacement, stress/ductility	Response/failure modes Peak and residual displacements, stress/ductility
Cellular quay wall		Threshold limit Order of magnitude displacement	Wall displacement Stress/ductility	Response/failure modes Peak and residual displacements, stress/ductility
Crane		Threshold limit	Displacement Stress/ductility	Response/failure modes Peak and residual displacements, stress/ductility
Break water		Threshold limit	Displacement	Response/failure modes Peak and residual displacements

Table 5.7. Outputs from dynamic analysis.

Structure and geotechnical modelling		Structure modelling	
		Linear	Non-linear
Geotechnical modelling	Linear (Equivalent linear)	Peak response Displacement/stresses	Failure mode of structure Peak and residual displacement/ductility factor/stresses for structures (assuming there are no effects from residual displacement of soils)
	Non-linear	Failure mode due to soil movement Peak and residual displacement/stresses from soils movement (assuming structure remains elastic)	Failure mode of soil-structure systems Peak and residual displacement/ductility factor/stresses including effects from residual displacements of soils

TECHNICAL COMMENTARIES

TC1: Existing Codes and Guidelines

TECHNICAL COMMENTARY 1

Existing Codes and Guidelines

This technical commentary presents a brief review of the earthquake provisions described in the existing codes and guidelines used at the various ports in the world. While most of the existing codes and guidelines for port structures include earthquake provisions, the degree to which specific earthquake-resistant design methods are addressed in these documents varies considerably. The objective of this brief review is to present:
- a list of the primary codes and guidelines for seismic design of port structures around the world;
- the current status of design practice around the world in light of the new design strategy proposed in this book.

T1.1 LIST OF SEISMIC DESIGN CODES AND GUIDELINES FOR PORT STRUCTURES

The existing codes and guidelines reviewed in this technical commentary are as follows:
- Technical standards for Ports and Harbour Facilities in Japan (Ministry of Transport, Japan, 1999);
- Recommendations of the Committee for Waterfront Structures Harbours and Waterways (EAU, Germany, 1996);
- ROM 0.6: Acciones y efectos Sísmicos en las Obras Marítimas y Portuarias (Puertos del Estado, Madrid, Spain, 2000);
- US Navy seismic design guidelines (Ferritto, 1997a, b);
- Seismic Guidelines for Ports, American Society of Civil Engineers – Technical Council on Lifeline Earthquake Engineering, Ports Committee (Werner, 1998);
- California Marine Oil Terminal Standard (California State Lands Commission, USA, under development as of May 2000);
- European Prestandard, Eurocode 8-Design provisions for earthquake resistance of structures (CEN, 1994):
  - ➢ Part 1-1: General rules – Seismic actions and general requirements for structures, ENV-1998-1-1, May 1994;

➤ Part 5: Foundations, retaining structures and geotechnical aspects ENV 1998-5, June 1994;
– New Zealand Standards;
   ➤ NZS 4203 (1992) "General Structural Design Loadings for Buildings"
   ➤ NZS 3101 Part 1 (1995) "The Design of Concrete Structures"
   ➤ NZS 3403 Part 1 (1997) "Steel Structures Standard"
   ➤ NZS 3403 Part 2 (1997) "Commentary to the Steel Structures Standard"
   ➤ Transit New Zealand (TNZ) Bridge Design Manual.

T1.2   REVIEWED ASPECTS OF CODES AND GUIDELINES

In order to identify the current status of design practice relative to the proposed design strategy in this book, existing codes and guidelines for port structures were reviewed with respect to the following aspects:
– Methodology:
   ➤ Is the methodology a single or dual level approach?
   ➤ What is the general statement on the acceptable level of damage?
– Damage criteria:
   ➤ What is the acceptable displacement/deformation for retaining structures in design?
   ➤ What is the acceptable ductility/strain limit for pile-supported structures in design?
– Analytical procedure:
   ➤ What factor is adopted for specifying the effective seismic coefficient as a fraction of peak ground acceleration for retaining structure design?
   ➤ What types of response spectra are used for pile-supported structures in design?

An overview of the existing codes and guidelines reviewed with respect to the aforementioned aspects is shown in Table T1.1. Additional specifics, to backup the descriptions in Table T1.1, are presented below.

(1) Japanese design (Port and Harbour Research Institute, 1997; Ministry of Transport, Japan, 1999).

The dual level approach has been adopted for structures of Special Class of Importance. However, the single level approach is adopted for structures of Class A, B, and C of Importance.

For structures of Special Class of Importance, the performance level is specified as follows:

     For L1: Minor or no damage/little or no loss of serviceability.
     For L2: Minor or little damage/little or short-term loss of serviceability.

For retaining structures of Special Class of Importance, the criteria for structural damage and criteria regarding serviceability are shown in Tables T1.2 and T1.3.

The seismic coefficient for use in retaining structures is defined as follows for Special Class structures.

$$k_h = \frac{a_{max}}{g} \qquad (a_{max} \leq 0.2g)$$
$$k_h = \frac{1}{3}\left(\frac{a_{max}}{g}\right)^{1/3} (a_{max} > 0.2g) \qquad\qquad \text{(T1.1)}$$

For Class B structures (designed with importance factor of 1.0), the code-specified seismic coefficients are about 60% of those given by Eqn. (T1.1).

For a pile-supported wharf with vertical piles, analysis is performed based on a simplified procedure and pushover method. The ductility limits for use in the simplified procedure (see Technical Commentary 7) for L1 earthquake motion are specified as shown in Table T1.4. Pushover analysis is performed for Special Class structures and the strain limits prescribed are:

Level 1 motion: equivalent elastic,
Level 2 motion: $\varepsilon_{max} = 0.44t_p/D_p$ for the embedded portion.

Pile-supported wharves with vertical steel piles are designed using the response spectra shown in Fig. T1.1 for L1 motion. These spectra have been computed based on 2D soil-structure interaction analysis for typical pile-supported wharf cross sections. For L2 motion, time history analysis should be performed and the results should meet the ductility limits for L2 earthquake motion shown in Table T1.5.

Comprehensive guidelines are shown on liquefaction potential assessment and implementation of remedial measures (Port and Harbour Research Institute, 1997).

(2) Spanish design (ROM0.6, 2000)
Dual earthquake levels must be considered in the design, but only Level 2 earthquake motion is considered for low seismic design acceleration.

Analysis procedures specified in ROM ranges from the simpler methods for the lower design acceleration to the more sophisticated methods for the higher design acceleration.

In most cases, liquefaction is one of the main aspects to be considered. In addition, the codes specifies the criteria with respect to:
• Gravity walls, breakwaters: overall stability, post-earthquake displacements.
• Sheet pile walls, piled decks: structural behaviour.

Pile-supported wharves/piers with vertical steel piles are designed using the ductility factors given as follows:

Wharves: Pile cap: $\mu = $ up to 5.0   Embedded portion: $\mu = 2.0$
Piers: Pile cap: $\mu = $ up to 4.0   Embedded portion: $\mu = 1.5$

Table T1.1. Overview of existing codes and guidelines for port structures.

Codes/Guidelines	Methodology	Performance level	Displacement	Ductility/strain limit	Analytical procedure (α factor)	Analytical procedure (Response spectra)	Remarks
	Single or dual level approach	(Acceptable damage level)	Displacement for retaining structures	Ductility/strain limit for pile-supported wharf/pier	The factor α for specifying the effective seismic coefficient as a fraction of a_{max}/g	Response spectra	
Japanese (1997, 1999)	Dual level for Special Class Single level for Class A, B, C	For Special Class: No damage for L1; Minor damage and little or short-term loss of serviceability for L2	See Tables T1.2 & T1.3	Simplified procedure uses displacement ductility factors. Pushover analysis uses strain limit	See Eqn. (T1.1)	L1: See Fig. T1.1 L2: Time history analysis [no response spectra specified]	These codes and guidelines have been specifically established for port structures
Spanish (2000)	Dual level approach: Seismic input in each level depends on the importance and life time of structure	Little or short-term loss of serviceability for L1: No collapse for L2	Not Specified	Ductility factors for pile head and embedded portions. Different for wharf and pier	α = 0.7	See Eqn. (T1.2)	
German (1996)	Single level approach. More detailed analysis is required if damage can endanger humane live	Not specified	Not specified	Not specified	Not specified; α to be taken from national code	Not specified	

		Dual Method L1 75 year L2 500 year Events	Serviceable under L1 and repairable under L2	Soil limits specified for wharf dikes	Ductility limits specified for piles	N/A	Linear or Nonlinear Dynamic Analysis	Time History
U S A	US Navy (1997)	Dual Method L1 75 year L2 500 year Events	Serviceable under L1 and repairable under L2	Soil limits specified for wharf dikes	Ductility limits specified for piles	N/A	Linear or Nonlinear Dynamic Analysis	
	ASCE-TCLEE (1998)	Dual level approach: exposure time and ground motion are based on the importance and design life	Critical structures: Serviceable under L1 and repairable under L2. Other structures: Repairable for L1	Site-specific based on serviceability and effect on adjacent structures	Ductility limits specified for piles	See Eqn. (T1.1)	a) response spectrum methods b) time history analysis c) pushover analysis	
	California Marine Oil Terminals (2000)	Dual Method L1 75 year L2 500 year Events	Serviceable under L1 and repairable under L2	Soil deformation limits for soil structure interaction	Strain limits for piles under limit conditions	N/A B) Multi-mode	A) Single Mode C) Pushover D) Inelastic	
Eurocode (1994)		Dual level approach	No reference to Port Structures	Not specified	Not specified. A guidance is given for the design of plastic hinging	Design ground acceleration fraction for a_{design} is specified	See Eqn. (T1.3)	These codes refer to buildings and other civil engineering works with no direct reference to port structures
New Zealand (1992–1997)		Dual method	The Ultimate Limit state peak response spectra is given	Not specified	Both are specified	Design Spectra are used	Based on seismic zone, subsoil condition, structure period and ductility factor	

Table T1.2. Structural damage criteria in Japanese standard.

Type of retaining walls	Water depth	
	Shallower than 7.5 m	Deeper than 7.5 m
Gravity quay wall • No repair needed for operation • Partial operation allowed	Horizontal displacement 0 to 0.2 m 0.2 to 0.5 m	Horizontal displacement 0 to 0.3 m 0.3 to 1.0 m
Sheet pile quay wall • No repair needed for operation • Partial operation allowed	Horizontal displacement 0 to 0.2 m 0 to 0.3 m	Horizontal displacement 0 to 0.3 m 0.3 to 0.5 m

Table T1.3. Serviceability criteria in Japanese ports.

Main body of retaining structure	Upper limit of settlement	0.2 to 0.3 m
	Upper limit of tilting	3–5°
	Upper limit of differential horizontal displacement	0.2 to 0.3 m
Apron	Upper limit of differential settlement on apron	0.3 to 1.0 m
	Upper limit of differential settlement between apron and background	0.3 to 0.7 m
	Upper limit of tilting	3–5% [towards sea] 5% [towards land]

Table T1.4. Ductility factor for L1 motion in the simplified analysis.

Special Class	$\mu = 1.0$
Class A	$\mu = 1.3$
Class B	$\mu = 1.6$
Class C	$\mu = 2.3$

Design response spectra are specified as follows:

For $T = 0$ to T_A: $S_A/a_{design} = 1 + 1.5\, T/T_A$

For $T = T_A$ to T_B: $S_A/a_{design} = 2.5$ (T1.2)

For $T > T_B$: $S_A/a_{design} = KC/T$

where $a_{design} = 0.04$ to $0.26g$, $T_A = KC/10$, $T_B = KC/2.5$

$K = 1.0$ to 1.5 (depends on type of seismicity)

$C = 1.0$ to 2.0 (depends on ground conditions)

(3) German design (EAU, 1996)

The German recommendations for quay walls describe a single level approach with a conventional seismic earth pressure based on the Mononobe-Okabe equation (Mononobe, 1924; Okabe, 1924). Pressures are assumed to increase linearly with depth, although it is accepted [R125] that tests have shown this not to be the case. The soil-water system is assumed to move as one body, and no water hydrodynamic suction in front of the wall is taken into account. The requirements for accuracy of the calculations are correspondingly more stringent when earthquake

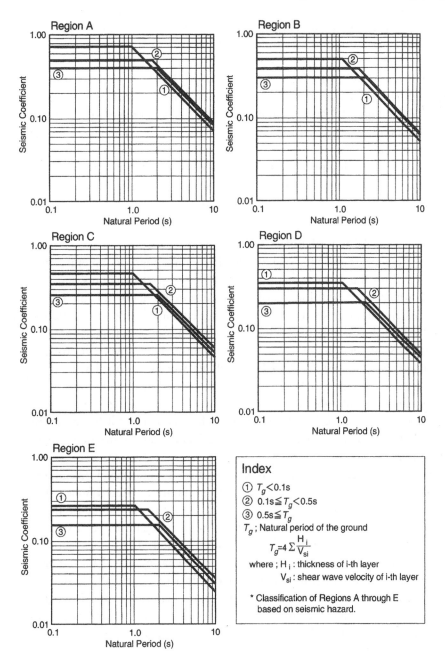

Fig. T1.1. Response spectra for L1 earthquake for pile-supported wharves in Japan.

Table T1.5. Ductility factor for L2 motion in the simplified analysis.

Special Class	$\mu = 1.25 + 62.5(t_p/D_p)$, but $\mu \leq 1.25$
	t_p = pile wall thickness
	D_p = pile diameter

damage can endanger human lives or cause destruction of supply facilities or the like which are of importance to the population [R124].

(4) US Navy/California State design (Ferritto, 1997a, b; Ferritto et al., 1999)

The US Navy code describes a dual level design and a performance level that is serviceable under L1 and repairable under L2. The damage criteria are deformation limits for wharf dikes and ductility limits for piles. The procedure requires a linear or nonlinear dynamic analysis. California State design is similar to the US Navy design in principle, but the ductility requirements are much more in detail and specified by strain limits. This procedure is still under development and will be finalized by 2001 in the form of regulatory guidance.

(5) ASCE-TCLEE guidelines (Werner, 1998)

This reference represents one of the first guidelines documents, specifically developed for the seismic analysis and design of port structures and facilities in North America. In addition to comprehensive treatment of seismic hazards, seismic design and analysis in North American engineering practice, and guidelines for specific port components, this document presents pertinent information on seismic risk reduction, and emergency response and recovery at ports. It should be noted that the ASCE-TCLEE guidelines were developed with the primary objective of providing a framework for the establishment of improved seismic risk evaluation and reduction procedures for ports in the United States. The recommendations outlined in the guidelines document are not to be interpreted as codes, nor are they intended to supercede local code requirements that may be applicable.

The ASCE-TCLEE seismic guidelines present a dual level design, consistent with the current state of practice at major ports in the United States. The multi-level design approach has been adopted at numerous ports in the form of two-level, and occasionally three-level design procedures. An example of the two-level approach as applied in the western United States is as follows:

Level 1. Under this first level of design, Operating Level Earthquake (OLE) ground motions are established, that have a 50-percent probability of exceedance in 50 years (corresponding to an average return period of about 72 years). Under this level of shaking, the structure is designed so that operations are not interrupted and any damage that occurs will be repairable in a short time (possibly less than 6 months).

Level 2. Under this second level of design, more severe Contingency Level Earthquake (CLE) ground motions are established that have a 10-percent probability of being exceeded in 50 years (consistent with most building codes and corresponding to an average return period of about 475 years). Under this level of shaking, the structure is designed so as to undergo damage that is controlled, economically repairable, and is not a threat to life safety.

It should be noted that the exposure times adopted for the Level 1 and Level 2 events in this example application, may vary regionally due to variations in the rate of seismicity, the type of facility, and economic considerations.

The damage criteria outlined in the guideline document are presented in the form of general performance-based recommendations. As such, the recommendations address the evaluation and mitigation of liquefaction hazards and ground failures, deformation limits for retaining structures, earth structures and other waterfront components, and ductility limits for piles. In order to evaluate these earthquake-induced loads and associated deformations, pseudo-static methods of analysis must often be supplemented with linear and/or nonlinear dynamic analysis, the level of analytical sophistication being a function of the intensity of the ground motions, the anticipated soil behaviour, and the complexity of the structure. General guidance on the level of analysis required for a variety of geotechnical and structural applications is provided (including dynamic soil-structure interaction analyses).

(6) Eurocode design (CEN, 1994)
The methodology of the Eurocode 8 describes in general, a dual level approach but in low seismicity zones [$a_{design} \leq 0.1g$] and for 'well defined' structures in seismic zones with small [$a_{design} \leq 0.04g$], a single level approach can be sufficient. The mentioned performance levels are:
- 'No collapse' requirement: retain structural integrity and a residual bearing capacity.
- 'Damage limitation' requirement: no damage and associated limitations of use, the costs of which would be disproportionately high compared with the cost of the structure itself.

Damage criteria in terms of maximum displacements and/or ductility levels are not specified. For piles, it is stated that they shall be designed to remain elastic. When this is not feasible, guidance is given for the design of potential plastic hinging and the region which it shall cover.

For the analytical procedure, the design ground acceleration a_{design} tends to coincide with the actual peak acceleration for moderate to high magnitude earthquakes in cases of medium to long source-to-site distances, which are characterised [on firm ground] by a broad and approximately uniform frequency spectrum, while a_{design} will be more or less reduced relative to the actual peak for near field, low magnitude events. The a_{design} corresponds to a reference period of 475 years, or as specified by the National Authority.

For retaining structures $k_h = a_{design} / r_{EC}g$ and $k_v = 0.5k_h$
where

$r_{EC} = 2$ for free gravity walls with acceptable displacements (mm) $\leq 300a_{design}/g$

$r_{EC} = 1.5$ as above with displacements (mm) $\leq 200a_{design}/g$

$r_{EC} = 1$ for the rest of the retaining structures

For retaining structures higher than ≈ 10 m, a refined estimate of a_{design} can be obtained by a free field one-dimensional analysis of vertically propagating waves.

For a linear analysis design of pile-supported structures, design spectra are defined as:

$$0 \leq T \leq T_B: \quad S_A(T) = \alpha \cdot S \cdot \left[1 + \frac{T}{T_B} \cdot \left(\frac{\beta_o}{q} - 1 \right) \right]$$

$$T_B \leq T \leq T_C: S_A(T) = \alpha \cdot S \cdot \frac{\beta_o}{q}$$

$$T_C \leq T \leq T_D: S_A(T) \begin{cases} = \alpha \cdot S \cdot \dfrac{\beta_o}{q} \cdot \left[\dfrac{T_C}{T} \right]^{k_{d1}} \\ \geq [0.20] \cdot \alpha \end{cases}$$

$$T_D \leq T: \quad S_A(T) \begin{cases} = \alpha \cdot S \cdot \dfrac{\beta_o}{q} \cdot \left[\dfrac{T_C}{T} \right]^{k_{d1}} \cdot \left[\dfrac{T_D}{T} \right]^{k_{d2}} \\ \geq [0.20] \cdot \alpha \end{cases}$$

(T1.3)

where α: design ground acceleration; β_o: spectral acceleration amplification factor for 5% damping, q: damping correction factor with reference value $q = 1$ for 5% viscous damping.

Values of the parameters S, k_{d1}, and k_{d2} are given depending on the subsoil class specified by shear wave velocity.

(7) New Zealand design

The seismic design loading for wharf structures is based on those within the N.Z. Standard NZS 4203: 1992 'General Structural Design Loadings for Buildings'. The requirements in this standard relate to an earthquake that has a 10% probability of being exceeded in a 50 year design life (or for a seismic event with a return period of 475 years). The specific material properties inclusive of damage criteria are set out in the concrete and steel standards. The concrete and steel standards are standards that have been fully developed to cater to the science of building design in NZ. These standards do not specifically include earth retaining or wharf quay type of structures, but they do form the basis of the design.

There is a recognition by New Zealand designers of the need to refer to NZ bridge design and/or relevant local or overseas standards to ensure that wharf structures are adequately detailed to address seismic cracking, variation in pile stiffness, effects on the structures of lateral spreading and/or liquefaction of the adjacent reclamation, pile ductility, and the diaphragm action of the deck to transfer loads between shorter stiff piles and longer more flexible piles. The torsional effects on the structure from variations in pile length and stiffness are also recognised.

Both the ultimate limit state and the serviceability limit state are applicable within the standards. Damage criteria used will depend on the form of construction and also on the material used. The design spectra are defined within the material standard and the response spectra are based on the seismic zone, the subsoil description, the fundamental period of oscillation of the structure and the structural ductility factor.

There is scope within the New Zealand design standards to vary performance requirements to cater for specific client requirements regarding design life. For example, definition of performance requirements to restrict concrete crushing and residual crack widths after design level and ultimate level earthquakes are not prescribed, but are recognised as an issue to be considered due to the rapid deterioration of exposed reinforcing in the highly corrosive port environment once any seismic cracking has occurred. Assumptions related to any predicted additional maintenance following a seismic event, do need to be canvassed by the designer as part of the design process.

T1.3   SEISMIC DESIGN PRACTICE FOR PORT STRUCTURES AROUND THE WORLD

The seismic design codes and guidelines reviewed in the previous section suggest the following.
- Since the mid-1990s, the dual level approach has been accepted as the new common basis in seismic design. The conventional, single level approach remains to be used for low design accelerations or structures with low seismic capacity demands;
- in the dual level approach, acceptable damage levels are typically defined as:
  - ➤ no damage or serviceable under L1 motion;
  - ➤ repairable or short-term loss of serviceability under L2 motion;
- damage criteria to specify limit conditions in engineering terms, including wall displacement and pile ductility demand, have been under development and further studies may be needed to establish them;
- analytical procedures adopted for design, including the factor for specifying the effective seismic coefficient for retaining structures and the response spectra used for pile-supported structures, vary depending on the region;
- liquefaction is considered a primary design factor for port structures.

The current state of seismic design practice may be in a rapidly changing transitional state, from the conventional design toward the performance-based design, the latter of which is fully discussed and illustrated in this book.

It must be emphasized that there is a significant difference between the design approach recommended in these guidelines and the conventional design concept. In the conventional design concept, especially when referring to the simplified analysis, an equivalent seismic coefficient is used as an input parameter representing adequately the ensemble of ground motions, and a factor of safety is applied to determine the dimensions and properties of the structure. In the proposed approach, the design is based on the seismic performance of the structure evaluated appropriately through response analysis for a variety of input earthquake motions. The ensemble of the seismic responses, rather than the ensemble of input motions, is used as a basis for accomplishing the design in the proposed guidelines. For each response analysis, input parameters most appropriate are those well defined in terms of applied mechanics, such as a peak ground acceleration for the simplified analysis and/or an equivalent parameter clearly defined in terms of peak ground acceleration. Consequently, no factor of safety should be applied to input data used in seismic analysis for evaluating the threshold level of the structure. This important distinction between conventional design and performance-based design should also be borne in mind when interpreting the design guidelines included in the various seismic codes.

TC2: Case Histories

TECHNICAL COMMENTARY 2

Case Histories

Selection of well-documented case histories of damage to port structures (1980–1999) is presented for the events listed in Table T2.1. The sequential numbers 1 through 29 shown in this table may be used as indexes in referring to the relevant case histories shown in Tables T2.2 through T2.30 in this technical commentary. These case histories illustrate various seismic effects on port structures.

T2.1 DAMAGE TO GRAVITY QUAY WALLS

As discussed in Section 4.1 in the Main Text, a gravity quay wall is made of a caisson or other gravity retaining structure placed on seabed. Stability against earth pressures from the backfill soil behind the wall is maintained by the mass of the wall and friction at the bottom of the wall. For gravity quay walls on firm foundations, typical failure modes during earthquakes are seaward displacements and tilting (Tables T2.3 through T2.5). For a loose backfill or natural loose sandy foundation, the failure modes involve overall deformation of the foundation beneath the wall, resulting in large seaward displacement, tilting and settlements (Table T2.7). A wall with a relatively small width to height ratio, typically less than about 0.75, exhibits a predominant tilting failure mode rather than horizontal displacements, resulting, in the extreme case, in the collapse of the wall (Table T2.2).

The evidence of damage to gravity quay walls (Tables T2.2 through T2.9) suggests that:

1) most damage to gravity quay walls is often associated with significant deformation of a soft or liquefiable soil deposit, and, hence, if liquefaction is an issue, implementing appropriate remediation measures against liquefaction may be an effective approach to attaining significantly better seismic performance;

2) most failures of gravity quay walls in practice result from excessive deformations, not catastrophic collapses, and, therefore, design methods based on displacements and ultimate stress states are desirable for defining the comprehensive seismic performance; and

3) overturning/collapse of concrete block type walls could occur when tilting is excessive, and this type of wall needs careful consideration in specifying damage criteria regarding the overturning/collapse mode.

T2.2   DAMAGE TO SHEET PILE QUAY WALLS

As discussed in Section 4.2 in the Main Text, a sheet pile quay wall is composed of interlocking sheet piles, tie-rods, and anchors. The wall is supported at the upper part by anchors and the lower part by embedment in foundation soil. Typical failure modes during earthquakes depend on structural and geotechnical conditions, as shown in Tables T2.10 through T2.15. The evidence of damage to sheet pile quay walls suggests that:

1) most damage to sheet pile quay walls is often associated with significant earth pressures from a soft or liquefiable backfill soil (Tables T2.10 and T2.12), and, hence, if liquefaction is an issue, implementing appropriate remediation measures against liquefaction may be an effective approach to attaining significantly better seismic performance (Tables T2.11 and T2.13);

2) if the anchor is embedded in competent soil, most failures of sheet pile quay walls occur at the sheet pile wall due to excessive bending moment (Tables T2.10 and T2.12), whereas, if the anchor is embedded in liquefiable soil, failures occur at the anchor, resulting in overall seaward movement and tilting of the sheet pile wall and anchor system; and

3) structural damage to a sheet pile quay wall is governed by stress states rather than displacements. It is therefore important to determine the preferred sequence and degrees of ultimate states that may occur in the composite sheet pile quay wall system.

T2.3   DAMAGE TO PILE-SUPPORTED WHARVES

As discussed in Section 4.3 in the Main Text, a pile-supported wharf is composed of a deck supported by a substructure consisting of piles and a rock dike/slope. The unsupported pile length above the dike/slope surface is variable. When rockfill suitable for construction of the dike is uneconomical, a gravity or sheet pile retaining structure is also constructed to replace a portion of the dike. The seismic response of pile-supported wharves is influenced to a great degree by complex soil-structure interaction during ground shaking. Typical failure modes during earthquakes depend on the magnitude of the inertia force relative to the ground displacement. In particular, the failure associated with the ground displacement shown in Tables T2.16, T2.17, T2.20, and T2.22 suggests that specific design consideration is required in the geotechnical aspect of seismic design.

Among various structural components in the pile-supported wharves, special attention should be paid to the application of batter piles. Batter piles remain the most efficient structural component for resisting lateral loads due to mooring, berthing and crane operation. However, the batter pile-deck system results in a much more rigid frame than one with vertical piles. As a result, large stress concentrations and shear failures of concrete batter piles have been observed during earthquakes (Tables T2.16 and T2.18). This mode of pile failure demonstrates that the structural design of batter piles used in regions of high seismicity must account for displacement demand and appropriate ductility.

Structural damage to a pile-supported wharf is governed by stress/strain states rather than displacements. It is important to determine the preferred sequence and degrees of ultimate states to occur in the composite pile-supported wharf system. Most pile failures are associated with liquefaction of soil which can result in buckling of the pile, loss of pile friction capacity, or development of pile cracking and hinging. Hinging may be at the cap location or at an interface between soil layers of differing lateral stiffness (Table T2.20). Damage to piles in the soil may also occur about 1 to 3 pile diameters below grade in non-liquefying soils due to the dynamic response of the wharf (see Technical Commentary 5).

T2.4   DAMAGE TO CELLULAR QUAY WALLS

As discussed in Section 4.4 in the Main Text, a cellular quay wall is made of a steel plate or sheet pile cell with sand or other fill. Resistance against inertia forces and earth pressures is provided by the fill friction at the bottom surface of the cell for a non-embedded cell, or by the resistance of the foundation subsoil at the cell embedment. Typical failure modes during earthquakes depend on cell embedment and geotechnical conditions, resulting in varying degrees of seaward displacement, settlement and tilting. In addition to the gross deformation, warping of the cellular cross section can occur (Table T2.23). Such performance of cellular quay walls should be considered in the seismic design criteria.

T2.5   DAMAGE TO CRANES

As discussed in Section 4.5 in the Main Text, a crane consists of an upper structure for handling cargoes, and a supporting structure for holding in place and transporting the upper structure. The crane is generally made of a steel frame. The supporting structure is either a rigid frame type or a hinged leg type, and rests on rails through the wheels.

A crane at rest is fixed to rails or to a quay wall with clamps or anchors. The strengths of these clamps or anchors provides the upper limit resistance against external forces acting on the crane. A crane in operation, however, is not supported with clamps or anchors. The lateral resistance of the crane against external forces during operation is only due to friction and resistance from the wheel flanges.

Typical failure modes during earthquakes are derailment of wheels, detachment or pull-out of the vehicle, rupture of clamps and anchors, buckling, and overturning. Widening of a span between the legs due to the deformation of the quay wall results in derailment or buckling of the legs (Table T2.26). Conversely, narrowing of a leg span can also occur due to the rocking response of the crane (Table T2.24). This is due to an alternating action of the horizontal component of resisting forces from the quay wall during a rocking type response involving uplifting of one side of the legs. The rocking response can also result in the derailment, leading to tilting. If the tilting becomes excessively large, overturning of the crane may result (Tables T2.25 and T2.27). The overturning typically occurs toward land because of the location of the center of gravity of the cranes at rest, and differential settlement on the crane foundation associated with the deformation of quay walls.

Crane rails are often directly supported either by a portion of a retaining wall, or by the deck of a pile-supported wharf. When the width of the gravity wall is small, or the quay wall is a sheet pile or cellular type, a separate foundation often consisting of piles is provided to support the rails. In order to achieve desirable seismic performance of quay walls with cranes, special consideration is required for the rail foundation. One such consideration could include providing a dedicated and cross-tied upper structure to support the rails.

T2.6   DAMAGE TO BREAKWATERS

As discussed in Section 4.6 in the Main Text, a breakwater is usually made of a rubble mound, a caisson, or a combination of both placed on the seabed. Stability against a horizontal external load is maintained by shear resistance of the rubble, friction at the bottom of the caisson, and with associated resistance to overturning and bearing capacity failure. Typical failure modes during earthquakes are settlement associated with significant foundation deformation beneath the caisson/rubble foundation (Table T2.28). It should be noted that even with moderate earthquakes, failure can happen due to extremely soft clay deposits and large seismic amplification within and beneath the rubble mound (Table T2.29).

Damage to breakwaters from a tsunami typically results in the caissons being washed off or tilting into the rubble mound (Table T2.30). If the performance objective of a breakwater includes preventive measures against a tsunami, appropriate design criteria need to be established with respect to both the earthquake motion and tsunamis.

Table T2.1. Selected case histories of damage to port structures (1980–1999).

No.	Port structure		Earthquake year (magnitude*)	Port	
1	Gravity	Block	1985 (M_s = 7.8)	San Antonio Port	Chile
2	quay		1986 (M_s = 6.2)	Kalamata Port	Greece
3	wall		1989 (M = 6.0)	Port of Algiers	Algeria
4		Caisson	1993 (M_J = 7.8)	Kushiro Port	Japan
5			1993 (M_J = 7.8)	Kushiro Port	Japan
6			1995 (M_J = 7.2)	Kobe Port	Japan
7		Block	1999 (M_w = 7.4)	Derince Port	Turkey
8		Caisson	1999 (M_s = 7.7)	Taichung Port	Taiwan
9	Sheet pile quay wall		1983 (M_J = 7.7)	Akita Port	Japan
10			1983 (M_J = 7.7)	Akita Port	Japan
11			1993 (M_J = 7.8)	Kushiro Port	Japan
12			1993 (M_J = 7.8)	Kushiro Port	Japan
13			1993 (M_J = 7.8)	Hakodate Port	Japan
14			1993 (M_s = 8.1)	Commercial Port, Guam	USA
15	Pile-	RC pile**	1989 (M_s = 7.1)	Port of Oakland	USA
16	supported		1990 (M_s = 7.8)	Port of San Fernando	Philippines
17	wharf	PC pile**	1994 (M_s = 6.8)	Port of Los Angeles	USA
18			1994 (M_s = 6.8)	Port of Los Angeles	USA
19		Steel pile	1995 (M_J = 7.2)	Kobe Port	Japan
20		PC pile	1995 (M_s = 7.2)	Port of Eilat	Israel
21	Dolphin	Steel pile	1995 (M_J = 7.2)	Kobe, Sumiyoshihama	Japan
22	Cellular quay wall		1995 (M_J = 7.2)	Kobe Port	Japan
23	Crane		1983 (M_J = 7.7)	Akita Port	Japan
24			1985 (M_s = 7.8)	San Antonio Port	Chile
25			1995 (M_J = 7.2)	Kobe Port	Japan
26			1999 (M_w = 7.4)	Derince Port	Turkey
27	Break-	Composite	1995 (M_J = 7.2)	Kobe Port	Japan
28	water	Rubble	1984 (M = 4.5)	Patras Port	Greece
29		Composite	1993*** (M_J = 7.8)	Okushiri Port	Japan

* For definition of magnitude, refer to Technical Commentary 3.
** RC: Reinforced concrete; PC: Prestressed concrete.
*** Damage due to Tsunami.

Table T2.2. Case history No. 1: San Antonio Port, Chile.

Structural and seismic conditions	Structure and location: Berths 1 & 2 Height: + 4.9 m Water depth: – 9.6 m Construction (completed): 1918–1935 Earthquake: March 3, 1985 Chile earthquake Magnitude: $M_s = 7.8$ Peak ground surface acceleration (PGA): 0.67 g PGA evaluated based on strong motion earthquake records.

Cross section: Block quay wall

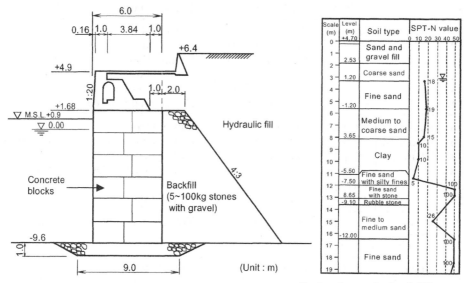

Boring log at the backfill

Features of damage/comments

Collapse of block type quay wall over 60% of wharf length (452 m) due to strong earthquake motion and backfill liquefaction.

References: Tsuchida et al. (1986); Wyllie et al. (1986)

Table T2.3. Case history No. 2: Kalamata Port, Greece.

Structural and seismic conditions	Height: + 2.1 m Water depth: − 9.5 m Earthquake: September 13, 1986 Kalamata earthquake Magnitude: $M_s = 6.2$ Peak ground surface acceleration (PGA): 0.2 to 0.3 g Peak ground surface velocity (PGV): 0.4 m/s PGA/PGV evaluated based on 1D & 2D equivalent linear analysis.
Seismic performance	Horizontal displacement: 0.15 ± 0.05 m Tilt: 4 to 5 degrees

Cross section: Block quay wall

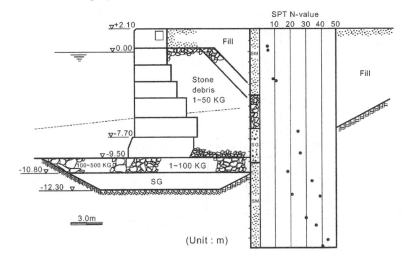

Index: SM = silty sand, SG = gravely sand

Features of damage/comments

Seaward displacement with tilt of block type quay wall constructed on firm foundation. The backfill settled 20 cm behind the wall with no settlement at a distance of 30 to 40 meters away. The quay wall remained serviceable.

Reference: Pitilakis and Moutsakis (1989)

Table T2.4. Case history No. 3: Port of Algiers, Algeria.

Structural and seismic conditions	Structure and location: Quay No. 34 Height: + 2.2 m Water depth: – 10.0 m Earthquake: October 29, 1989 Chenoua earthquake Magnitude: $M = 6.0$, Epicenter located 60 km to the west of Algiers
Seismic performance	Horizontal displacement: 0.5 m Vertical displacement: 0.3 m

Cross section: Block quay wall

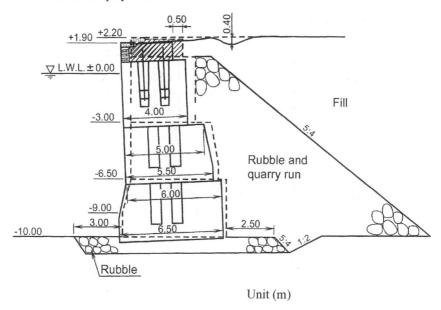

Unit (m)

Features of damage/comments

During this earthquake, 8,000 buildings were destroyed; there were ground ruptures, landslides, and the coastal road was closed by falling rock. No additional detailed information is available on the earthquake.

Reference: Manja (1999)

Table T2.5. Case history No. 4: Kushiro Port, Japan.

Structural and seismic conditions	Structure and location: West No. 1, West Port District Height: + 3.0 m Water depth: − 9.0 m Design seismic coefficient: $k_h = 0.20$ Construction (completed): 1975 Earthquake: January 15, 1993 Kushiro-Oki earthquake Magnitude: $M_J = 7.8$ (JMA) Peak ground surface acceleration (PGA): 0.47 g Peak ground surface velocity (PGV): 0.63 m/s PGA/PGV evaluated based on strong motion earthquake records.
Seismic performance	Horizontal displacement: 0.75 m Vertical displacement: 0.2 m Tilt: 2%

Cross section: Caisson quay wall

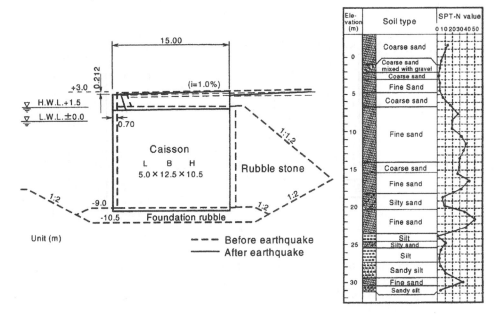

Features of damage/comments

Seaward displacement of caisson type quay wall constructed on firm foundation. Liquefaction of backfill.

Reference: Iai et al. (1994)

Table T2.6. Case history No. 5: Kushiro Port, Japan.

Structural and seismic conditions	Structure and location: East Quay, Kita Wharf, East Port District Height: + 2.5 m Water depth: – 8.5 m Design seismic coefficient: $k_h = 0.15$ Construction (completed): 1950 Earthquake: January 15, 1993 Kushiro-Oki earthquake Magnitude: $M_J = 7.8$ (JMA) Peak ground surface acceleration (PGA): 0.47 g Peak ground surface velocity (PGV): 0.63 m/s PGA/PGV evaluated based on strong motion earthquake records.
Seismic performance	Horizontal displacement: 1.9 m Vertical displacement: 0.2 to 0.5 m

Cross section: Caisson quay wall

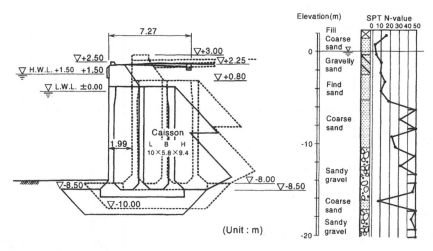

(Unit : m)

Features of damage/comments

Seaward displacement of caisson type quay wall constructed on firm foundation. Liquefaction of backfill. The displacement was larger presumably because of a smaller width to height ratio than the case history shown in the previous page (i.e. West No. 1 quay wall, West Port District, Kushiro Port, Japan).

Reference: Iai et al. (1994)

Table T2.7. Case history No. 6: Kobe Port, Japan.

Structural and seismic conditions	Structure and location: Rokko Island, RC5 Height: + 4.0 m, Water depth: – 14.0 m Design seismic coefficient: $k_h = 0.15$ Construction (completed): 1989 Earthquake: January 17, 1995 Hyogoken-Nambu earthquake Magnitude: $M_J = 7.2$ (JMA) Peak ground surface acceleration (PGA): 0.53 g Peak ground surface velocity (PGV): 1.06 m/s PGA/PGV evaluated based on strong motion earthquake records.
Seismic performance	Horizontal displacement: 4.2 to 5.2 m Vertical displacement: 1.5 to 2.2 m Tilt: 4 to 5°

Cross section: Caisson quay wall

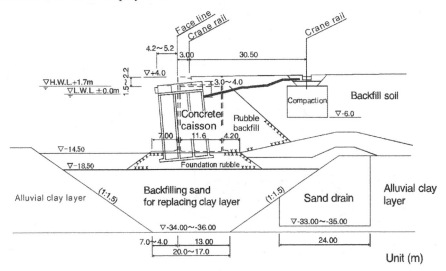

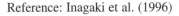

Features of damage/comments

Loose foundation soil below the wall induced significant seaward displacement and tilting, associated with the deformation in the foundation soil. Horizontal displacement of caisson walls is closely correlated with thickness of loose foundation deposit as shown in the right figure.

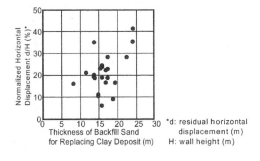

Reference: Inagaki et al. (1996)

Table T2.8. Case history No. 7: Derince Port, Turkey.

Structural and seismic conditions	Structure and location: No. 6 quay wall Height: + 2.25 m Water depth: – 12.0 m Earthquake: August 17, 1999 Kocaeli earthquake Magnitude: $M_w = 7.4$ (USGS) Peak ground surface acceleration (PGA): 0.2 to 0.25 g PGA evaluated based on strong motion earthquake records.
Seismic performance	Horizontal displacement: 0.7 m

Cross section: Block quay wall

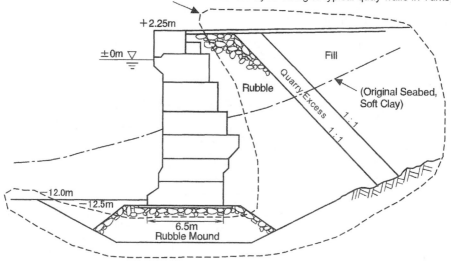

Inside broken lines : Presumed by referring to typical quay walls in Turkey

Features of damage/comments

The concrete blocks are put in oblique to the horizontal in Turkey ports as shown in the figure right (a front view of a block type quay wall in Turkey)

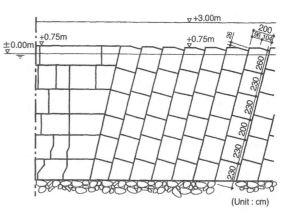

Reference: Sugano and Iai (1999)

Table T2.9. Case history No. 8: Taichung Port, Taiwan.

Structural and seismic conditions	Structure and location: No. 1 Quay Wall Height: + 6.2 m Water depth: − 13.0 m Design seismic coefficient: $k_h = 0.15$ Construction (completed): 1974 Earthquake: September 21, 1999 Ji-Ji earthquake Magnitude: $M_s = 7.7$ Peak ground surface acceleration (PGA): 0.16 g PGA evaluated based on strong motion earthquake records.
Seismic performance	Horizontal displacement: 1.5 m Vertical displacement: 0.1 m

Cross section: Caisson quay wall

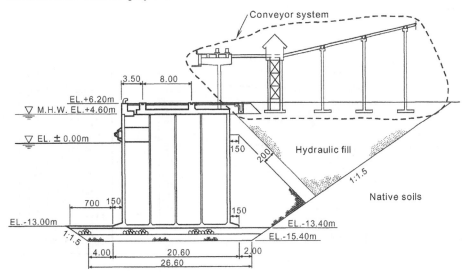

Features of damage/comments

Displacement of caisson quay wall constructed on firm foundation. Liquefaction of backfill. Settlement at backfill was 1m maximum.

Reference: Sugano et al. (1996)

Table T2.10. Case history No. 9: Akita Port, Japan.

Structural and seismic conditions	Structure and location: Ohama No. 2 Wharf Height: + 2.0 m Water depth: – 10.0 m Design seismic coefficient: $k_h = 0.10$ Construction (completed): 1976 Earthquake: May 26, 1983 Nihonkai-Chubu earthquake Magnitude: $M_J = 7.7$ (JMA) Peak ground surface acceleration (PGA): 0.24 g Peak ground surface velocity (PGV): 0.32 m/s PGA/PGV evaluated based on strong motion earthquake records.
Seismic performance	Horizontal displacement: 2.0 m (at the top of the wall) Vertical displacement: 0.3 to 1.3 m

Cross section: Sheet pile quay wall

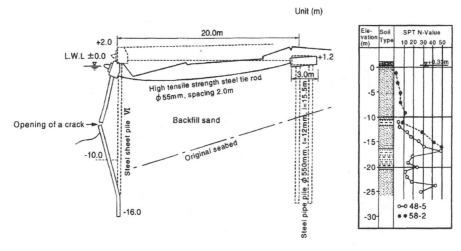

Features of damage/comments

The sheet pile wall suffered significant damage due to liquefaction of backfill.

Reference: Iai and Kameoka (1993)

Table T2.11. Case history No. 10: Akita Port, Japan.

Structural and seismic conditions	Structure and location: Ohama No. 1 Wharf Height: + 2.0 m Water depth: – 10.0 m Design seismic coefficient: $k_h = 0.10$ Construction (completed): 1970 Earthquake: May 26, 1983 Nihonkai-Chubu earthquake Magnitude: $M_J = 7.7$ (JMA) Peak ground surface acceleration (PGA): 0.24 g Peak ground surface velocity (PGV): 0.32 m/s PGA/PGV evaluated based on strong motion earthquake records.
Seismic performance	Horizontal displacement: 0 m (no damage) Vertical displacement: 0 m (no damage)

Cross section: Sheet pile quay wall

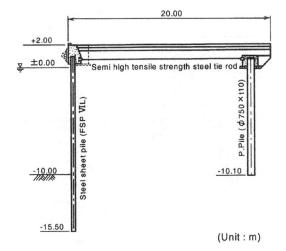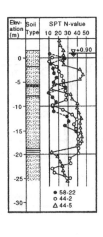

(Unit : m)

Features of damage/comments

Although the cross section of this sheet pile quay wall is similar to that of Ohama No. 2 Wharf (shown in the previous page), this quay wall suffered no damage. No signs of liquefaction were evident at this quay wall. Occurrence or non-occurrence of backfill liquefaction resulted in the significant difference of seismic performance of sheet pile quay walls.

Reference: Iai and Kameoka (1993)

Table T2.12. Case history No. 11: Kushiro Port, Japan.

Structural and seismic conditions	Structure and location: South Quay Wall, Fishery Port, East Port District Height: + 2.7 m Water depth: − 7.5 m Design seismic coefficient: $k_h = 0.20$ Construction (completed): 1980 Earthquake: January 15, 1993 Kushiro-Oki earthquake Magnitude: $M_J = 7.8$ (JMA) Peak ground surface acceleration (PGA): 0.47 g Peak ground surface velocity (PGV): 0.63 m/s PGA/PGV evaluated based on strong motion earthquake records.
Seismic performance	Horizontal displacement: 0.4 m (at the top of the wall) Vertical displacement: 0.1 to 0.3 m

Cross section: Sheet pile quay wall

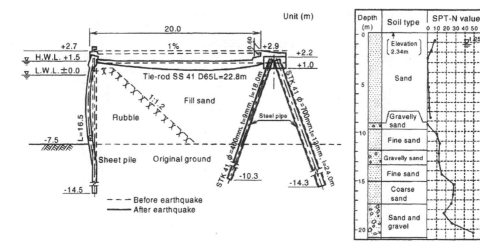

Features of damage/comments

Liquefaction of backfill resulted in opening of a crack in the sheet pile wall at an elevation of − 4 m.

Reference: Iai et al. (1994)

Table T2.13. Case history No. 12: Kushiro Port, Japan.

Structural and seismic conditions	Structure and location: South Quay, West No. 1 Wharf, West Port District Height: + 3.0 m Water depth: – 12.0 m Design seismic coefficient: $k_h = 0.20$ Construction (completed): 1975 Earthquake: January 15, 1993 Kushiro-Oki earthquake Magnitude: $M_J = 7.8$ (JMA) Peak ground surface acceleration (PGA): 0.47 g Peak ground surface velocity (PGV): 0.63 m/s PGA/PGV evaluated based on strong motion earthquake records.
Seismic performance	Horizontal displacement: 0 m (no damage) Vertical displacement: 0 m (no damage) Tilt: 0° (no damage)

Cross section: Sheet pile quay wall

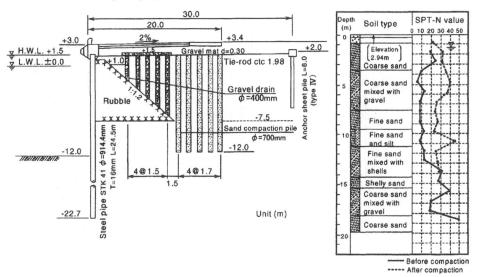

Features of damage/comments

The quay walls with remedial measures against liquefaction suffered no damage.

Reference: Iai et al. (1994)

Table T2.14. Case history No. 13: Hakodate Port, Japan.

Structural and seismic conditions	Structure and location: No. 6 Quay Wall, Benten District Height: + 2.3 m Water depth: − 8.0 m Design seismic coefficient: $k_h = 0.15$ Construction (completed): 1973 Earthquake: July 12, 1993 Hokkaido-Nansei-Oki earthquake Magnitude: $M_J = 7.8$ (JMA) Peak ground surface acceleration (PGA): 0.12 g Peak ground surface velocity (PGV): 0.33 m/s PGA/PGV evaluated based on strong motion earthquake records.
Seismic performance	Horizontal displacement: 5.2 m Vertical displacement: 1.6 m Tilt: 15°

Cross section: Sheet pile quay wall

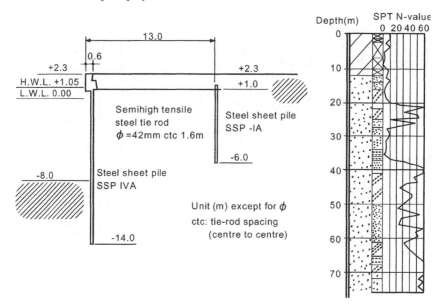

Features of damage/comments

Despite the small PGA, in the order of 0.1 g, significant deformation occurred due to liquefaction of backfill as well as foundation soil.

Reference: Inatomi et al. (1994)

Table T2.15. Case history No. 14: Commercial Port, Guam, USA.

Structural and seismic conditions	Structure and location: Berths F3 through F6 Height: + 2.7 m Water depth: – 10.5 m Earthquake: August 8, 1993 Guam earthquake Magnitude: $M_s = 8.1$ Peak ground surface acceleration (PGA): 0.15 to 0.25 g PGA evaluated based on strong motion earthquake records.
Seismic performance	Horizontal displacement: 0.6 m

Cross section: Sheet pile quay wall

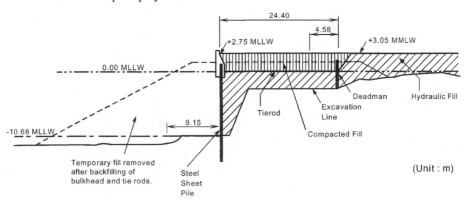

Features of damage/comments

This quay wall was constructed using a three-step dredging sequence. The last step is depicted in the figure.

References: Comartin (1995); Mejia and Yeung (1995); Vazifdar and Kaldveer (1995)

Table T2.16. Case history No. 15: Port of Oakland, USA.

Structural and seismic conditions	Structure and location: 7th Street Terminal wharf Height: + 4.5 m Water depth: − 11.0 m Earthquake: October 17, 1989 Loma Prieta earthquake Magnitude: $M_s = 7.1$ Peak ground surface acceleration (PGA): 0.22 to 0.45 g PGA evaluated based on strong motion earthquake records.
Seismic performance	Horizontal displacement: 0.1 m
Cross section:	Pile-supported wharf (41cm square section reinforced concrete piles)

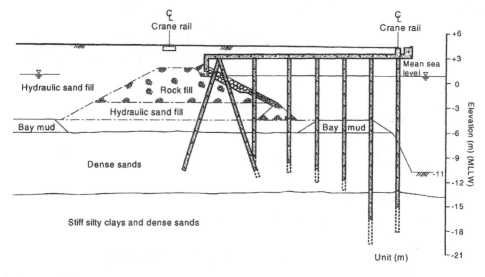

Features of damage/comments

Tensile failures at or near the tops of most of the rearmost line of batter piles (inclined inboard). Evidence of liquefaction of the backfill. Hydraulic sand fill below the dike (i.e. rock fill) is presumed to be the major geotechnical cause of damage.

References: Benuska (1990); Egan et al. (1992)

Table T2.17.   Case history No. 16: Port of San Fernando, Philippines.

Structural and seismic conditions	Structure and location: Government Pier No. 1 Earthquake: July 16, 1990 Luzon, Philippines, earthquake Magnitude: $M_s = 7.8$
Seismic performance	Horizontal displacement: Displacement along the longitudinal direction of the pier toward the sea. Many gaps opened in the longitudinal direction of the concrete deck, including a 0.7 m gap. Summation of all the gaps amounted to about 1.5 m at the seaside end of the pier.

Cross section: Pile-supported pier (square reinforced concrete piles)
200 m long, 19 m wide pier that consists of a concrete deck supported by vertical and battered concrete piles.

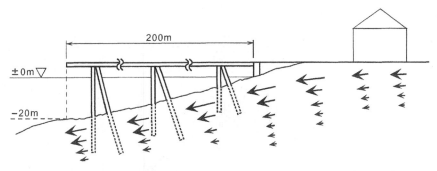

Schematic figure. Arrows indicate presumed lateral deformation of the foundation ground.

Features of damage/comments

Extensive opening of cracks on the pier deck. Cracks and chipping at pile caps, exposing reinforcing steel bars, which were either bent or cut.
Damage was presumably caused by lateral deformation of the foundation ground as illustrated above.

Reference: Japan Society of Civil Engineers (1990)

114   PIANC

Table T2.18. Case history No. 17: Los Angeles Port, USA.

Structural and seismic conditions	Structure and location: B127, APL Terminal Height: + 4.5 m Water depth: − 10.5 m Construction (completed): 1960s Earthquake: January 17, 1994 Northridge earthquake Magnitude: $M_s = 6.8$ Peak ground surface acceleration (PGA): 0.25 g PGA evaluated based on strong motion earthquake record.

Cross section: Pile-supported wharf (46 cm prestressed octagonal concrete piles)

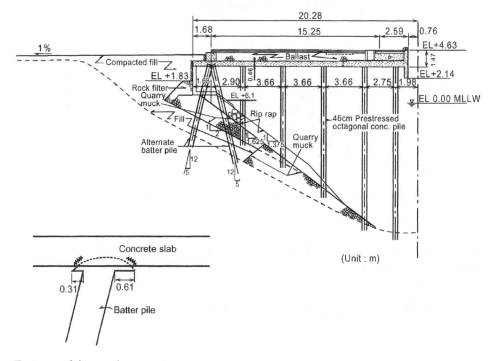

Features of damage/comments

Pull-out or chipping of concrete off batter piles at pile cap.

References: Hall (1995); Iai and Tsuchida (1997)

Table T2.19. Case history No. 18: Los Angeles Port, USA.

Structural and seismic conditions	Structure and location: B126(New), APL Terminal Height: + 4.5 m Water depth: – 13.5 m Design seismic coefficient: k_h = 0.12, 0.04 in normal direction Construction (completed): 1982 Earthquake:January 17, 1994 Northridge earthquake Magnitude: M_s = 6.8 Peak ground surface acceleration (PGA): 0.25 g PGA evaluated based on strong motion earthquake record.
Seismic performance	Horizontal displacement: 0.1 m

Cross section: Pile-supported wharf (61 cm prestressed octagonal concrete piles)

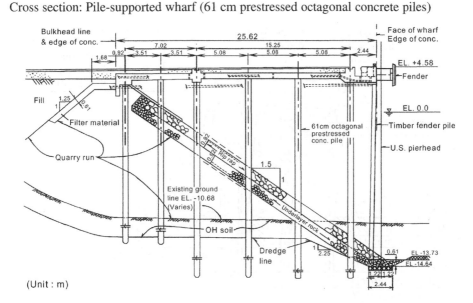

(Unit : m)

Features of damage/comments

Liquefaction of backfill.
Hairline crack at pile cap of most landward row of piles.
Horizontal displacement of 8 cm relative to B126(Old) at an expansion joint caused damage to crane rails.

References: Hall (1995); Iai and Tsuchida (1997)

Table T2.20. Case history No. 19: Kobe Port, Japan.

Structural and seismic conditions	Structure and location: Takahama Wharf Height: + 4.0 m Water depth: − 7.5 m Design seismic coefficient: $k_h = 0.15$ Earthquake: January 17, 1995 Hyogoken-Nambu earthquake Magnitude: $M_J = 7.2$ (JMA) Peak ground surface acceleration (PGA): 0.53 g Peak ground surface velocity (PGV): 1.06 m/s PGA/PGV evaluated based on strong motion earthquake records.
Seismic performance	Horizontal displacement: 1.5–1.7 m Tilt: 3%

Cross section: Pile-supported wharf ($D_p = 700$ mm, $t_p = 10, 12, 14$ mm, steel piles)

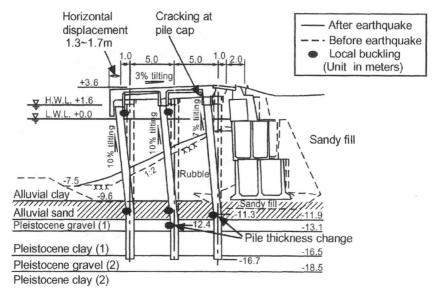

Features of damage/comments

Buckling of steel piles was found not only at the pile caps but also in the embedded portion. The damage was associated with significant deformation of alluvial sand layer below the rubble mound/dike.

Reference: Iai (1998b)

Table T2.21. Case history No. 20: Port of Eilat, Israel.

Structural and seismic conditions	Structure and location: Main Wharf Height: + 2.5 m Water depth: – 10.5 m Earthquake: November 22, 1995 Earthquake Magnitude: $M_s = 7.2$ The port was located more than 100 km away from the epicenter.

Cross section: Pile-supported wharf (46 cm octagonal prestressed concrete pile)

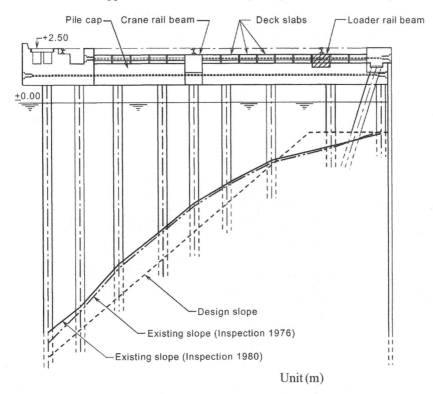

Unit (m)

Features of damage/comments

5 to 15 mm displacements at the expansion joints. No damage to piles.

Reference: Buslov (1996)

Table T2.22. Case history No. 21: Kobe, Japan.

Structural and seismic conditions	Structure and location: Sumiyoshihama district Height: + 5.5 m, Water depth: – 12.5 m Design seismic coefficient: $k_h = 0.15$ Construction (completed): 1982 Earthquake: January 17, 1995 Hyogoken-Nambu earthquake Magnitude: $M_J = 7.2$ (JMA) Peak ground surface acceleration (PGA): 0.53 g Peak ground surface velocity (PGV): 1.06 m/s PGA/PGV evaluated based on strong motion earthquake records.
Seismic performance	Horizontal displacement: 0.1 to 0.6 m landward Tilt: 2 to 5° landward

Cross section: Pile-supported dolphin (steel piles) 160 m long and 12.5 m wide

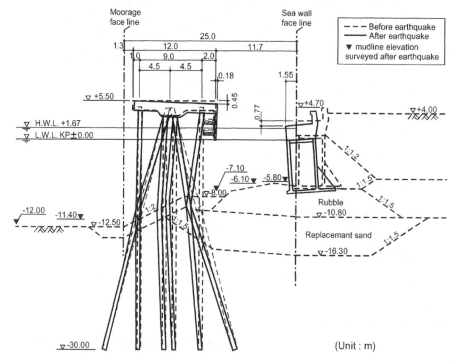

Features of damage/comments

Steel piles buckled near the foundation. Lateral deformation was about 1.0 m at the buckling point.
Retaining wall: horizontal displacement 1.5–1.7 m seaward, vertical displacement 0.6 to 1.0m, tilt: 3 to 4° seaward.
Damage to the pile-supported jetty was not only due to the seismic inertia force in the deck but also due to lateral displacement of foundation soil toward the sea.

Reference: Nishizawa et al. (1998)

Table T2.23. Case history No. 22: Kobe Port, Japan.

Structural and seismic conditions	Structure and location: Maya Wharf No. 1 Height: + 4.0 m, Water depth: – 12.0 m Design seismic coefficient: $k_h = 0.15$ Construction (completed): 1964 Earthquake: January 17, 1995 Hyogoken-Nambu earthquake Magnitude: $M_J = 7.2$ (JMA) Peak ground surface acceleration (PGA): 0.53 g Peak ground surface velocity (PGV): 1.06 m/s PGA/PGV evaluated based on strong motion earthquake records.
Seismic performance	Horizontal displacement: 1.3 to 2.9 m Vertical displacement: 0.6 to 1.3 m Tilt: 11° (max)

Cross section: Steel plate cofferdam type

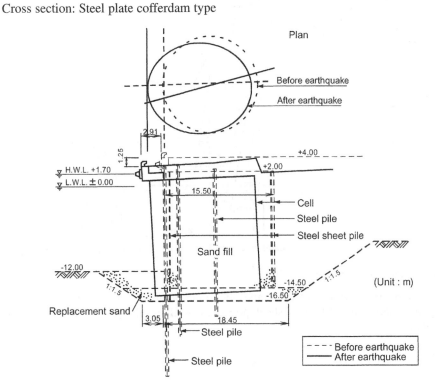

Features of damage/comments

Note the warping of the cellular cross section as shown above.

Reference: Sugano et al. (1998)

Table T2.24. Case history No. 23: Akita Port, Japan.

Structural and seismic conditions	Structure and location: Nakajima No. 2 Wharf (sheet pile quay wall) Earthquake: May 26, 1983 Nihonkai-Chubu earthquake Magnitude: $M_J = 7.7$ (JMA) Peak ground surface acceleration (PGA): 0.24 g Peak ground surface velocity (PGV): 0.32 m/s PGA/PGV evaluated based on strong motion earthquake records.

The crane derailed and inclined 20° landward.

Cross section: Crane on sheet pile quay wall

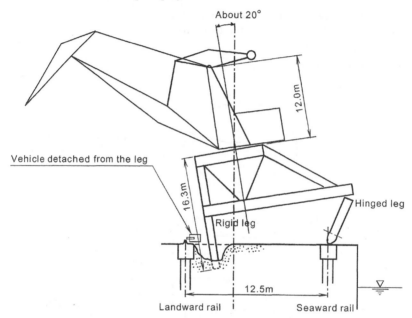

Features of damage/comments

Due to the rocking response of the crane, the crane derailed. The derailed legs were landed on liquefied backfill of a sheet pile quay wall, resulting in a 20° landward inclination.

Reference: Tsuchida et al. (1985)

Table T2.25. Case history No. 24: San Antonio Port, Chile.

Structural and seismic conditions	Structure and location: Berths 1 & 2 Earthquake: March 3, 1985 Chile earthquake Magnitude: $M_s = 7.8$ Peak ground surface acceleration (PGA): 0.67 g PGA evaluated based on strong motion earthquake records.

Cross section: Crane on gravity quay wall (rail span = 5.25 m)

Features of damage/comments

Tilting and overturning of cranes due to rocking response and differential settlement at the liquefied backfill.

Reference: Wyllie et al. (1986)

Table T2.26. Case history No. 25: Kobe Port, Japan.

Structural and seismic conditions	Structure and location: Rokko Island, RC5 (Gravity quay wall) Earthquake: January 17, 1995 Hyogoken-Nambu earthquake Magnitude: $M_J = 7.2$ (JMA) Peak ground surface acceleration (PGA): 0.53 g Peak ground surface velocity (PGV): 1.06 m/s PGA/PGV evaluated based on strong motion earthquake records.

Cross section: Container crane on gravity quay wall

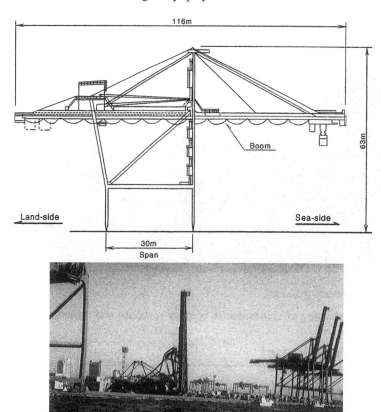

Features of damage/comments

Buckling at upper portion of crane legs not only due to inertia force but also due to widening of the rail span associated with the movement of the caisson wall toward the sea.

References: Tanaka and Inatomi (1996); Scott (1997)

Table T2.27. Case history No. 26: Derince Port, Turkey.

Structural and seismic conditions	Structure and location: No. 6 Quay Wall (Gravity quay wall) Earthquake: August 17, 1999 Kocaeli earthquake Magnitude: $M_w = 7.4$ (USGS) Peak ground surface acceleration (PGA): 0.2 to 0.25 g PGA evaluated based on strong motion earthquake records.

Cross section: Crane on gravity quay wall (rail span = 5.5 m)

Features of damage/comments

Cranes derailed due to rocking response. Tilted toward the landside due to differential settlement of backfill, resulted in collapse of one crane.

Reference: Sugano and Iai (1999)

Table T2.28. Case history No. 27: Kobe Port, Japan.

Structural and seismic conditions	Structure and location: Breakwater No. 7 Height: + 5.0 m Water depth: − 15.0 m Construction (completed): 1970–1972 Earthquake: January 17, 1995 Hyogoken-Nambu earthquake Magnitude: M_J = 7.2 (JMA) Peak ground surface acceleration (PGA): 0.53 g Peak ground surface velocity (PGV): 1.06 m/s PGA/PGV evaluated based on strong motion earthquake records.
Seismic performance	Horizontal displacement: 0.6 m Vertical displacement: 1.4 to 2.6 m

Cross section: Vertical composite type

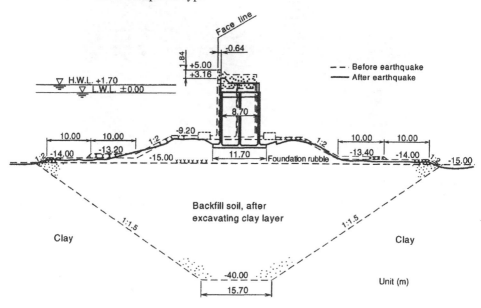

Features of damage/comments

Settlement associated with significant foundation deformation beneath the caisson/rubble foundation.

Reference: Iai et al. (1998)

Table T2.29. Case history No. 28: Patras Port, Greece.

Structural and seismic conditions	Structure and location: Southern extension Height: + 2.0 m Water depth: – 18.0 m Construction: 1984 (just completed) Earthquake: Late February of 1984 Earthquakes Magnitude: M = 3.5 to 4.5 Peak ground surface acceleration (PGA): 0.05 g PGA evaluated based on 1D equivalent linear analysis.
Seismic performance	Vertical displacement: 3.0 to 4.0 m

Cross section: Rubble mound type

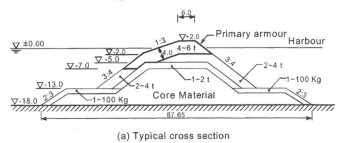

(a) Typical cross section

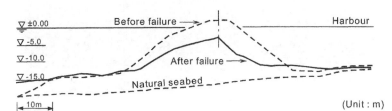

(b) Cross section before and after the failure

Features of damage/comments

Abrupt settlements of the breakwater (3.0 to 4.0 m) took place after a series of moderate earthquakes of magnitude 3.5 to 4.5. The damage was caused mainly by the large amplification (up to 2.5 times) of the accelerations within the mound.

The breakwater was constructed on a normally consolidated soft clay layer 30 to 38 m thick with SPT N-values ranging from 0 to 15 and undrained shear strength ranging from 5 to 40 kN/m^2, increasing with depth. It was underlain by a thick medium-stiff clay layer with SPT N-value greater than 42 and undrained shear strength ranging from 120 to 250 kN/m^2. The breakwater was constructed in two stages; the level of the mount reached 11 m from its foundation at the first stage, and after a period of 6 months, the level reached 18 m. Total settlement due to consolidation was about 3 m.

Reference: Memos and Protonotarios (1992)

Table T2.30. Case history No. 29: Okushiri Port, Japan.

Structural and seismic conditions	Structure and location: North breakwater, Section C Height: + 3.1 m Water depth: – 7.6 m Construction: (completed): 1988–1992 Earthquake: July 12, 1993 Hokkaido-Nansei-Oki earthquake Magnitude: $M_J = 7.8$ (JMA)

Cross section: Vertical composite type

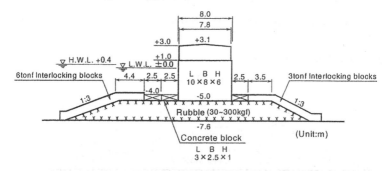

Features of damage/comments

Caissons were washed off or tilted into the rubble mound due to tsunami.

Reference: Inatomi et al. (1994)

TC3: Earthquake Motion

T3.1 Size of Earthquakes
    T3.1.1   Intensity
    T3.1.2   Magnitude
T3.2 Strong Ground Motion Parameters
T3.3 Seismic Source and Travel Path Effects
T3.4 Local Site Effects
T3.5 Seismic Hazard and Design Earthquake Motion
    T3.5.1   Seismotectonic approach
    T3.5.2   Direct approach based on seismic history
    T3.5.3   Design accelerograms
T3.6 Ground Motion Input for Seismic Analysis of Port Structures
    T3.6.1   Gravity quay walls
    T3.6.2   Sheet pile quay walls
    T3.6.3   Pile-supported wharves
    T3.6.4   Embankments and breakwaters

TECHNICAL COMMENTARY 3

Earthquake Motion

Earthquakes are complex natural phenomena, with their origin in the release of tectonic stress which has accumulated in the earth's crust. Their principal effects on port structures are caused by oscillatory ground movement, which depends on such factors as seismic source, travel path, and local site effects.

Each port, and to a certain extent each structure, requires a specific evaluation of the design parameters of ground motion. This technical commentary outlines the definitions of primary parameters, and makes recommendations pertaining to the basic data to be collected and the analytical procedures to be followed in the seismic design process.

T3.1 SIZE OF EARTHQUAKES

The basic parameters which characterise the size of earthquakes are intensity, magnitude and energy release.

T3.1.1 Intensity

Intensity is a measure of the destructiveness of the earthquake, as evidenced by human reaction and observed damage. It varies from one location to another, depending on the size of the earthquake, the focal distance and the local site conditions. Seismic damage and the corresponding intensity depend on characteristics of seismic motion, (acceleration, duration and frequency content) as well as the natural frequencies and vulnerability of the affected structures. Intensity is the best single parameter to define the destructiveness of an earthquake at a given site, but it cannot be used as input for dynamic analysis. In many cases, especially for historic earthquakes, it is the only parameter available for characterising the earthquake motion.

Several different seismic intensity scales have been adopted in different parts of the world. The Modified Mercalli (MM) Scale (Wood and Neumann, 1931; Richter, 1958) is widely used in North and South America. The early intensity scales of Mercalli or Rossi-Forel have been updated to include the twelve degrees (I to XII). The Medvedev-Sponheuer-Karnik (MSK) Scale (Medvedev and Sponheuer, 1969), recently revised to the European Macroseismic Scale (EMS) (Grunthal, 1998), is commonly used in Europe. These scales are roughly equivalent. Japan maintains the Japan Meteorological Agency (JMA) Scale, that consists of eight degrees (0 to VII). Relationships between the different seismic intensity scales can be found in the publication by ISSMGE(TC4) (1999).

Intensity is used to estimate the locations of historic earthquakes, establish the rate of occurrence of earthquakes as a function of their size, as well as to estimate the strength and attenuation characteristics of seismic ground motions. Based on intensities at different locations, a map of contours of equal intensity, called an isoseismal map, is plotted. Figure T3.1 shows examples of typical isoseismal maps. Note that the contours could be affected by population centres, regions of poor construction and areas of weak soils.

Maximum intensity in the epicentral region, I_o or I_e, is a measure of the size of an earthquake. The use of intensity for seismic hazard analysis, can be found in the publication by ISSMGE(TC4) (1999).

T3.1.2   Magnitude

Magnitude is a physical measure of the size of the earthquake, typically evaluated based on the recorded data. There are several scales based on the amplitude of seismograph records: the Richter local magnitude M_L, the surface wave magnitude M_S, the short-period body wave magnitude m_b, the long-period body wave magnitude m_B and the Japan Meteorological Agency magnitude M_J. Moment magnitude M_w is calculated from the seismic moment, which is a direct measure of the factors that produce the rupture along the fault.

The relationship between the various magnitude scales can be seen in Fig. T3.2. M_W is presently preferred by seismologists. The use of M_L for magnitudes up to 6 and M_S for magnitudes 6 to 8 does not significantly differ from the M_W values. Earthquakes with magnitude less than 3 are considered as microtremors, while those measuring up to 5 are considered minor earthquakes with little associated damage. Maximum recorded magnitude is about $M_W = 9.5$ (e.g. Chilean earthquake of 1960).

T3.2   STRONG GROUND MOTION PARAMETERS

For the seismic analysis envisaged in this book, earthquakes are characterised by the ground motions that they produce, which is usually described by means of one or several of the following parameters or functions:

- Peak Ground Horizontal Acceleration, PGA_H, or simply PGA, is the maximum absolute value reached by ground horizontal acceleration during the earthquake. It is also called peak acceleration or maximum acceleration.
- Peak Ground Vertical Acceleration, PGA_V, is the same as above, but for vertical movement.

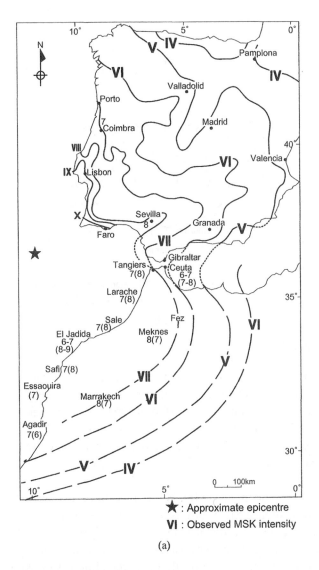

(a)

Fig. T3.1. Examples of isoseismal maps.
      (a) 1755 Lisbon earthquake. MSK scale (after Elmrabet et al., 1991).
      (b) 1964 Niigata earthquake. JMA scale (after Okamoto, 1973).
      (c) 1989 Loma Prieta earthquake, MM Scale (after Jones et al., 1995).

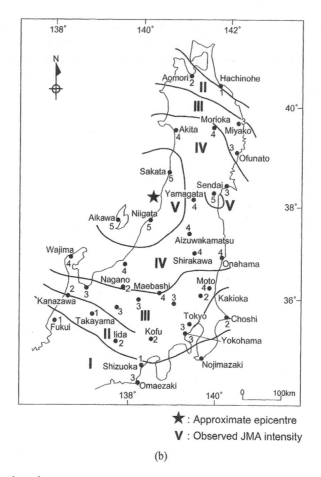

(b)

Fig. T3.1. Continued.

- Accelerogram $a(t)$ is the time history of acceleration. It can be horizontal or vertical. It is usually an irregular function of time. Figure T3.3 shows some examples of accelerograms from different types of earthquakes.
- Peak Ground Horizontal Velocity, PGV_H, or simply PGV, is the maximum horizontal component of the ground velocity during the earthquake.
- Peak Ground Vertical Velocity, PGV_V, is the same as above, but for vertical component.
- Peak Ground Displacement, PGD, is the maximum displacement.
- Acceleration Response Spectrum $S_A(T, D)$ represents the maximum acceleration (absolute value) of a linear single degree-of-freedom (SDOF) oscillator, with period T and damping $D\%$ of critical, when the earthquake motion is applied to its base. The SDOF oscillator is the simplest model of a structure. Thus the spectrum represents a good approximation of the response of the different structures when they are subjected to an earthquake. Similarly, there is a Velocity Response Spectrum, $S_V(T, D)$, and a Displacement Response

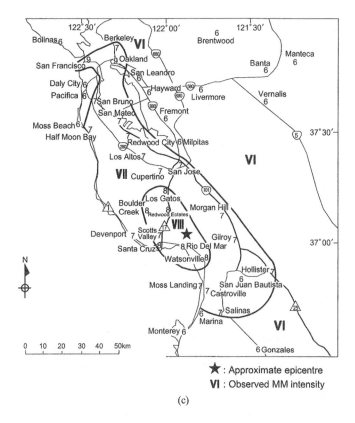

★ : Approximate epicentre
VI : Observed MM intensity

(c)

Fig. T3.1. Continued.

Spectrum $S_D(T, D)$, although, in these cases, the velocity and the displacement are not absolute but relative to the base. The response spectra usually refer to horizontal movement, but can also be defined for vertical movement. Figure T3.4 shows examples of response spectra of different types of earthquakes.

– Fourier Spectrum represents the decomposition of an accelerogram as the sum of a series of harmonic functions of variable amplitude, phase and frequency (see Appendix of Technical Commentary 4).

– Duration is a measure of the time between the beginning and the end of the ground motion. Two types of duration are used in practice:

> Bracketed Duration, associated with a given acceleration 'a' (usually 5% or 10% of 'g'), which is the time between the first and last occasions when acceleration exceeds value 'a'.

> Significant Duration, associated with a given proportion of seismic energy E_n (usually 90%), which is the time needed to build up between the $(100 - E_n)/2\%$ and the $(100 + E_n)/2\%$ of the total seismic energy in the accelerogram.

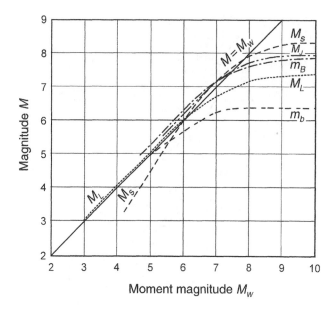

Fig. T3.2. Relationship between moment magnitude M_W and various magnitude scales: M_L (local magnitude), M_S (surface wave magnitude), m_b (short-period body wave magnitude), m_B (long-period body wave magnitude), and M_J (Japan Meteorological Agency magnitude) (after Heaton et al., 1982; as reported by Idriss, 1985).

– Equivalent number of uniform cycles, N_{eq}, is a parameter related to the duration of ground movement, used to approximate an irregular time history by N_{eq} cycles with uniform amplitude. It is useful in calculating resistance to liquefaction by some simplified methods. The parameter N_{eq} depends on the duration of the earthquake and on the relative amount of low and high peaks. It can be computed for each time history. Some authors (Seed and Idriss, 1982; Arango, 1996) have tabulated N_{eq} as a function of magnitude.

T3.3   SEISMIC SOURCE AND TRAVEL PATH EFFECTS

The tectonic mechanism in the seismic source zone, the source-to-site distance, and the attenuation characteristics of the motions along the travel path, influence the resulting ground motion at the site of interest. Depending on the position of the hypocentre and its relationship with the global tectonics processes, earthquakes can be classified as

➤ interplate
➤ intraplate

Interplate earthquakes originate in the great stress accumulated in contacts between tectonic plates (the principal fragments into which the earth crust is divided). Most destructive earthquakes, with magnitude above 7 to 8, belong to

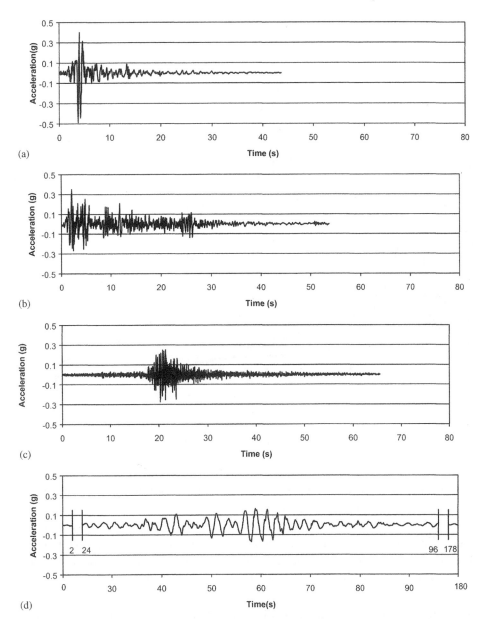

Fig. T3.3. Examples of accelerograms from different types of earthquakes and site conditions.
(a) Parkfield, CA, USA, 1966-06-28. A record of almost a single shock.
(b) El Centro, CA, USA, 1940-05-19. A record of moderate duration with a
wide band frequency content.
(c) Lima, Peru, 1966-10-17. A record with predominant high frequency
components.
(d) Mexico City, Mexico, 1985-09-19. A long duration record with predominant low frequency components.

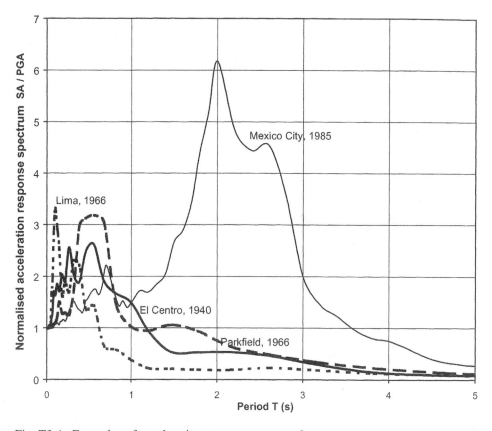

Fig. T3.4. Examples of acceleration response spectra shapes.

this category. The earthquakes which originate in the subduction zones are a particular type of interplate earthquakes.

Intraplate earthquakes originate in faults within a plate and away from its edges. The intraplate earthquakes typically do not reach such high magnitudes. Magnitudes above 7.5 are unusual, although large historic earthquakes have been documented in those regions (e.g. 1811–1812 New Madrid Earthquakes, USA).

As shown in Fig. T3.5, the type of faults causing the earthquakes are usually classified as reverse, normal, or strike-slip. Tectonic movement in reverse and normal faults occurs mainly in the direction of the dip, with a relatively high vertical component. Faults of strike-slip type are typically associated with horizontal movement. In general, reverse faults allow greater normal stress to be transmitted than normal faults and are capable of storing and, consequently, releasing in an earthquake a larger amount of energy. The type of fault also influences the frequency content of the vibratory motion and the intensity of the motions in the near field of the source (i.e. near-fault effects).

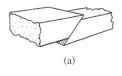

(a)

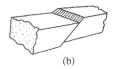

(b)

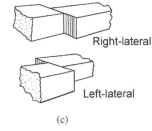

Right-lateral

Left-lateral

(c)

Fig. T3.5. Types of seismic faults.
(a) Reverse.
(b) Normal.
(c) Strike-slip.

Another conventional way of classifying earthquake motions is related to the distance between the hypocentre and the site where the earthquake is felt. High frequencies attenuate faster than low ones. Near field ground motions are traditionally distinguished from other types. The term "near field earthquake" is useful in characterising the earthquake motion close to the seismic source, as represented by their higher peak acceleration and greater content in high frequencies. However, a continuous variation of these parameters with epicentral distance is presently accepted.

In practice, the effects of seismic source and travel path are taken into account through magnitude and distance. The movements at the bedrock or at an outcropping rock has an amplitude that increases with magnitude and decreases with distance. Predominant periods are influenced by the same factors. Generally, the greater the magnitude or focal distance, the greater the predominant period. Figure T3.6 compares the response spectra of several Mexican earthquakes. All of the earthquake motions associated with these response spectra came from essentially the same source and were recorded at about the same distance. The variations in amplitude and frequency content, particularly in the long-period range, are apparent in Fig. T3.6. Figure T3.7 compares the median response spectra for strike-slip events of magnitude 7 at several distances. The decrease of spectral ordinates and the displacement of predominant periods towards long period range are also apparent in Fig. T3.7.

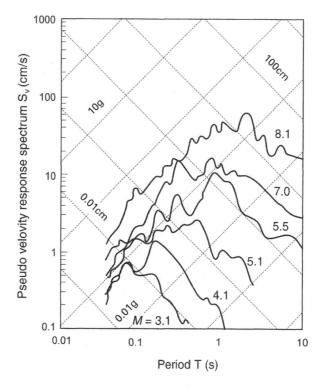

Fig. T3.6. Effects of magnitude on response spectra (after Anderson, 1991).

T3.4   LOCAL SITE EFFECTS

There are important differences in ground motion depending on the local site conditions. The main local factor that influences the movement of the surface during an earthquake is the presence of soil (especially when soft) over the bedrock. Effects of the topography (i.e. hill effects) and of the shape of the soil deposits (i.e. basin effects) have also been noted.

The effects of a soft soil layer is usually the amplification of the vibration, especially in some low frequencies. The main physical mechanism responsible for the motion amplification at soft soil sites is related to the resonance of the components of vibration with frequency near the natural frequency of the soil layer. The fundamental resonant period for shear wave, T_s, propagating through a soil deposit with a shear wave velocity of V_s with a thickness of H, is given by $T_s \approx 4H/V_s$. At the surface, the upgoing and downgoing waves of such a frequency are in phase, resulting in a double amplitude wave. The resonance amplification could be significantly modified by the damping and the non-linear properties of the soil.

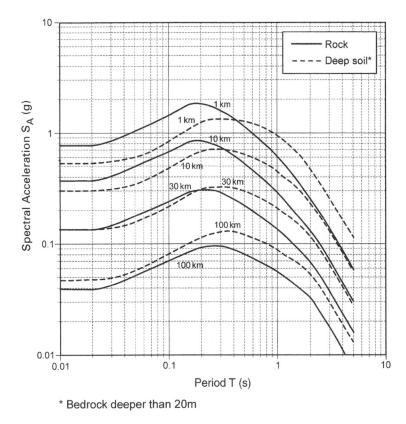

Fig. T3.7. Effects of distance on response spectra (after Abrahamson and Silva, 1997).

Figure T3.8 shows the differences in observed surface accelerations depending on local site conditions. For low and medium PGA, acceleration in the soft soil surface is larger than acceleration in the rock outcropping. However, for high acceleration in the bedrock, the shear strength of soil limits the level of the acceleration, resulting in the decrease in acceleration amplification, and eventually in the deamplification.

As shown in Fig. T3.9, the frequency content of the movement at the surface of a soil deposit is different from that in a nearby outcrop rock or in the bedrock. Typically, the predominant periods in rock is tenths of a second. Depending on the thickness and stiffness of the soil layer, the predominant period at the surface can be tenths of a second to a few seconds. Accordingly, the maximum seismic vibratory effects for structures founded on rock can occur for natural periods of tenths of a second (e.g. stiff buildings, gravity quay walls), while, at the surface of a soil deposit, the most affected structures can be those with a natural period of tenths of a second to a few seconds (e.g. high-rise buildings, offshore platforms, pile-supported wharves).

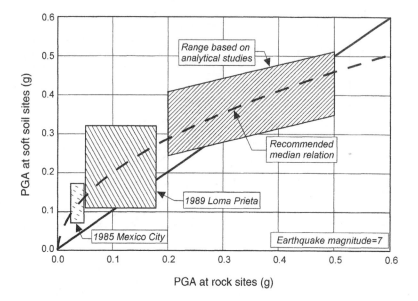

Fig. T3.8. Comparison of peak horizontal accelerations at soft soil sites with accelerations at rock sites (after Idriss, 1991).

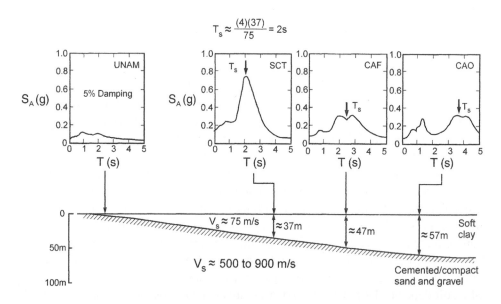

Fig. T3.9. Effects of soil conditions in recorded acceleration response spectra for several sites in Mexico City during the 1985 earthquake (after Dobry, 1991).

T3.5   SEISMIC HAZARD AND DESIGN EARTHQUAKE MOTION

Peak horizontal acceleration, peak horizontal velocity and response spectra ordinates are commonly used parameters to characterise the seismic hazard at a given site. There are two approaches to compute these parameters of earthquake motions for the seismic analysis of port structures. They are a seismotectonic approach and a direct approach based on seismic history.

T3.5.1   Seismotectonic approach

In the seismotectonic approach, there are two different, but complementary methods:
– Deterministic method: The largest earthquake(s) which could affect a given site is used. Commonly, the result of a deterministic analysis consists of the Maximum Credible Earthquake (MCE), the maximum earthquake that appears possible to occur under the known tectonic conditions, taking into account seismic history of the region.
– Probabilistic method: Several values of the earthquake motion parameters are used, usually acceleration or response spectra ordinates, associated with annual exceedance probability.

The procedures for deterministic and probabilistic analysis are shown schematically in Figs. T3.10 and T3.11. The process includes the following steps:

Step 1. Identification of active faults and other seismic sources.

In both the deterministic and probabilistic methods, all sources capable of producing significant ground motion at the site must be considered. The locations and other parameters of active, and potentially active seismic sources, should be identified.

In the probabilistic method, the temporal occurrence of earthquakes should also be characterized. In addition to hazards associated with the specific faults, broader seimotectonic provinces, i.e. regions with uniform tectonic and seismic conditions, are often defined. Recent earthquakes in several seismically active regions of the world demonstrate that the current state of knowledge of both the spatial and temporal occurrence of potentially damaging earthquakes can be incomplete. This uncertainty in the characterization of the seismic hazard is compounded in regions of low- to moderate-seismicity, and in areas where the seismic sources are not well understood. In light of the seismic hazard associated with unidentified sources, the inclusion of areal, or 'random', sources is warranted in most regions of the world. The distribution and rate of occurrence of earthquakes associated with areal sources are specified based on the nature of the seismotectonic province.

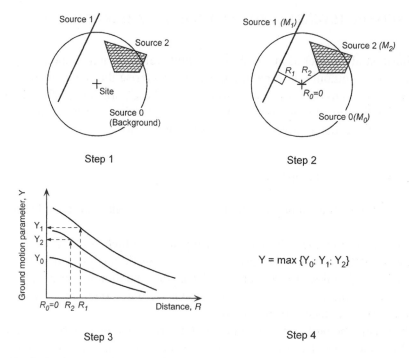

Fig. T3.10. Main steps of a deterministic seismic hazard analysis.

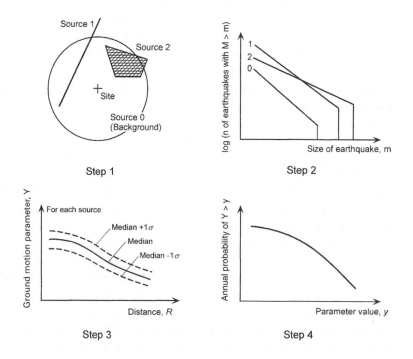

Fig. T3.11. Main steps of a probabilistic seismic hazard analysis.

The area studied should include seismic sources, both onshore and off-shore. The methodology to be applied includes the interpretation of:
– Topographic and bathymetric maps;
– Seismicity maps;
– Geophysical surveys;
– Repeated high precision geodetic measurements;
– Aerial photographs;
– Geomorphological data;
– Stratigraphic correlations;
– Paleoseismicity (i.e. geologic guidance for pre-historic earthquakes).

Step 2. Characterisation of each seismic source activity

– Step 2a. In the Deterministic Method:

The maximum earthquake (M_i) and the minimum distance (R_i) from the source to the site are defined. The term 'maximum earthquake' is not well defined. It is preferable to refer to the Maximum Credible Earthquake, MCE, the largest earthquake that appears capable of occurring in a seismic source. As pointed out by Wyss (1979), Fig. T3.12, dimensions of the fault segment, which could move at the same time, limit the size (magnitude) of the associated earthquakes. If causative faults are not known, the regional seismic history is the basic information which helps the engineer to decide the maximum credible earthquake. For R_i,

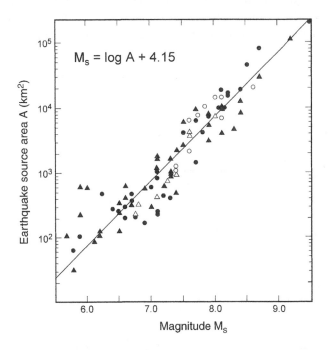

$$M_s = \log A + 4.15$$

Fig. T3.12. Data and regression for magnitude versus fault-rupture area (after Wyss, 1979).

epicentral, hypocentral or fault plane distance may be used, depending on the attenuation model considered in Step 3.

– Step 2b. In the Probabilistic Method:

The parameters of the earthquake occurrence statistics are defined, including the probability distribution of potential rupture locations within each source and the recurrence relationship. Commonly, it is assumed that all points within the source have the same probability of originating an earthquake. The recurrence relationship specifies the average rate at which an earthquake of a given size will be exceeded, and also, the maximum earthquake, the same as discussed in the deterministic method.

For modelling the occurrence of earthquakes of different magnitudes, the conventional exponential model (Gutenberg and Richter, 1944), or any of its variants are commonly used. They are based on the Gutenberg-Richter relationship, that relates magnitude (or intensity) M with the mean annual number of events, n, that exceeds magnitude (or intensity) M:

$$\log n = a - bM \tag{T3.1}$$

The coefficients a and b must be obtained by regression of the data of each seismic source. They could depend on the range of earthquake sizes used in the regression. If the seismic catalogue is incomplete for small earthquakes, as it is usual, only earthquakes with a size beyond a certain threshold level must be used.

Figure T3.13 shows the application of Gutenberg-Richter relationship to primary worldwide seismicity zones for earthquakes with magnitude $m > 6$. This

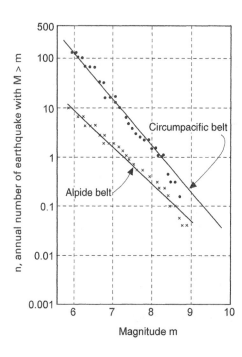

Fig. T3.13. Gutenberg-Richter recurrence law applied to worldwide seismicity (after Esteva, 1970).

model was first derived for a set of regional data, that included several seismic sources as in seismotectonic provinces in regions of low seismicity, but it would not be appropriate for the occurrence of large earthquakes in a fault. For those cases, the elastic rebound theory predicts the occurrence of a large 'characteristic' earthquake when the elastic stress in the fault exceeds the resistance of the rock. Characteristic earthquakes repeatedly occur with a segment of a fault having the maximum possible source dimensions.

The temporal model commonly used for the occurrence of earthquakes is a Poisson model, in which each earthquake is a memoryless event, independent of the previous and next earthquake at the same location. The elastic rebound theory suggests that the occurrence of characteristic earthquakes on a particular fault segment should not be independent of past seismicity. A renewal process better than a Poisson process would represent the occurrence of these large characteristic earthquakes. However, the Poisson model seems adequate, physically and observationally, to represent the smaller events, presumably due to local readjustment of strain, secondary to main elastic rebound cycles. A hybrid (Renewal-Poisson) model was proposed by Wu et al. (1995). Figure T3.14 illustrates this model. In this mode, small (to moderate) earthquakes follow a Poisson process in time and the exponential part of a characteristic distribution in magnitudes. Characteristic earthquakes follow a general renewal process in time and the non- exponential part of a characteristic distribution in magnitudes. Times and magnitudes of all earthquakes are assumed mutually independent. The terms m_l^P and m_u^P are the lower and upper bounds of the magnitudes of the small earthquakes; m_l^C and m_u^C are those of the characteristic earthquakes. The value of m_u^P is assumed less than or equal to m_l^C.

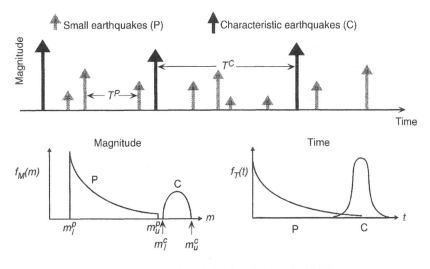

Fig. T3.14. Hybrid Poisson renewal model (after Wu et al., 1995).

Step 3. Determination of the attenuation relationship for the acceleration, response spectra ordinates or other parameters of interest.

– Step 3a. In the Deterministic Method:

One attenuation relationship is typically used that should adequately represent the attenuation for the region. If a statistical analysis has been done, a median relationship would not be enough. A relationship should be selected that corresponds to a conservative estimate of the design parameters.

– Step 3b. In the Probabilistic Method:

Ground motion parameters (e.g., PGA, PGV, spectral acceleration) are routinely estimated in practice based on the probabilistic evaluation of mean values obtained from the attenuation relationships. In regions of high seismicity, or in applications involving long exposure intervals that approach the return period for the largest earthquake(s) expected in the region, the standard deviation term established for the specific attenuation relationship being employed will often be used. The application of the standard deviation term in estimates of the ground motion parameters accounts for the probability of experiencing greater than mean motions during the period of interest. The mean attenuation function and the standard deviation should be computed through the statistical analysis of data from earthquakes of the same region or, at least, from earthquakes of similar tectonic environment, recorded in stations with travel paths and local ground conditions similar to those of the site of interest.

In recent years, it has become common practice to use specific attenuation relationships for each of the spectral ordinates of different periods. A general expression for an attenuation relationship is:

$$\log y_g = f_1\,(F_T) + f_2\,(M) + f_3\,(R) + f_4\,(S_T) + \varepsilon_\sigma \tag{T3.2}$$

where:

y_g = ground motion parameter or response spectrum ordinate

F_T = a set of discrete variables describing the fault type

M = magnitude

R = a measure of distance

S_T = a set of discrete variables describing the site subsoil conditions or a continuous variable depending on the average shear wave velocity in the deposit

f_i = functions, f_2 is often assumed linear in powers of M, f_3 depends on R, $\log R$ and, sometimes, M

ε_σ = a random error term with zero mean and σ standard deviation

This procedure allows the inclusion of fault type and distance effects and supplies appropriate response spectra for rock. However, results for soil sites are averages of values from different soil conditions and will not represent any particular site. In many cases, it may be preferable to first

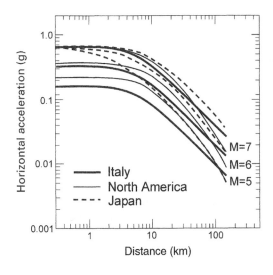

Fig. T3.15. Comparison of proposed PGA attenuation with distance for several regions of the world (after ISSMGE (TC4), 1999).

obtain the ground motion parameters in rock and then compute the seismic response at the ground surface.

Figure T3.15 shows the comparison among the attenuation of PGA_H from Japan, Italy and North America. Selected attenuation relations for spectral ordinates can be found in Abrahamson and Silva (1997), or in Sabetta and Pugliese (1996).

Step 4. Definition of the seismic hazard

Ground motion, primarily described by PGA and spectral ordinates must be defined.

Other parameters such as intensity or duration can also be obtained in a similar way.

– Step 4a. In the Deterministic Method:

Calculate the values of the motion parameters at the site from each source of the earthquakes and select the maximum.

– Step 4b. In the Probabilistic Method:

i) Calculate the annual number of occurrences of earthquakes from each source which produce, at the site, a given value of the PGA_H (or other earthquake motion parameters).

ii) Calculate the total annual number, n, of exceedance.

iii) Calculate the reciprocal of the mean annual rate of exceedance for the earthquake effect (PGA_H):

$$T_R = \frac{1}{n}$$

(T3.3)

This is the return period of earthquakes exceeding that PGA_H

iv) Calculate the probability $P_R(a_{max}, T_L)$ of the PGA_H being exceeded in the life T_L of the structure.

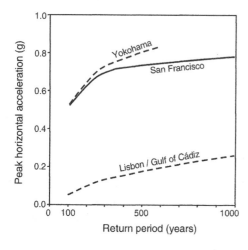

Fig. T3.16. Peak horizontal bedrock acceleration evaluated for different regions.

$$P_R(a_{max}, T_L) = 1 - (1 - n)^{T_L} \approx T_L n = \frac{T_L}{T_R} \quad (\approx \text{for } T_L/T_R \ll 1.0) \tag{T3.4}$$

The results of the analysis, sensitive to the details of the procedures used, reflect the seismic hazard of each site. As presented in Fig. T3.16, the seismic hazard is a variable in different parts of the world.

Step 5. If the attenuation relationships do not match with the local site conditions (e.g. if attenuation relationships for rock are used and there is a surface soil deposit), convert the motion parameters to the specific site conditions using empirical amplification ratios or numerical dynamic soil response models.

For preliminary studies, or when data are scarce, the methodology previously discussed is simplified. The traditional method, which is still used, consists of calculating the PGA_H by means of a probabilistic or a deterministic study and scaling a response spectrum to this PGA_H. In this method, care must be taken in the selection of the response spectrum shape, which depends on fault type, magnitude, distance to rupture zone and local site conditions. Figure T3.17 shows the shapes of elastic response spectra proposed by Bea (1997).

T3.5.2   Direct approach based on seismic history

If a particular earthquake controlling the seismic hazard is defined directly from the seismic history, and taking into account the tectonic characteristics of the region around the site, the seismic hazard analysis is significantly simplified. An example is the conventional practice in seismic design codes, where every time that a set of the most recent strong earthquake motions (larger than the previously

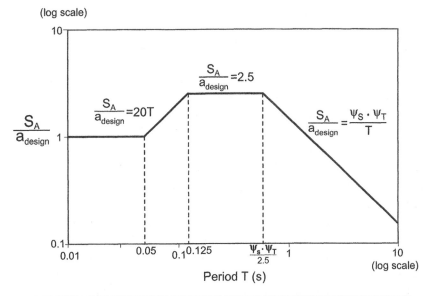

Fig. T3.17. Design response spectra shape proposed by Bea (1997) with modification.

Soil condition	Ψ_s	Tectonic condition	Ψ_T
A-Rock (Vs > 750 m/s)	1.0	**Type A** Shallow crustal faulting zones	1.0
B - Stiff to very stiff soils, gravels (Vs = 360 to 750 m/s)	1.2	**Type B** Deep subduction zones	0.8
C - Medium stiff to stiff clays (Vs = 180 to 360 m/s)	1.4	**Type C** Mixed shallow crustal & deep subduction zones	0.9
D - Soft to medium stiff clays (H = 1 to 20 m; Vs < 180 m/s)	2.0	**Type D** Intraplate zones	0.8
E - Conditions other than above	Site specific studies required	**Default value**	1.0

recorded ones) are obtained, the seismic code has been modified to incorporate the response spectra of the last recorded earthquake motions.

In this direct approach, the magnitude, the distance and the fault type of the design earthquake are defined in the simplified manner described above. Accelerograms recorded from earthquakes in these magnitude and distance ranges, with the similar fault type, propagation path and recording site conditions as the design earthquake, are then used to define response spectra at various percentile levels suitable for design.

T3.5.3   Design accelerograms

Depending on the type of the analysis, time histories might be defined. Either recorded or generated accelerograms may be adequate.

Recorded time histories must be selected from the collections of accelerograms caused by earthquakes whose fault type, magnitude and distance are similar to those of the design earthquake and can be scaled to the design PGA$_H$. Commonly it is necessary to use several accelerograms to cover the implicit contingency in its selection. When the strong ground motion parameters of the design earthquake are influenced by several seismic sources, it is not possible to find one recorded accelerogram that incorporates all the characteristics of the design motion, including all the spectral ordinates and duration. In such a case, several records of each earthquake type should be used.

Typical sources of digitised accelerograms can be found at:
- the National Geophysical Data Centre (http://www.ngdc.noaa.gov) in the USA; and
- the K-NET (http://www.K-NET.bosai.go.jp) in Japan (Kinoshita, 1998).

In particular, Port and Harbour Research Institute, Japan, have operated the strong motion recording network throughout port areas in Japan since 1964. The recorded strong motion data have been compiled as annual publication (e.g. Sato et al., 1999), together with occasional compilation and update of local site conditions at the strong motion recording stations (e.g. Ichii et al., 1999). Opening of a web site for easy access to these data are planned in near future.

For the spectral analysis of structures, it is important that the frequency content of the recorded time histories matches the spectral shape of the design earthquake. If the maximum recorded acceleration is different from the design PGA$_H$, the recorded accelerogram can be scaled to this value. In some cases, a slight modification of the time scale in the recorded accelerogram is also acceptable to achieve a better agreement between the frequency contents of the recorded and the design acceleration time histories. For liquefaction analysis, it is also important that the duration of a recorded accelerogram matches the design duration.

Artificial time histories must match the PGA$_H$, the duration and the response spectrum of the design earthquake.

Formulation of design earthquake motions for near source regions, including Level 2 earthquake motion, could require adequate modelling of fault rupture process and travel path effects. Practice-oriented guidelines proposed by the Japan Society of Civil Engineers are summarized by Ohmachi (1999).

The seismic hazard analysis discussed in Section T3.5.1 can be generalized by using earthquake magnitude and source-site distance as primary earthquake parameters (Kameda and Ishikawa, 1988). In this approach, the magnitude and distance are evaluated as conditional mean values, given that the ground motion intensity (e.g. PGA$_H$) exceeds the level specified by the annual probability of exceedance. Other ground motion parameters, including predominant period, ground motion duration, response spectral ordinates, can be determined as a function of these hazard-consistent magnitude and distance. This approach also forms the basis for establishing the link between probabilistic seismic hazard analysis and the deterministic scenario earthquakes, and makes it possible to associate the scenario earthquakes with the specific risk level (Ishikawa and Kameda, 1991).

T3.6   GROUND MOTION INPUT FOR SEISMIC ANALYSIS OF PORT STRUCTURES

As discussed in Chapter 5 in the Main Text, the seismic input to be used for analysis and design depends on the type of port structure and on the type of analysis being performed. For dynamic analysis of soil-structure interaction, seismic motions at the bedrock is required. If the seismic data consist of free field ground surface motions, it is necessary to compute the subsurface/bedrock motion by numerical deconvolution techniques. For simplified analysis and simplified dynamic analysis, the seismic input depends on the configuration of the structure.

In some cases, the effects of elevation difference between the landward ground surface and the seabed can be an important issue, that should be considered in the development of the response spectra and other earthquake motion parameters. There is not a criteria well established in design practice, and this issue needs careful studies if higher reliability is required on the results of the simplified and simplified dynamic analyses.

T3.6.1   Gravity quay walls

Simplified analysis of gravity quay walls assumes a uniform field of seismic acceleration. Therefore, the design acceleration can be the average acceleration over the height of the caisson and the backfill soil wedge areas. To estimate the average acceleration, one-dimensional site response analysis can be performed using the geotechnical configuration at B and C locations shown in Fig. T3.18.

Simplified dynamic analyses include the Newmark-type rigid block sliding and the simplified chart approaches based on the parametric studies (see Technical Commentary 7). For the Newmark-type analysis, a time history of acceleration averaged over the caisson and the backfill soil wedge can be used. For simplicity, the time history at the elevation of the bottom of the caisson scaled to the average of the peak acceleration over the sliding mass is often used.

For simplified chart approaches based on the parametric studies, the input acceleration must be representative of the one used to develop the chart.

T3.6.2   Sheet pile quay walls

Simplified and simplified dynamic analyses of sheet pile quay walls assumes uniform field of seismic acceleration on the active and passive soil wedges.

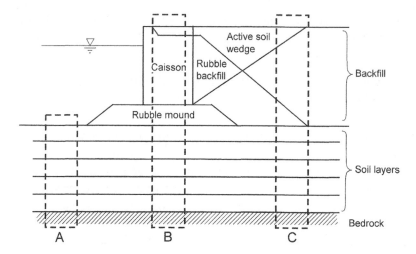

Fig. T3.18. Earthquake motion input for seismic analysis of gravity quay walls.

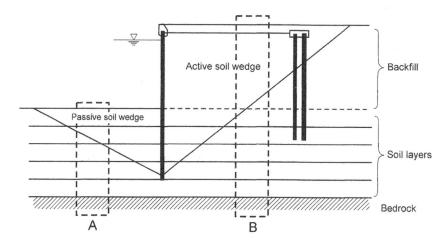

Fig. T3.19. Earthquake motion input for seismic analysis of sheet pile quay walls.

Therefore, the design acceleration can be the average over the active and passive soil wedges. One-dimensional site response analysis can be performed using the geotechnical configuration either in front of or behind the wall in locations A and B shown in Fig. T3.19. There may be an option to adopt different accelerations for the active and passive soil wedges.

T3.6.3   Pile-supported wharves

For simplified and simplified dynamic analyses of pile-supported wharves, the equivalent fixity pile model is often adopted as shown in Fig. T3.20. It is typical

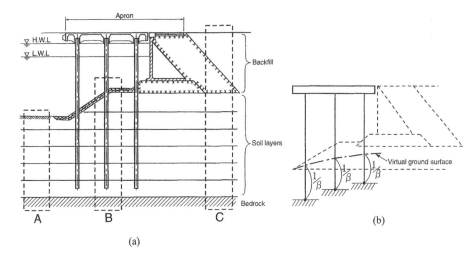

Fig. T3.20. Earthquake motion input for seismic analysis of pile-supported wharves.
(a) Pile-supported wharf.
(b) Modelling by a frame structure having equivalent fixity lengths.

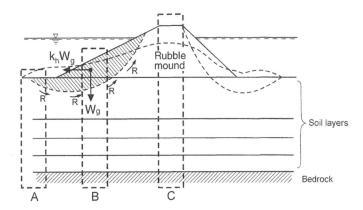

Fig. T3.21. Earthquake motion input for seismic analysis of embankments and break-waters.

to obtain the seismic action at the base of the equivalent structure by an one-dimensional site response analysis of the soil column in the middle portion of dike/slope below the deck, in location B in Fig. T3.20a.

T3.6.4   Embankments and breakwaters

Simplified and simplified dynamic analyses of embankments and breakwaters, often assumes a sliding failure block as shown in Fig. T3.21. In this case, average acceleration over the sliding block may be adequate for sliding block analysis. It can be obtained by a one-dimensional response analysis of a wedge shear beam or averaging the results of the one-dimensional site response analysis of soil columns A and B, shown in Fig. T3.21.

TC4: Geotechnical Characterisation

T4.1   Mechanical Behaviour of Soil under Cyclic Loads
    T4.1.1   Soil behaviour in pre-failure conditions
    T4.1.2   Soil behaviour at failure
T4.2   Measurement of Soil Parameters
    T4.2.1   In situ penetration tests
    T4.2.2   In situ geophysical tests
    T4.2.3   Laboratory tests
    T4.2.4   Combined use of in situ and laboratory tests
T4.3   Evaluation of Pre-Failure Soil Properties through Empirical Correlations
    T4.3.1   Small strain stiffness
    T4.3.2   Small strain damping
    T4.3.3   Non-linear pre-failure behaviour
T4.4   Assessment of Liquefaction Potential
    T4.4.1   Liquefaction potential assessment using field data
    T4.4.2   Liquefaction potential assessment combining field and laboratory data
T4.5   Geotechnical Characterisation Strategy with Reference to the Analysis Requirements
Appendix: Interpretation of Geophysical Tests
    A.1   Time domain analysis
    A.2   Frequency domain analysis
    A.3   Spectral analysis and propagation velocity

TECHNICAL COMMENTARY 4

Geotechnical Characterisation

This technical commentary addresses characterisation of subsoil and earth-fill for site response, liquefaction and soil-structure interaction analysis. The subsoil refers to the natural soil deposit interacting with a structure foundation, and extending from the ground surface to the bedrock as illustrated in Fig. T4.1. The subsoil can significantly modify the seismic waves propagating from the bedrock, and affect the behaviour of a port structure under earthquake loading. The earth-fill is the man-made deposit interacting with the structural surfaces of a port structure. The earth-fill can be either a backfill (for quay walls as shown in Fig. T4.1) or a foundation layer (for quay walls and breakwaters). In some cases, e.g. for rubble mound breakwaters, the earth-fill can be the structural material. Appropriate characterisation of subsoil and earth-fill is important for site response and seismic analysis of port structures.

This technical commentary begins with reviewing the soil behaviour under cyclic loads in Section T4.1, followed by the relevant field and laboratory tests in Section T4.2. Empirical correlations are addressed with respect to deformation properties in Section T4.3, and procedures for liquefaction potential assessment are summarized in Section T4.4. The technical commentary concludes with the recommended strategy for geotechnical characterisation in Section T4.5, with an Appendix on the interpretation of geophysical tests. The contents of this technical commentary reflect the state-of-the-practice in the geotechnical characterisation around the world.

T4.1 MECHANICAL BEHAVIOUR OF SOIL UNDER CYCLIC LOADS

The effects of earthquake loads on a soil element for port structures can be represented by a complex time history of shear stress $\tau(t)$. The pre-existing quasi-static stress state due to gravity further affects the soil element behaviour. Depending on the level of earthquake motions considered for design, the shear strain level induced in the soil can vary. According to the strain level considered,

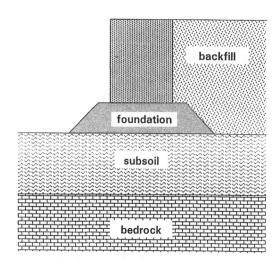

Fig. T4.1. Schematic view of subsoil and backfill in a typical quay wall.

soil behaviour may be characterised with increasing complexity, and accordingly specified with different geotechnical parameters and variables. The soil behaviour may be broadly classified into pre-failure and failure conditions, depending on the strain level. The relevant soil behaviour is reviewed in the following subsections.

T4.1.1   Soil behaviour in pre-failure conditions

The level of shear strain γ in a competent horizontally layered site is usually less than 1% for low and moderate earthquake shaking. Even at such low to moderate strain amplitudes, the soil exhibits non-linear behaviour, typically specified by two parameters; the equivalent shear modulus G and equivalent damping ratio D (Richart et al., 1970). These parameters define the characteristics of a shear stress-strain loop of a soil element undergoing a cyclic simple shear as shown in Fig. T4.2.

The equivalent parameters vary with increasing cyclic strain amplitude γ, as schematically shown in Fig. T4.3. With increasing strain amplitude, the non-linear, irreversible behaviour of a soil becomes more pronounced, resulting in the decrease in the equivalent shear modulus G and the increase in the equivalent damping ratio D (Fig. T4.3(a)). In the following, the pre-failure soil behaviour is described with reference to three strain levels: small strains (Fig. T4.3(b)), medium strains (Fig. T4.3(c)), and large strains (Fig. T4.3(d)).

(1) Small strains
Figure T4.3(a) shows that the equivalent shear modulus G of a soil initially takes a maximum value (G_0 or G_{max}), after that it decays with increasing shear strain level. Although soil non-linearity starts from a very low strain level, it is common practice to assume $G(\gamma)$ constant and equal to G_0 if the shear strain level is lower

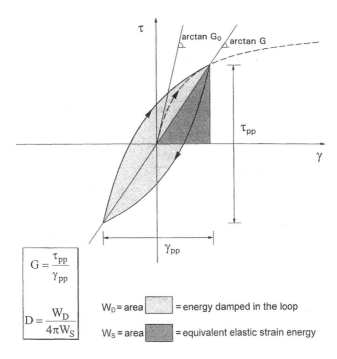

Fig. T4.2. Typical shear stress-strain loop and definitions of the equivalent parameters G and D.

than a threshold level designated by linear threshold γ_l (Vucetic, 1994). The equivalent damping ratio D takes its minimum constant value D_0 in the same strain range. Within this strain range, the τ-γ loops are approximately elliptical, and soil response can be adequately represented by a linear model.

For clean granular soils, even at the highest loading rates, dissipative effects are negligibly small ($D_0 \cong 0$) and the τ-γ relationship is practically linear (Fig. T4.3(b)). Therefore, an elastic model can be appropriate for the analyses.

Fine-grained soils can exhibit rate-dependent behaviour even in such a small strain level; the equivalent parameters G_0 and D_0 increase with higher loading rate (Shibuya et al., 1995; d'Onofrio et al., 1999a). This rate-dependency may be represented by a visco-elastic model.

(2) Medium strains

Beyond the linear threshold γ_l, soil behaviour begins to show a marked non-linearity and, as shown in Fig. T4.3(a), the equivalent shear modulus G gradually decreases with increasing γ. The shape of the τ-γ loop changes, taking a more pronounced spindle shape: the equivalent shear modulus $G(\gamma)$ decreases and the energy absorbed per cycle (i.e. $D(\gamma)$) increases.

In the medium strain level, the soil behaviour is practically unaffected by the previous loading history; regardless of the previous loading history, the same

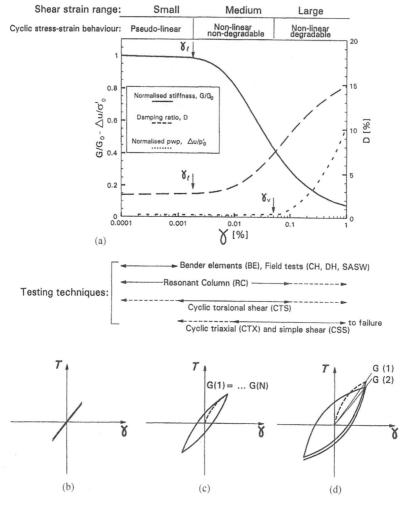

Fig. T4.3. Variation of the equivalent shear modulus (*G*) and damping ratio (*D*) with the
cyclic shear strain amplitude.
(a) Typical $G(\gamma)$, $D(\gamma)$ curves and related experimental techniques.
(b) Typical stress-strain loop at small-strains.
(c) Typical stress-strain loop at medium strains.
(d) Typical stress-strain loop at large strains.

stress-strain loop is obtained for a loading cycle at the same strain level
(Fig. T4.3(c)). Within this strain range, the cyclic soil behaviour can be termed
as 'stable': cyclic loads neither produce residual volumetric strains, ε_v, in
drained conditions, nor appreciable pore-water pressure build-up, Δu, in
undrained conditions. Therefore, the saturated soil behaviour can be represented by an equivalent one-phase medium, and total stress analysis may be still
applicable.

In practice, the stress-strain response can be analytically described by either an equivalent-linear or a non-linear model. An equivalent-linear model is typically specified by the parameters, G and D, referring to the current strain level γ. Alternatively, a non-linear elasto-plastic model, with adequate kinematic hardening rules, may represent the stress-strain relationship throughout irregular sequence of loading. If the kinematic hardening rule is based on Masing's rule (Hardin and Drnevich, 1972), appropriate modifications need to be introduced in order to avoid undesirable effects, including ratcheting/drifting of hysteresis loops (Pyke, 1979).

The equivalent linear model is typically adopted for site response analyses in the frequency domain, using amplification (or transfer) functions, with iterations on soil parameters to account for strain dependent properties (e.g. Schnabel et al., 1972). The non-linear elasto-plastic approach is typically implemented in time-domain analyses and performed by means of step-by-step time integration, with progressive updating of tangential soil properties (e.g. Finn et al., 1977).

(3) Large strains

In the large strain levels, beyond the so-called volumetric threshold γ_v (which is typically greater than γ_l by one to two orders of magnitude), the soil undergoes irreversible micro-structural changes. The effects of repeated cyclic loading become evident with respect to the shear-volumetric coupling, inducing pore-water pressure build-up for undrained cyclic simple shear loads.

In such conditions, a progressive increase in the shear strain amplitude at constant cyclic stress or, conversely, a reduction in stress at constant strain amplitude can be observed (Fig. T4.3(d)). As a consequence, the equivalent soil parameters G and D progressively change with the number of loading cycles. As an analogy to 'fatigue behaviour' for solid materials, these changes may be defined by cyclic degradation indexes, δ_G and δ_D, defined as

$$\delta_G = \frac{G(N)}{G(1)} \qquad \delta_D = \frac{D(N)}{D(1)} \tag{T4.1}$$

A corresponding degradation parameter (Idriss et al., 1978) may be defined for the shear modulus G

$$t_G = -\frac{\log \delta_G}{\log N} \tag{T4.2}$$

At large strain levels, cyclic soil behaviour can no longer be described by the same conventional $G(\gamma)$ and $D(\gamma)$ relationships used for equivalent linear analysis. A total stress analysis may be still appropriate, provided that the variation of modulus and damping with the number of cycles is appropriately accounted for, such as through the degradation parameters defined above. Alternatively, the constitutive models based on an effective stress approach (e.g. elasto-plastic with kinematic hardening) accounting for shear-volumetric coupling, cyclic degradation and stress-strain behaviour approximating failure may be required (Sagaseta et al., 1991). The analyses of a saturated soil need to be carried out by coupled two-phase effective stress approaches (e.g. Ishihara and Towhata, 1982; Zienkiewicz and Bettess, 1982), which may incorporate the description of states close to failure, as it will be more extensively discussed in Section T4.1.2.

The above conceptual framework associating strain levels with constitutive models and types of analysis is shown in Table T4.1 (adapted after Ishihara, 1982).

T4.1.2 Soil behaviour at failure

As discussed in Section T4.1.1, large strain behaviour of soil is invariably associated with marked cyclic degradation effects, resulting in significant irrecoverable strains and/or dynamic pore-water pressures. Beyond the large strain level near or at failure state, further excessive inelastic deformations and significant loss of shear strength of the soil may occur, leading to significant damage to port structures.

Permanent displacements and failure of saturated soils are a consequence of the progressive pore pressure increase at large strains, beyond the volumetric threshold γ_v. This build-up takes place in all granular soils, either in loose or dense states, but the pattern of behaviour is quite different from each other: Loose

Table T4.1. Relationship between strain levels-soil models-methods of analysis.

Soil properties (parameters)	Shear strain amplitude, g [%] (indicative values)					
	10^{-4} Small	10^{-3} Medium	10^{-2}	10^{-1}	1 Large	10 Failure
Stiffness (G_0, G)	━━					
Damping (D_0, D)	▪▪▪▪▪▪━━━━━━━━━━━━━━━━━━━━━━━━━━━━━━━━					
Pore pressure build-up (Δu)				━━━━━━━━━━━━━━━━		
Irrecoverable strains (ε_v^p, ε_s^p)				━━━━━━━━━━━━━━━━		
Cyclic degradation (δ_G, δ_D)				━━━━━━━━━━━━━━━━		
Mechanical model	Linear visco-elastic		Lin. eq. visco-elastic Elasto-plastic		Non-linear	
Method of analysis	Linear Total stresses		Linear equivalent Total stresses		Incremental Effective stresses	

sand is typically contractive, and may abruptly lose its strength when, after some cycles of shaking, the pore-water pressure build-up becomes equal to the initial effective stress, i.e. $u = \sigma'_0$ (Fig. T4.4(a)). This phenomenon is called liquefaction. Dense sand, on the contrary, is typically dilatant, and behaves more stable and recovers its resistance even if it is momentarily lost. This behaviour is called cyclic mobility. This behaviour is shown in Fig. T4.4(b), in which the abrupt increase in shear strain typical of liquefaction is not recognised.

Under constant amplitude cyclic shear stress, gradual build-up in pore pressure causes the effective stress path to approach near the origin of the failure envelope

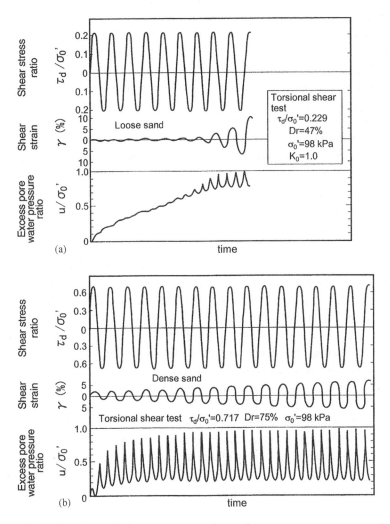

Fig. T4.4. Time histories of shear stress, strain, and pore water pressure ratio (after Ishihara, 1985).
(a) Loose sand.
(b) Dense sand.

for loose sand (Fig. T4.5). Once this stress state is reached (either in extension/compression cycles or simple shear cycles), the shear strain amplitude significantly increases and the slope of the peak-to-peak line of the hysteresis loops becomes nearly horizontal, leading to a shear modulus nearly zero. The cyclic mobility behaviour of dense sands (Fig. T4.5(b)) is similar in that the stress path gradually approaches the failure envelope, but the changes in shear strain amplitude and stress-strain loop are less distinctive: after many cycles of shaking, the shear strain reaches a limiting value and the stress-strain loops take a hysteretic shape of hard-spring type.

An alternative interpretation of this phenomenon may be given in terms of the stress path position relative to the phase transformation lines as shown in Fig. T4.6 (Ishihara, 1985). If the stress path is confined within the zone between the phase transformation lines and the horizontal axis, the degradation of the soil is very small (low to medium pore-water pressures and shear strain levels). Once the stress path goes beyond the phase transformation lines, the pore pressures increase rapidly, and large shear strains appear and continue to grow. When the stress point reaches the failure envelope, a 'steady' (or 'critical') state may be reached, at which the soil continuously deforms at constant volume.

In engineering practice, onset of liquefaction is characterised in two ways, depending on the mechanical variable used. If the pore-water pressure build-up, u, is used for defining the liquefaction, the onset of liquefaction is characterised by

$$r_u = \frac{u}{\sigma'_{v0}} = 1 \tag{T4.3}$$

where r_u is the pore-water pressure ratio which, for horizontally layered ground, is typically defined with the vertical effective stress σ'_{v0}. An alternative definition assumes that liquefaction is triggered when the shear strain amplitude reaches a limiting value; the value of 5% is typically adopted.

The difference in the definition of liquefaction is typically minor for loose clean sand. However, for dense sand or soils other than clean sands, differences may arise. Liquefaction resistance is typically summarized as the cyclic shear stress ratio, τ/σ'_{v0}, plotted against the number of loading cycles to cause liquefaction, N_c, as shown in Fig. T4.7.

Analysis involving the liquefaction phenomenon is typically based on effective stress modelling. For horizontally layered ground, volume change and excess pore-water pressure may be interrelated as suggested by Martin et al. (1975). In undrained conditions (i.e. no volume change), the tendency of the soil skeleton to contract is compensated by the elastic rebound with effective stress decrease (Fig. T4.8). This is written as

$$u = K_u \varepsilon_{vd} \tag{T4.4}$$

where ε_{vd} is the volumetric densification of the soil (inelastic volumetric strain) and K_u is the rebound bulk modulus.

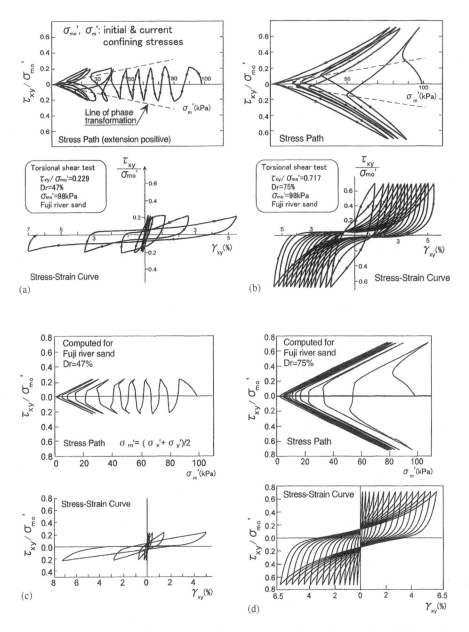

Fig. T4.5. Measured and computed cyclic behaviour of sand (see Fig. T4.4 for time histories).
   (a) Loose sand (measured) (after Ishihara, 1985).
   (b) Dense sand (measured) (after Ishihara, 1985).
   (c) Loose sand (computed) (after Iai et al., 1992).
   (d) Dense sand (computed) (after Iai et al., 1992).

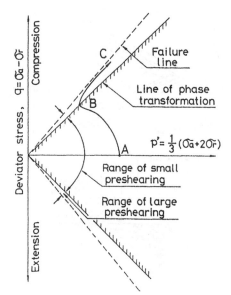

Fig. T4.6. Phase transformation lines in the stress plane (after Ishihara and Towhata, 1980).

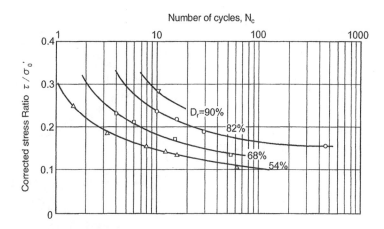

Fig. T4.7 Typical relationship between cyclic stress ratio and number of cycles to liquefaction (after De Alba et al., 1976).

If drainage is permitted in a soil layer, the rate of pore pressure generation, $\partial u/\partial t$, is balanced with the rate of pore pressure dissipation, due to consolidation of the layer. The governing equation is:

$$\frac{\partial u_d}{\partial t} = c_v \frac{\partial^2 u_d}{\partial z^2} + \frac{\partial u}{\partial t} \tag{T4.5}$$

in which c_v is the coefficient of consolidation and u_d stands for the pore-water pressure at an elevation z in the drained layer. The effects of pore-water pressure

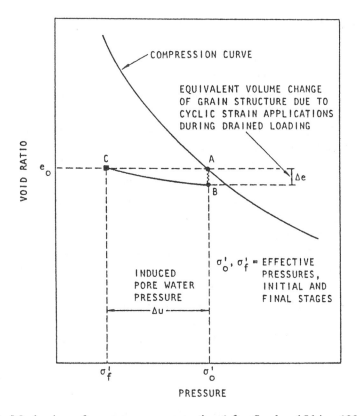

Fig. T4.8. Mechanism of pore pressure generation (after Seed and Idriss, 1982).

redistribution within the soil deposit play an important role in the development of pore-water pressures. For highly permeable deposits, pore-water pressure dissipation can occur at a faster rate than pore-water pressure generation, thereby preventing the sand from liquefying.

For more complicated situations (e.g. involving a thin sand layer sandwiched between impervious materials, irregular stratigraphy or complex soil-structure systems), more sophisticated analyses of liquefaction, based on inelastic two-phase medium approaches, may be needed. In such cases, the constitutive model should incorporate all the fundamental features of the liquefaction phenomenon, including strain-softening of the soil skeleton, hysteresis of stress-strain loops, anisotropy and coupling between the solid and fluid phases. This will lead to non-linear effective stress analyses based on elasto-plastic models generalized for two- or three-dimensional analysis.

A suitable elasto-plastic model in effective stresses should be simple, numerically robust, yet sophisticated enough to reproduce the essential features of soil behaviour described in the beginning of this subsection. The essential requirements for such a constitutive model should ideally include the ability to:

- follow the stress path close to the shear failure line during cyclic loading of a dense saturated sand, such as shown in the upper plot in Fig. T4.5(b);
- reproduce the hysteresis loop of a hardening spring type, such as shown in the lower plot in Fig. T4.5(b);
- reproduce the progressive increase in the shear strain amplitude, such as shown in the lower plot in Fig. T4.5(a, b);
- analyse the cyclic behaviour of an anisotropically consolidated soil.

An example which meets these requirements is the constitutive model based on a multiple shear mechanism (Towhata and Ishihara, 1985) among others. This model was coded into a computer code (FLIP, see Iai et al., 1992) and found to be capable of simulating the behaviour of sand discussed earlier and as shown in Fig. T4.5(c, d). While these models rely on a fundamental physical understanding of liquefaction phenomena, they still require careful calibration and field validation studies in order to routinely apply them in practice (Blázquez, 1995). Nevertheless, research is very active in this field and a growing number of computer programs, e.g. DIANA (Zienkiewicz et al., 1999), DYNAFLOW (Prevost, 1981), GEFDYN (Aubry et al., 1985), TARA (Finn et al., 1986), among others, have been made available for numerical analysis.

T4.2   MEASUREMENT OF SOIL PARAMETERS

The experimental determination of the cyclic soil properties defined in Section T4.1, requires the use of conventional and/or ad-hoc testing procedures, which can be grouped into the following main categories:

- in-situ penetration tests;
- in-situ geophysical tests;
- laboratory tests;

Table T4.2 shows each test type and its applicability for a specific range of strains and frequencies designated, and whether or not the test allows for direct or indirect determination of soil properties.

T4.2.1   In-situ penetration tests

The static (CPT) and dynamic (SPT) penetration tests are used for determining parameters for site response analysis, soil-structure interaction analysis and liquefaction assessment of both subsoil and earth-fills. In particular:

- strength parameters: the shear strength of soils is typically estimated from CPT/SPT data based on semi-empirical background knowledge and

Table T4.2. Field and laboratory investigation techniques and related parameters.

Test class		Test type stress state	Consolidation	Shear strain γ [%]	Frequency f [Hz]	Stiffness	Damping	Strength C	F
Field	Penetration	SPT	Lithostatic	—	—	$N \rightarrow V_S \rightarrow G_0$	—	ϕ'	—
		CPT	Lithostatic	—	—	$q_c \rightarrow V_S \rightarrow G_0$	—	ϕ'	c_u
	Geophysical	Down-Hole	Lithostatic	$<10^{-3}$	10–100	$V_S \rightarrow G_0$	possible	—	—
		Cross-Hole	Lithostatic	$<10^{-3}$	10–100	$V_S \rightarrow G_0$	—	—	—
		SASW	Lithostatic	$<10^{-3}$	10–100	$V_R \rightarrow V_S \rightarrow G_0$	—	—	—
Laboratory	Cyclic	Triaxial	Axisymmetric	$>10^{-2}$	0.01–1	$q:\varepsilon_a \rightarrow E \rightarrow G$	$W_D/W_S \rightarrow D$	$q/\sigma'_r : N_c$	—
		Simple shear	Axisymmetric	$>10^{-2}$	0.01–1	$\tau:\gamma \rightarrow G$	$W_D/W_S \rightarrow D$	$\tau/\sigma'_v : N_c$	—
		Torsional shear	Axisymmetric or true triaxial	10^{-4}–1	0.01–1	$\tau:\gamma \rightarrow G_0, G$	$W_D/W_S \rightarrow D$	—	—
	Dynamic	Resonant column	Axisymmetric or true triaxial	10^{-4}–1	> 10	$f_r \rightarrow G_0, G$	H.p., R.f. $\rightarrow D$	—	—
		Bender elements	Axisymmetric	$<10^{-3}$	> 100	$V_S \rightarrow G_0$	—	—	—

Legend: V_S = shear wave velocity; V_R = Rayleigh wave velocity; f_r = resonant frequency; H.p. = half-power method; R.f. = resonance factor method; N = SPT blow count; q_c = CPT tip resistance; ϕ' = friction angle in effective stresses; c_u = undrained shear strength; C = coarse-grained, F = fine-grained soils; q/σ'_r = deviator/radial stress ratio; τ/σ'_v = shear/vertical stress ratio; N_c = number of cycles.

experience. The relevant parameters are traditionally specified in terms of either effective (coarse grained soils) or total stresses (fine-grained soils);

• shear-wave velocity profiles: regional and worldwide experience has led to the proposals of a large number of correlations between penetration resistance (q_c for CPT, N for SPT) and shear-wave velocity (V_S) or initial shear modulus (G_0). Examples of selected correlations based on CPT tip resistance and SPT blow count are given in Table T4.3;

• liquefaction resistance: as discussed later in Section T4.4, liquefaction charts based on field correlation of penetration test data are widely used in design practice, as often specified in codes and guidelines (e.g. ISSMGE(TC4), 1999). However, the penetration test procedures are different depending on the regional practice. Table T4.4 shows the variability of the SPT procedures and energy efficiency in various regions around the world. As an attempt to reduce the variability, Seed et al. (1985) made recommendations for North-American practice as shown in Table T4.5. If the energy efficiency of a particular procedure is different from that adopted in the proposed empirical correlation, the measured penetration test data should be corrected accordingly (see Section T4.4).

Table T4.3. Examples of regional correlations between penetration resistance and shear modulus or shear wave velocity.

Source	Expression	Coefficients	Soil
Rix and Stokoe (1991)	$G_0 = k(q_c)^a(\sigma'_{v0})^b$	$k = 1634$ $a = 0.25,\ b = 0.375$	Quartz sand
Mayne and Rix (1993)	$G_0 = k(q_c)^a(e)^b$	$k = 406$ $a = 0.695,\ b = -1.13$	Clay
Imai (1977)	$V_S = aN^b$	$a = 102,\ b = 0.29$ $a = 81,\ b = 0.33$ $a = 114,\ b = 0.29$ $a = 97,\ b = 0.32$	Holocene clays Holocene sands Pleistocene clays Pleistocene sands
Ohta and Goto (1978)	$V_S = 69\alpha\beta z^{0.2}N^{0.17}$	$\alpha = 1.0$ $\alpha = 1.3$ $\beta = 1.00$ $\beta = 1.09$ $\beta = 1.07$ $\beta = 1.14$ $\beta = 1.15$ $\beta = 1.45$	Holocene Pleistocene Clay Fine sand Medium sand Coarse sand Gravelly sand Gravel
Baldi et al. (1988)	$V_S = Kz^{0.16}N^{0.10}$	$K = f(\text{soil})$	Holocene soils

Key: G_0, q_c, σ'_{v0}, in kPa; N = SPT blow count at 60% maximum energy delivered; z = depth, in m; V_S = shear wave velocity in m/s.

Table T4.4. Variability of regional SPT procedures (Seed et al., 1985; Skempton, 1986; after ISSMGE (TC4), 1999).

Country	Hammer type	Hammer release	Estimated rod energy ER_m(%)	Correction factor for 60% rod energy
Japan[a]	Donut	Tombi	78	1.30
	Donut[b]	Rope and pulley with special throw release	67	1.12
				1.2 (average)
U.S.A.	Safety[b]	Rope and pulley	60	1.00
	Donut	Rope and pulley	45	0.75
Argentina	Donut[b]	Rope and pulley	45	0.75
China	Donut[b]	Free-fall[c]	60	1.00
	Donut	Rope and pulley	50	0.83
U.K.	Pilcon	Trip	60	1.00
	Old standard	Rope and pulley	60	1.00

[a]Japanese SPT results have additional corrections for borehole diameter and frequency effects.
[b]Prevalent method in each country today.
[c]Pilcon type hammers develop an energy ratio of about 60%.

Table T4.5. Recommended SPT procedure for use in liquefaction assessment (Seed et al., 1985; as reported ISSMGE (TC4), 1999).

Borehole	4 to 5 in. diameter rotary borehole with bentonite drilling mud for borehole stability
Drill bit	Upward deflection of drilling mud (tricone of baffled drag bit)
Sampler	O.D. = 2.00 in.; I.D. = 1.38 in. – Constant (i.e., no room for liners in barrel)
Drill rods	A or AW for depth less than 50 ft; N or NW for greater depths
Energy delivered to sampler	2520 in.-lbs.(60% of theoretical maximum)
Blowcount rate	30–40 blows per minute
Penetration resistance count	Measures over range of 6~18 in. of penetration into the ground

T4.2.2   In-situ geophysical tests

In-situ geophysical tests permit direct measurement of shear-wave velocity V_S, from which the initial shear stiffness G_0 can be derived through the well-known relationship

$$G_0 = \rho V_S^2 \qquad (T4.6)$$

where ρ denotes the bulk soil density.

As shown in Fig. T4.3(a) and Table T4.2, the strain levels attained by the geo-physical testing are never higher than 0.001% (i.e. within the linear range), and the frequencies used are usually higher than those typical of earthquakes.

The field geophysical tests can be classified into surface and borehole tests, with obvious implications in terms of cost of test execution proportional to drilling and related operations.

(1) Surface tests

The well-known seismic reflection and refraction surveys give only a rough delineation of major units of soil layering, for instance allowing for the localisa-tion of a bedrock. A description of the testing procedure and its interpretation is reported in most textbooks (e.g. Kramer, 1996) and will not be presented here. Worth remembering though is that these geophysical tests are most commonly adopted for undersea investigations.

Higher resolution and degree of reliability among the surface geophysical tests has been achieved by the Spectral Analysis of Surface Waves (SASW) method (Stokoe et al., 1994). This test is based on the analysis of the propagation of Rayleigh waves generated at an appropriate distance by applying a vertical pulse on the surface of a subsoil (Fig. T4.9). The receivers are located symmetrically with reference to the vertical profile investigated. The depth of subsoil surveyed by this method is in proportion to the wave length $\lambda_R = V_R/f$ (V_R, f = velocity and frequency of surface waves). By decomposing transient surface waves into Fourier components and computing their phase velocity, V_{ph} (see Appendix), it is possible to describe an experimental dispersion curve (V_{ph}:λ) representative of the subsoil shear stiffness profile. A numerical inversion procedure then needs to be used, by assuming a layered wave velocity profile that most closely fits to the experimental dispersion curve.

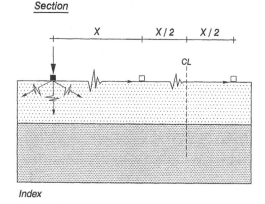

Fig. T4.9. Experimental setup for SASW tests.

Even though the SASW data reduction procedure may be cumbersome, this technique is operationally inexpensive, since it does not require drilling of a bore-hole. In addition, its execution has been successfully tested underwater (e.g. Luke and Stokoe, 1998).

(2) Borehole tests

The borehole tests can be classified according to the position of the impact source. In the Down-Hole (DH) technique, a horizontal impact source is placed on the surface and one or more receiver geophones are installed in the borehole. The receivers are then lowered as shown in Fig. T4.10. Shear wave pulses with predominant horizontal components (SH waves) are generated from the surface and recorded at every geophone location. In the conventional test interpretation practice, the shear-wave velocity V_S results from the ratio between the inclined ray path length and the first arrival time at the receiver (see Appendix).

The Cross-Hole (CH) test is carried out by drilling two (or more than two) boreholes, in which a vertical impact source and one (or more) receiver geo-phones are installed (Fig. T4.11). Shear-wave velocity V_S is given by the ratio between the source-receiver distance and the first arrival time at the receiver (see Appendix). The damping ratio (D_0) can be evaluated by using two receiving bore-holes (Mok et al., 1988).

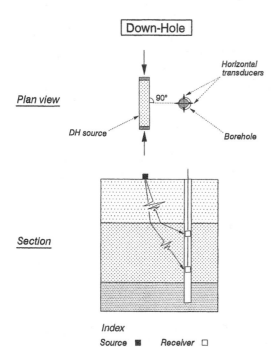

Fig. T4.10. Experimental setup for Down-Hole tests.

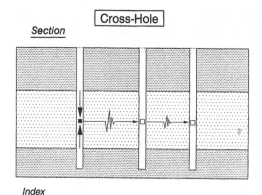

Fig. T4.11. Experimental setup for Cross-Hole tests.

An updated version of the above techniques, the so-called Seismic Cone (SC) utilises a cone penetrometer tip as a receiver, to avoid drilling of the corresponding borehole (Robertson et al., 1985). The seismic cone enables measurement of both the shear-wave velocity and the tip resistance at the same depth, thus permitting calibration of direct correlations between q_C and V_S. This technique has been successfully utilised in sea-bottom investigations, since it overcomes some of the disadvantages (e.g. site accessibility and deployment) of other geophysical tests such as cross-hole and down-hole.

The data recorded in borehole tests are analysed according to well-established procedures, including manual picking of first waveform arrivals and numerical Fourier analysis of the digitised records. More details are given in the Appendix, which is located at the end of this technical commentary.

For most offshore geophysical tests carried out in the sea floor, the following issues should be considered:
1) the boreholes have to be cased with either steel or plastic, which limits the type of sondes that have to be used;
2) the high water content of sea floor deposits can create problems in the interpretation of the logs, eventually 'masking' some of the responses to geophysical logging;
3) practice has shown that the sondes have to be run very slowly through the sea floor material to ensure that an adequate response is recorded.

T4.2.3   Laboratory tests

Laboratory tests can characterise soil behaviour over a wide strain range which, according to the selected technique, spans from γ as low as 0.0001% up to the non-linear range and towards failure (see Fig. T4.3(a) and Table T4.2).

The laboratory testing techniques can be classified into:

- cyclic tests, in which inertia forces in the soil element are negligible, and the mechanical response is characterised by direct measurement of stress and strain values;
- dynamic tests, in which the response needs to be interpreted by means of a dynamic analysis of equilibrium of the soil element.

(1) Cyclic tests

a) Cyclic triaxial test (CTX)

The cyclic triaxial test (CTX) involves the application of cyclic compression-extension loads on a cylindrical specimen (Fig. T4.12(a)), controlling stress or strain amplitudes. There are some studies that suggest that strain-controlled tests may be preferable to the frequently used stress-controlled tests because they may be less sensitive to experimental factors (Dobry et al., 1979, 1980; Talaganov, 1986). The most up-to-date versions of mechanical and hydraulic loading systems allow the application of arbitrary stress and strain paths to cylindrical specimens. Since the principal stress directions alternatively switch between horizontal and vertical, this testing pattern cannot fully reproduce the ideal simple shear loading conditions acting on a soil element during an earthquake.

In the CTX, the axial stiffness and damping parameters can be directly obtained by analysing the deviator stress-axial strain loops (see Fig. T4.12(b) and Table T4.2). CTX tests are traditionally oriented to analyse cyclic failure behaviour, described by the relationship between the deviatoric/radial effective stress ratio (q/σ'_r) and the number of cycles at liquefaction (N_C). However, experimental limitations (such as equipment compliance and resolution in the axial strain

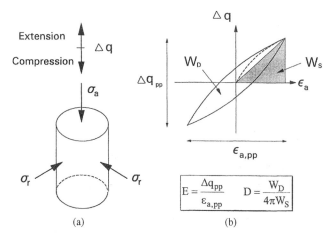

Fig. T4.12. Cyclic triaxial test.
(a) Loads on soil specimen.
(b) Interpretation.

measurement) may limit the reliability of CTX tests for the characterisation of pre-failure behaviour at small and medium strain levels.

Recent research developments in advanced triaxial equipments (e.g. Tatsuoka et al., 1997) showed that shear strains as low as 0.001% can be resolved in static (e.g. Goto et al., 1991) and cyclic (e.g. Olivares, 1996) triaxial tests by using local displacement transducers (LDT). However, these techniques have not yet been commonly adopted in design practice.

b) Cyclic simple shear test (CSS)

The cyclic simple shear test (CSS) is performed by applying a cyclic tangential load on the boundaries of a short-height specimen (Fig. T4.13(a)), the shape of which can be either prismatic (Cambridge type) or cylindrical (NGI and SGI types, shown in Fig. T4.13(a)). The non-compliant cell walls permit K_0 consolidation state before the application of the cyclic shearing, which is usually strain-controlled.

The stiffness and damping parameters can be directly obtained by interpreting the shear stress-strain loops (Fig. T4.13(b)) and defined in Fig. T4.2. Since the state of shear stress and strain within the specimen can be far from uniform, the reliability of measurements can be much reduced at small and medium strains. As in the CTX tests, this laboratory technique can be used to investigate liquefaction potential, described by the relationship between the shear/vertical effective stress ratio (τ/σ'_v) and the number of cycles at liquefaction (N_C).

c) Cyclic torsional shear test (CTS)

The cyclic torsional shear test (CTS) is performed by applying a cyclic torque to the upper base of a cylindrical slender specimen (Fig. T4.14(a)). The possibility

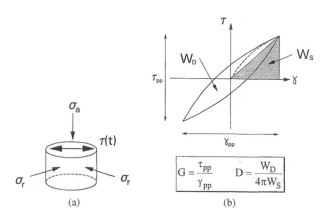

Fig. T4.13. Cyclic simple shear test.
  (a) Loads on soil specimen.
  (b) Interpretation.

of using hollow specimens, and independently controlling inner and outer confining pressures (Fig. T4.14(a)), permits the application of triaxial consolidation prior to the application of shear loads. However, testing on natural soils is practically feasible only for solid specimens (Fig. T4.14(b)). The test is normally stress-controlled, with torsional loads applied by means of several kinds of devices, more frequently electro-magnetic (e.g. Isenhower, 1979; d'Onofrio et al., 1999b) or electro-mechanical (e.g. Bolton and Wilson, 1989; Alarcon et al., 1986).

As in the CSS test, the stiffness and damping parameters can be directly obtained by interpreting the shear stress-strain loops according to their definitions (Fig. T4.14(c)). Due to the high degree of uniformity and to the elevated resolution in the measurement of shear stress and strains in the specimen, the reliability of pre-failure equivalent parameters evaluated (G and D) is expected to be high at small and medium strains. Normally, with electro-magnetic devices, it is not possible to bring a specimen to failure state; in such a case, the CTS test is not suitable for assessing liquefaction potential.

(2) Dynamic tests
a) Resonant column test (RC)
The resonant column test (RC) is currently performed with essentially the same electro-magnetic driving devices adopted for CTS tests (Fig. T4.15(a)).

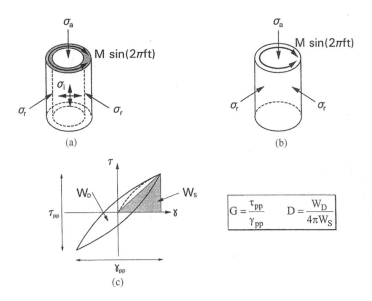

Fig. T4.14. Cyclic torsional shear test.
     (a) Loads on hollow cylinder specimen.
     (b) Loads on solid cylinder specimen.
     (c) Interpretation.

A cylindrical specimen is loaded with a constant amplitude torque with variable frequency until the resonance of the assembly of the soil specimen and the driving system is achieved (Fig. T4.15(b)).

The soil parameters are obtained from the interpretation of the frequency response curve (Drnevich et al., 1978): the shear-wave velocity V_S (hence, the shear modulus G) is obtained from the measurement of the resonant frequency f_r, and the damping ratio D from the half-power frequencies f_1 and f_2. The damping ratio can also be measured from the acquisition of the free vibration decay curve (Fig. T4.15(c)). However, with the RC test, as with the CTS test, it is not possible to bring a specimen to failure. Consequently, the test is only suitable for investigating deformation properties.

b) Bender elements (BE)

The bender elements (BE) are active and passive transducers, made with piezo-ceramics or piezocrystals (Dyvik and Madshus, 1985), which can be installed at the top and at the bottom of a specimen mounted in a triaxial cell (Fig. T4.16(a)). Recently, this technique has been successfully introduced in large-scale triaxial equipment that was sized to test coarse-grained soils such as gravels and rockfills (Modoni et al., 2000).

When the active (transmitter) transducer is excited, it generates a single pulse shear-wave that propagates along the specimen length and is recorded at the other end by the passive (receiver) transducer. The direct interpretation of the test is

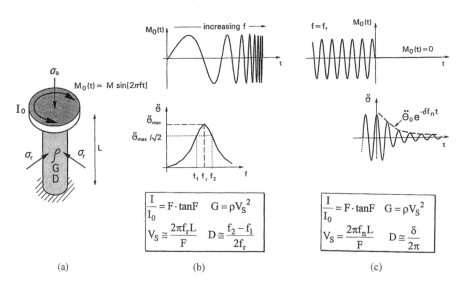

(a)   (b)   (c)

Fig. T4.15. Resonant column test.
       (a) Loads on soil specimen.
       (b) Interpretation of resonance test.
       (c) Interpretation of free vibration decay test.

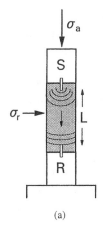

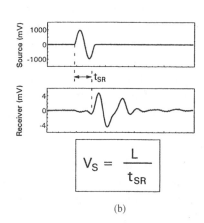

(a)   (b)

Fig. T4.16. Bender element test.
   (a) Loads on soil specimen.
   (b) Interpretation.

therefore similar to that of the geophysical field tests. The visual analysis of time records of vibration amplitudes permits the measurement of shear-wave velocity V_S, from which the small-strain shear modulus G_0 is calculated through Eqn. (T4.6). Wave propagation analysis in the frequency domain (such as that adopted for in-situ tests) is also recommended for a more accurate investigation (Viggiani and Atkinson, 1995).

T4.2.4 Combined use of in-situ and laboratory tests

It is important to remember in the geotechnical design practice, that field and laboratory investigation techniques have their own advantages and disadvantages.

In-situ geophysical tests can measure soil properties in their natural state. This is achieved by interpreting vibration records, based on the theory of propagation of seismic waves within a layered half-space. However, these methods do not allow the measurements of non-linear soil properties since the vibration energy level of geophysical tests is typically not strong enough to allow the development of large strains.

Laboratory tests allow the measurements of soil behaviour in a strain range, much wider than the in-situ tests. The laboratory characterisation of an earth-fill material before its placement may be achieved through a reliable reproduction of the in-situ conditions by means of appropriate specimen preparation, compaction and re-consolidation procedures. However, for natural subsoil, the laboratory results are often affected by soil disturbance during sampling. Other disadvantages include limits in reproducing in-situ stress states, and scaling effects due to the limited size of soil specimen tested for seismic excitation. The sample disturbance can significantly affect the test results, especially for underwater sed-iments. Care should be taken for choosing the most appropriate sampling tech-nique among those available in local practice; regional recommendations are available on the suitable sampling methods depending on the soil type and den-sity (e.g. CEN, 1997; Port and Harbour Research Institute, 1997)

In view of these advantages and disadvantages, a combined use of in-situ and laboratory test results may be recommended as follows.

(1) Pre-failure conditions

The shear modulus and damping ratio at small-strain, G_0 and D_0, are sensitive to the sampling procedure. In the medium to large strain ranges, the normalised equivalent shear modulus, G/G_0, is relatively insensitive to sampling disturbance. Therefore, the design strain dependent soil stiffness may be best evaluated through

$$G(\gamma) = (G_0)_{\text{field}} \cdot \left[\frac{G(\gamma)}{G_0} \right]_{\text{lab}} = (G_0)_{\text{field}} \cdot [\bar{G}(\gamma)]_{\text{lab}} \tag{T4.7}$$

In Eqn. (T4.7), $(G_0)_{\text{field}}$ is preferably determined from the shear-wave velocity measured in-situ, while the strain-dependent normalised shear modulus G/G_0 should be obtained by the laboratory tests.

The $D(\gamma)$ relationship can be expressed as

$$D(\gamma) = (D_0) + \bar{D}(\gamma) \tag{T4.8}$$

and it is typically determined from laboratory tests only. If D_0 is measured in-situ (e.g. from CH tests performed with at least two receiver boreholes), then Eqn. (T4.8) should read:

$$D(\gamma) = (D_0)_{\text{field}} + [\bar{D}(\gamma)]_{\text{lab}} \tag{T4.9}$$

where $[\bar{D}(\gamma)]$ represents the laboratory measured damping ratio scaled to the rel-evant small-strain value D_0.

(2) Failure conditions

With reference to failure conditions, the following laboratory tests have been developed to characterise soil properties:
- direct liquefaction tests (stress or strain-controlled), with pore-pressure recording on undrained saturated samples;
- indirect liquefaction tests (constant volume tests), with densification recording on dry soil samples.

By cyclic laboratory tests, a considerable saving in testing effort may be achieved, since the pre-failure properties of the soil may be obtained along with the cyclic strength characteristics. However, it is most often the case that these laboratory test results are strongly affected by many testing factors, frequently leading to low reliability in the test results. Figure T4.17 shows examples of the effects on the cyclic strength curves of experimental factors such as compaction procedures (Fig. T4.17(a)) and re-consolidation techniques (Fig. T4.17(b)).

For the reasons stated above, in-situ estimators of liquefaction susceptibility (by means of static and dynamic penetration resistance, or V_S values) are currently used (see Section T4.4). Each one of these field measurements has advantages and disadvantages, and their suitability is also soil-dependent (Table T4.6).

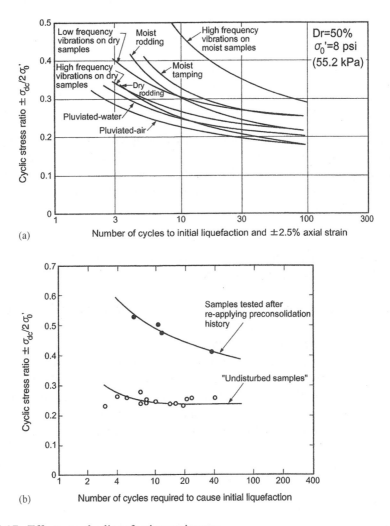

Fig. T4.17. Effects on the liquefaction resistance.
    (a) Effects of sample preparation techniques (after Mulilis et al., 1975).
    (b) Effects of re-consolidation techniques (after Mori et al., 1978).

Table T4.6. Comparative performance of various field tests for assessment of liquefaction potential (Youd and Idriss, 1997).

Feature	Test Type		
	SPT	CPT	V_S
Number of test measurements at liquefaction sites	Abundant	Abundant	Limited
Type of stress-strain behaviour influencing test	Partially drained, large strain	Drained, large strain	Small strain
Quality control and repeatability	Poor to good	Very good	Good
Detection of variability of soil deposits	Good	Very good	Fair
Soil types in which test is recommended	Non-gravel	Non-gravel	All
Test provides sample of soil	Yes	No	No
Test measures index or engineering property	Index	Index	Engineering property

To summarise, site and laboratory tests are complementary. By combining the test results of both investigations, an adequate description of soil behaviour for seismic design may be achieved. The most appropriate experimental strategy may be chosen in accordance with the analysis requirements; this point will be discussed in Section T4.5.

T4.3   EVALUATION OF PRE-FAILURE SOIL PROPERTIES THROUGH EMPIRICAL CORRELATIONS

Conventional practice in geotechnical characterisation suggests that, whenever a soil property is directly measured or indirectly evaluated, it must be reviewed in a light of pre-existing test data and experience. In compiling the knowledge gathered on natural soils, pre-failure soil properties are often correlated with grain size, micro- and macro-structure, and geologic age. In earthquake geotechnical design practice, empirical correlations with index properties are commonly referred to, and often suggested by relevant codes or guidelines (e.g. EPRI, 1993; PHRI, 1997). Some of these useful empirical correlations are summarised and discussed below.

T4.3.1   Small-strain stiffness

The small-strain shear modulus G_0 can be expressed as a function of the state variables (mean effective stress p', void ratio e, and overconsolidation ratio OCR) as (Hardin, 1978)

$$\frac{G_0}{p_a} = S \left(\frac{p'}{p_a} \right)^n f(e) OCR^m \tag{T4.10}$$

where p_a = atmospheric pressure, $f(e)$ is a decreasing function of e, and S, n, m are dimensionless parameters. The function $f(e)$ varies depending on the soil types. The most common relationship for $f(e)$ is that given by Hardin and Black (1968)

$$f(e) = \frac{(2.973 - e)^2}{1 + e} \tag{T4.11}$$

Taking this function as a reference for many natural soils, Mancuso et al. (1997) obtained the variation of S and n with I_P shown in Fig. T4.18. For non-plastic granular materials, the parameter S varies from 200 to 400; for fine-grained soils, it gradually decreases with increasing plasticity index I_P (Fig. T4.18(a)). The parameter n increases with I_P from an initial range 0.4–0.6 (gravels and sands) to about unity for high plasticity clays (Fig. T4.18(b)). The plot showing the dependency of the third parameter, m, on I_P (Fig. T4.18(c)) was originally reported by Hardin and Black (1969).

For normally consolidated soils ($OCR = 1$), e and p' are correlated with each other (i.e. the normal compression line); therefore, the dependency of G_0 on the state variables can be simplified as a power function of the mean effective stress p' as follows (Rampello et al., 1994)

$$\frac{G_0}{p_a} = S \left(\frac{p'}{p_a} \right)^n \tag{T4.12}$$

The small-strain stiffness data on marine clays were collected from existing literature by d'Onofrio and Silvestri (2001), compiled as shown in Fig. T4.19. The stiffness coefficient S decreases with I_P (Fig. T4.20(a)) in the similar manner as shown in Fig. T4.18(a), but with different meaning because of the different formulation of Eqn. (T4.12) with respect to Eqn. (T4.10). The trend for n in Fig. T4.19(b) is almost the same as in Fig. T4.18(b).

The small-strain stiffness G_0 can be also expressed as a function of soil density ρ and shear-wave velocity V_S, through Eqn. (T4.6). The case histories indicate that, at moderate depths, the typical values of ρ and V_S for the majority of soils are those reported in Table T4.7. These values can be useful for obtaining a rough estimate of G_0.

T4.3.2   Small-strain damping

The laboratory tests show that, irrespective of the soil type, the damping ratio at small-strains, D_0, decreases with the mean effective stress, p'. Based on the review of literature data, the influence of soil type on D_0 is reported by Mancuso et al. (1997) as follows:

- for granular soils (sands, gravels, rockfills), the range of variation of D_0 with stress state and history is very narrow and approaches zero;

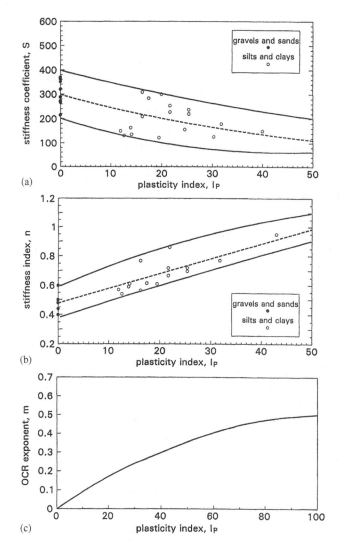

(a)

(b)

(c)

Fig. T4.18. Variation of stiffness parameters with plasticity index I_P for natural soils (after Mancuso et al., 1997).
(a) Parameter S.
(b) Parameter n.
(c) Parameter m.

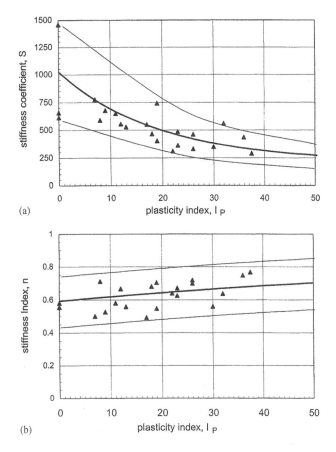

Fig. T4.19. Soil stiffness parameter variation with plasticity index I_P for normally consolidated marine clays (after d'Onofrio and Silvestri, 2001).
   (a) Parameter S.
   (b) Parameter n.

- for fine-grained natural soils, the average values of D_0 and the rate of decrease
  with stress level become higher going from stiff to soft clays; also, at the same
  stress state and history, D_0 increases with plasticity index;
- the values of D_0 of medium/fine-grained compacted materials are often higher than those typical of natural clays, due to the relatively young age of the
  man-made soil deposits.

For the same normally consolidated marine soils analysed to obtain the correlation in Fig. T4.19, d'Onofrio and Silvestri (2001) used the simple power function
relationship

$$D_0 = Q\left(\frac{p'}{p_a}\right)^{-j} \tag{T4.13}$$

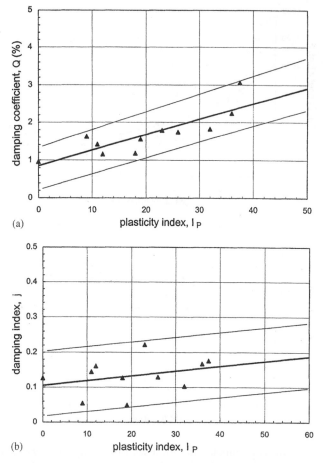

Fig. T4.20. Soil damping parameters variation with plasticity index I_P for normally consolidated marine clays (after d'Onofrio and Silvestri, 2001).
(a) Parameter Q.
(b) Parameter j .

to relate the decrease of the small-strain damping to the increase of mean effective stress. The correlation with the plasticity index I_P of the damping parameters
Q (representing the value of D_0, expressed in %, at a $p' = p_a$) and j, shown in Fig.
T4.20 was obtained. The plots confirm that both the values of D_0 and its rate of
decrease with p' increase with soil plasticity.

T4.3.3   Non-linear pre-failure behaviour

Several correlations have been proposed in the literature for specifying the strain-
dependent properties $G(\gamma)$ and $D(\gamma)$ using the plasticity index I_P.

Table T4.7. Typical ranges of soil density and shear wave velocity.

Soils	Typical ranges	
	ρ (t/m^3)	V_S (m/s)
Organic soils, peats	1.0 to 1.3	<100
Very soft clays	1.4 to 1.7	
Loose volcanic silts/sands	1.0 to 1.7	100 to 200
Soft clays, loose sands	1.6 to 1.9	
Dense/cemented volcanic silts/sands	1.2 to 1.9	200 to 400
Stiff clays, dense sands	1.8 to 2.1	
Cemented sands and gravels	1.9 to 2.2	400 to 800
Intact rocks	2.0 to 3.0	>800

Based on the test data found in literatures, including their own laboratory test data, Vucetic and Dobry (1991) report the average $\bar{G}(\gamma)$ and $D(\gamma)$ curves shown in Fig. T4.21. Both plots show that the degree of non-linearity (hence, the energy dissipation) increases with decreasing plasticity. The granular soils with $I_P = 0$, show non-linearity over a linear threshold strain γ_1 of about 0.001%. Note that the curves relevant to the damping ratio are not scaled to the small-strain value, D_0, and that, according to the observations in Section T4.3.2, such small-strain value should increase with I_P and decrease with p'.

The strain dependent curves for Japanese marine clays have been proposed by Zen et al. (1987) as shown in Figs. T4.22 and T4.23. The effects of variations in the state variables (e, p', OCR) on the shape of normalised curves were found to be negligible for fine-grained high plasticity soils (Fig. T4.22(a)). On the other hand, the increase in stress level can significantly shift to the right the $\bar{G}(\gamma)$ relationship for a medium to low plasticity soil (Fig. T4.22(b, c)). In contrast to the soils compiled by Vucetic and Dobry (1991) (Fig. T4.21), the Japanese marine clays do not appear to exhibit dependency on plasticity for $I_P > 30$. Again, the degree of non-linearity is seen to increase with decreasing plasticity. The influence of state and consolidation variables on the increase of damping ratio with shear strain was not clearly established, and the average $D(\gamma)$ curves given in Fig. T4.23 were suggested.

The linear threshold strain γ_1, beyond which soil non-linearity is observed, can be also correlated with the plasticity index I_P. Figure T4.24(a) shows that, depending on particle grading and microstructure, γ_1 may vary from 0.0001% to more than 0.01%. The values of γ_1 for granular soils increase with decreasing particle size; those for volcanic sands and silts are higher with respect to those typical of other non-plastic soils, due to the stronger inter-particle bonds provided by cementation and/or particle interlocking. The degree of non-linearity of compacted fine-grained soils may be moderately affected by plasticity (Mancuso et al., 1995), whereas that of natural clays can be strongly affected by

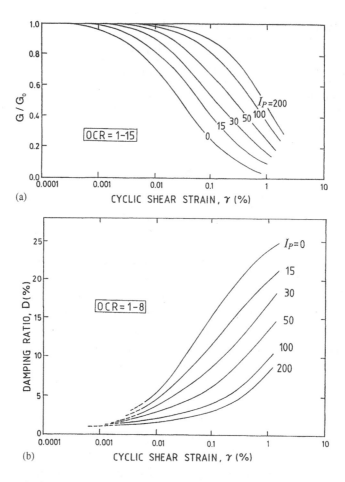

Fig. T4.21. Strain dependent stiffness and damping (after Vucetic and Dobry, 1991).
    (a) Normalised equivalent shear modulus $\overline{G}(\gamma)$.
    (b) Equivalent damping $D(\gamma)$.

macro-structure. For instance, the linear threshold level of a fissured clay can be much lower than that of homogeneous reconstituted samples of the same soil (see Fig. T4.24(a)).

The volumetric threshold level γ_v, beyond which soil dilatancy is observed, increases with I_P in a similar way as that of γ_l, and is about 1 to 2 orders higher than γ_l as shown in Fig. T4.24(b).

The average dependency of the degradation index t_G (defined in Section T4.1.1) on the strain level is shown in Fig. T4.25 for normally consolidated soils with different plasticity (Vucetic, 1994). A significant degradation occurs when the strain exceeds the volumetric threshold γ_v (increasing with I_P); thereafter, t_G increases with the strain level. At a given strain amplitude, the rate of degradation of shear modulus decreases with soil plasticity.

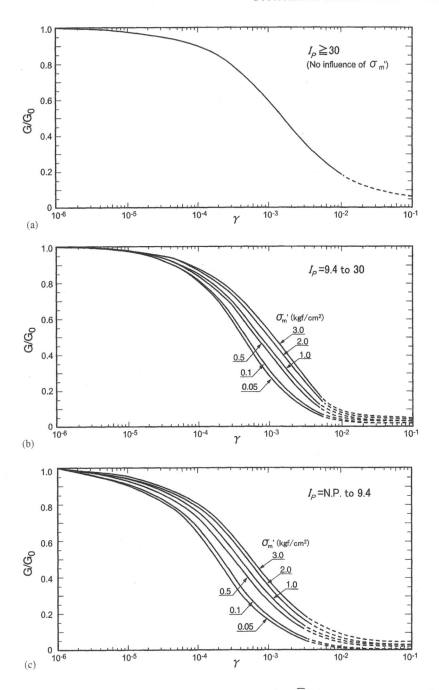

Fig. T4.22. Normalised equivalent shear modulus stiffness $\overline{G}(\gamma)$ of Japanese soils (after
Zen et al., 1987).
(a) Soils with plasticity index $I_P \geq 30$.
(b) Soils with plasticity index $9.4 \leq I_P \leq 30.191$.
(c) Soils with plasticity index $I_P \geq 9.4$ including non-plastic soils.

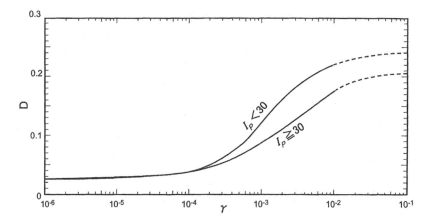

Fig. T4.23. Equivalent damping ratio $D(\gamma)$ of Japanese soils (after Zen et al., 1987).

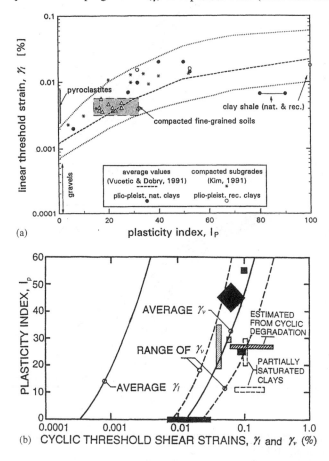

Fig. T4.24. Influence of grain size and microstructure on the threshold strains.
(a) Linear threshold shear strain γ_l (after Mancuso et al., 1995).
(b) Volumetric threshold shear strain γ_v (after Vucetic, 1994).

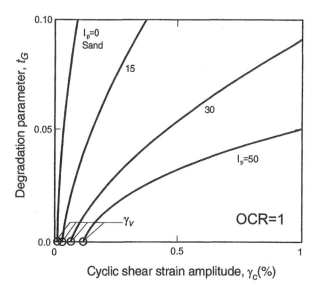

Fig. T4.25. Influence of plasticity index on strain-dependent parameter t_G (after Vucetic, 1992).

T4.4   ASSESSMENT OF LIQUEFACTION POTENTIAL

Studies on liquefaction potential assessment have been primarily based on laboratory tests. These have lead to a development of conventional test procedures that approximately represent equivalent field stress conditions. For example, irregular time histories of earthquake motions have been often approximated by equivalent uniform cyclic loads in the laboratory (Seed et al., 1975; Annaki and Lee, 1977). Hence, soil liquefaction is characterised by cyclic resistance curves (see Fig. T4.7) and pore-water pressure time histories (see Fig. T4.4).

Despite the further developments in laboratory liquefaction test procedures (Seed, 1979; Ishihara, 1985), the laboratory-based approach has been known to be sensitive to many factors and uncertainties. The liquefaction resistance can be quite sensitive to physical properties of soils, including fines content and degree of cementation. They are also highly sensitive to small variations of the degree of saturation in the soil specimen (Fig. T4.26). Thus, the laboratory tests are often not performed in routine design practice; liquefaction potential is typically evaluated based on field investigation results (Section T4.4.1). If required, the liquefaction potential assessment can be improved by combining additional information obtained from laboratory test (Section T4.4.2).

T4.4.1   Liquefaction potential assessment using field data

Assessment of liquefaction potential is typically accomplished based on simplified procedures correlating undrained cyclic resistance at in-situ conditions with field performance (i.e. case histories of liquefaction observed during earthquakes). These procedures are available in seismic codes and technical recommendations (e.g. CEN, 1994; Youd and Idriss, 1997; ISSMGE(TC4), 1999). In these procedures, two variables need to be evaluated: the seismic load (cyclic stress ratio, *CSR*) and the capacity of the soil to withstand liquefaction (cyclic resistance ratio, *CRR*). The soil will be then determined liquefiable if *CSR* > *CRR*.

(1) Evaluation of seismic load
There are several ways to define the cyclic stress ratio *CSR*. The most common expression of CSR is given by the following equation (Seed and Idriss, 1971):

$$CSR = \frac{\tau_{av}}{\sigma'_{v0}} = 0.65 \frac{a_{max(surface)}}{g} \frac{\sigma_{v0}}{\sigma'_{v0}} r_d$$

(T4.14)

where $a_{max(surface)}$ is the peak horizontal acceleration at the ground surface, g is the acceleration of gravity, σ_{v0} and σ'_{v0} are respectively, the total and effective vertical overburden stresses. The coefficient r_d is a stress reduction factor at the depth of interest, which accounts for the deformability of the soil profile. The range of variation of r_d with depth, as proposed by Seed and Idriss (1971), is shown in Fig. T4.27. Analytical expressions to approximate the mean $r_d(z)$ profile are suggested by the consensus made at the workshop on evaluation of liquefaction resistance of soils at the National Center for Earthquake Engineering Research, hereafter called 'NCEER workshop' (Youd and Idriss, 1997). As indicated in the figure, the validity of these profiles becomes uncertain at depths greater than 15 m. When possible and applicable, it is suggested to assess the value assumed for r_d through the site response analysis.

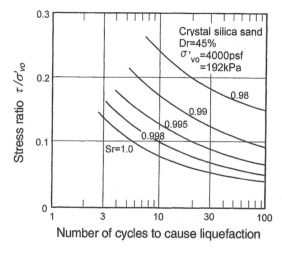

Fig. T4.26. Influence of the degree of saturation on the liquefaction resistance (after Martin et al., 1978).

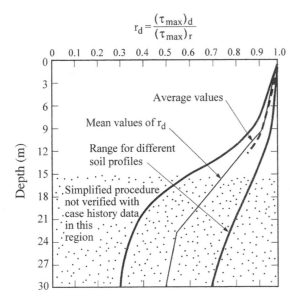

$$r_d = \frac{(\tau_{max})_d}{(\tau_{max})_r}$$

Fig. T4.27. Variation of the stress reduction factor r_d with depth within soil profile (after Youd and Idriss, 1997; adapted from Seed and Idriss, 1971).

(2) Correction and normalisation of field data

The liquefaction resistance *CRR* is typically evaluated using empirical charts to be discussed in the subsection (3). In these charts, the in-situ test data (such as SPT *N*-values, cone penetration resistance, or the shear-wave velocity) should be normalised to account for variability in the reference depth and corrected to a reference standard testing procedure, as shown below.

As mentioned in Section T4.2.1 and shown in Table T4.4, the energy efficiency of SPT procedures are different in the various regions in the world. The same table suggests broadly that energy efficiencies, ER_m are 67 to 78% for Japanese correlations and about 60% or even less for most other countries. If the energy efficiency of a particular procedure, ER_m is different from that adopted in the proposed empirical correlation, the measured blow count N_m should be scaled as follows

$$N_{ER} = \frac{ER_m}{ER} N_m \qquad (T4.15)$$

where *ER* and N_{ER} denote the energy efficiency and blow count used for establishing the correlation between the cyclic strength and the SPT value. For example, the measured SPT blow count, N_m, in North American practice is typically normalised to 60% energy efficiency as follows (Seed et al., 1985)

$$(N_1)_{60} = C_N \frac{ER_m}{60} N_m \qquad (T4.16)$$

where $(N_1)_{60}$ is the reference blow count (at an overburden effective stress of 1 atmosphere (= 1kgf/cm^2 = 98 kPa) with 60% energy ratio, C_N a stress correction factor depending also on relative density (Fig. T4.28(a)).

A more general expression has been proposed by the NCEER workshop (Youd and Idriss, 1997), namely

$$(N_1)_{60} = C_N C_E C_B C_R C_S N_m \tag{T4.17}$$

where the correction factor C_N accounts for overburden stress, C_E for hammer energy ratio ER_m, C_B for borehole diameter, C_R for rod length and C_S for samplers with or without liners. The suggested ranges of values for these factors are listed in Table T4.8. The expression of the stress correction factor C_N in Table T4.8 refers to the overburden effective stress normalised with respect to the atmospheric pressure p_a (Liao and Whitman, 1986). An alternative expression for the correction factor C_N may be given by

$$C_N = \frac{1.7}{\sigma'_{v0} + 0.7} \tag{T4.18}$$

where σ'_{v0} is expressed in kgf/cm^2 (Meyerhof, 1957).

A similar normalisation procedure should be applied to the measured tip resistance q_c, to obtain the reference value q_{c1} through the expression

$$q_{c1} = C_q \frac{q_c}{p_a} \tag{T4.19}$$

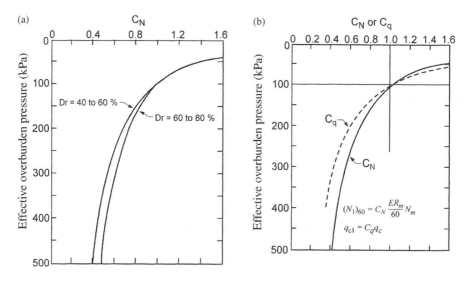

Fig. T4.28. Stress correction factors for SPT and CPT.
        (a) C_N for SPT blow count (after Seed et al., 1985).
        (b) C_q for CPT tip resistance (after Seed et al., 1983).

Table T4.8. Correction factors for SPT blow count (Robertson and Wride, 1997).

Factor	Equipment variable	Correction	Value
Overburden stress		C_N	$(p_a/\sigma'_{v0})^{0.5}$ $C_N \leq 2$
Energy ratio	Donut hammer Safety hammer Automatic-trip donut-type hammer	C_E	0.5 to 1.0 0.7 to 1.2 0.8 to 1.3
Borehole diameter	65 mm to 115 mm 150 mm 200 mm	C_B	1.0 1.05 1.15
Rod length	3 m to 4 m 4 m to 6 m 6 m to 10 m 10 m to 30 m > 30 m	C_R	0.75 0.85 0.95 1.0 > 1.0
Sampling method	Standard sampler Sampler without liners	C_S	1.0 1.1 to 1.3

The curve indicated by Seed et al. (1983) for C_q is shown in Fig. T4.28(b); NCEER workshop (Youd and Idriss, 1997) suggested to use the generalized expression

$$C_q = \left(\frac{p_a}{\sigma'_{v0}} \right)^n \qquad\qquad (T4.20)$$

with n dependent on grain characteristics of the soil, and ranging between 0.5 (clean sands) and 1.0 (clayey soils).

The measured shear-wave velocity V_S can be normalised to a reference effective overburden stress through the expression

$$V_{S1} = V_S \left(\frac{p_a}{\sigma'_{v0}} \right)^{1/n} \qquad\qquad (T4.21)$$

where n can be 3 (Tokimatsu et al., 1991) or 4 (Finn, 1991; Robertson et al., 1992).

(3) Use of liquefaction charts

A liquefaction chart is usually expressed as a curve in a plane defined with the cyclic stress ratio on the ordinate and the selected field estimator of liquefaction resistance on the abscissa. The curve limits the boundary between the experimental points representing observed liquefaction cases (above) and non-liquefied cases (below). Therefore, it empirically represents the locus of the *CRR*, i.e. the minimum *CSR* required to produce liquefaction of a soil layer characterised by the reference field property.

Figures T4.29 through T4.31 show the criteria obtained as the consensus at the NCEER workshop (Youd and Idriss, 1997) for evaluation of liquefaction resistance, respectively, based on normalised and corrected values of SPT blow count $(N_1)_{60}$, CPT tip resistance q_{c1N}, and shear-wave velocity V_{S1}. The curves in these figures are given for earthquakes with moment magnitude $M_W = 7.5$, on gently sloping ground (slopes lower than about 6%) with low overburden pressures (depths less than about 15m). The values of *CRR* for moment magnitudes other than 7.5 need to be corrected by a scaling factor, C_M, decreasing with the design earthquake magnitude. Table T4.9 shows the values originally suggested by Seed and Idriss (1982), together with the upper and lower bounds suggested

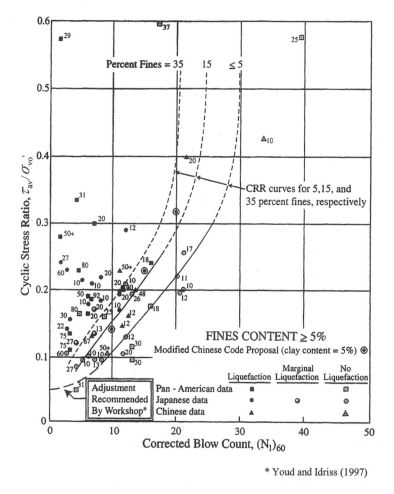

* Youd and Idriss (1997)

Fig. T4.29. Evaluation of *CRR* from SPT *N*-value (Youd and Idriss, 1997, modified after Seed et al., 1985).

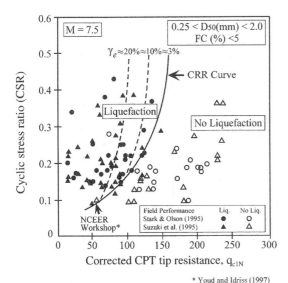

Fig. T4.30. Evaluation of *CRR* from CPT data (Robertson and Wride, 1997).

as a consensus made in the NCEER workshop (Youd and Idriss, 1997) on the basis of more updated contributions by several researchers.

In the charts in Figs. T4.29 and T4.31, additional curves relative to soils other than clean sands are suggested, referring to variable fines contents, F_C, defined as the percentage finer than 0.075 mm. Several correction procedures were suggested to take into account in a general way, the increase of *CRR* with increased fines content percentage (Seed et al., 1985). For the application of the SPT-based method on soils with non-plastic fines ($I_P < 5$), Robertson and Wride (1997) suggested to convert the calculated $(N_1)_{60}$ to an equivalent value for clean sand, $(N_1)_{60cs}$

$$(N_1)_{60cs} = K_s(N_1)_{60cs} \tag{T4.22}$$

$$K_s = 1 + 0.025(F_c - 5) \tag{T4.23}$$

where F_c is the fines content. To date, a further correction factor (larger than K_S) for soil with plastic fines has not been adopted by the geotechnical engineering community due to the insufficient availability of data.

Similarly, for the application of the CPT-based method to sands with fines, Robertson and Wride (1997) suggested to convert the normalised tip resistance, q_{c1N}, to an equivalent value for clean sand, $(q_{c1N})_{cs}$

$$(q_{c1N})_{cs} = K_c q_{c1N} \tag{T4.24}$$

where the correction factor K_c is expressed by a function of the so-called 'soil behaviour type index', I_c, as described in Fig. T4.32(a). The index I_c can be estimated from the apparent fines content from the plot of Fig. T4.32(b), or calculated by means of the combination of measurements of both tip and sleeve resistance, with the procedure fully described by Robertson and Wride (1997).

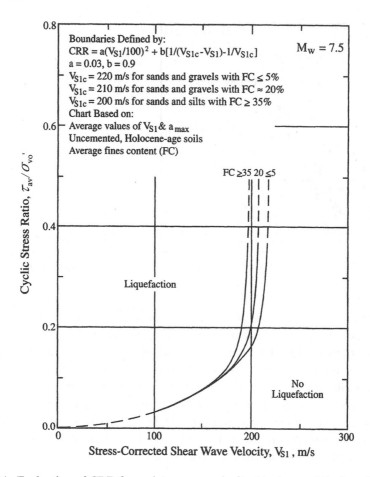

Fig. T4.31. Evaluation of CRR from shear wave velocity (Andrus and Stokoe, 1997).

Table T4.9. Magnitude scaling factors required to modify *CRR* for $M_W \neq 7.5$.

	Seed and Idriss (1982)	Youd and Idriss (1997)
Magnitude, M_W	C_M	C_M
5.5	1.43	2.20 to 2.80
6.0	1.32	1.76 to 2.10
6.5	1.19	1.44 to 1.60
7.0	1.08	1.19 to 1.25
7.5	1.00	1.00
8.0	0.94	0.84
8.5	0.89	0.72

More sophisticated correction factors to account for layer thickness, high over-consolidation states, static shear stress for sloping ground, and age of deposit, have begun to appear in various publications, but they still belong to the research domain. The topic was extensively discussed during the NCEER workshop

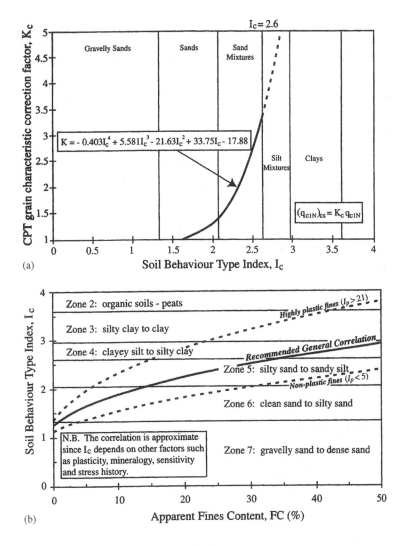

Fig. T4.32. Correction factors for CPT resistance (Robertson and Wride, 1997).
    (a) Correction factor K_c.
    (b) Soil behaviour index I_C.

(Youd and Idriss, 1997); nevertheless, most of these latter corrections appear of minor importance for conditions specific to port structures.

T4.4.2   Liquefaction potential assessment combining field and laboratory data

A methodology combining field and laboratory data has been developed and extensively used in standard design practice in Japan (Port and Harbour Research Institute, 1997; Yamazaki et al., 1998; Ministry of Transport, Japan, 1999). In this procedure, liquefaction potential assessment takes place in two steps. The first step uses the grain size distribution of soil and standard penetration test blow counts (SPT N-value) for a soil layer. If the results of this first step are close to the borderline between liquefaction and non-liquefaction conditions, then the first step can be supplemented with the second step, cyclic triaxial tests.

(1) Grain size distribution and SPT N-value (1st step)

Initially, a soil is classified according to the grain size distribution zones shown in Fig. T4.33. The soil is considered non-liquefiable if the grain size distribution falls outside of the liquefaction possibility zone in Fig. T4.33. For the soil with grain size distribution curve falling inside the liquefaction possibility zone, the liquefaction potential is evaluated based on the following procedure.

a) Equivalent N-value

An equivalent N-value, N_{65}, is calculated by correcting for the effective overburden pressure of 65 kPa using the following expression

$$N_{65} = \frac{N_m - 0.019 \, (\sigma'_v - 65)}{0.0041 \, (\sigma'_v - 65) + 1.0}$$

(T4.25)

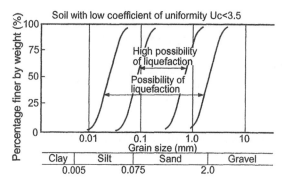

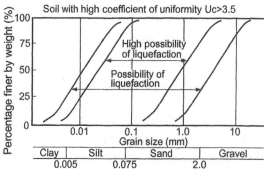

Fig. T4.33. Grain-size distribution of soil having the possibility of liquefaction (Ministry of Transport, Japan, 1999).

where N_m = measured SPT N-value of a soil layer (obtained with typical energy efficiency in Japan), σ'_v = effective overburden pressure of a soil layer (kPa), calculated with respect to the ground surface elevation at the time of the standard penetration test.

b) Equivalent acceleration

An equivalent acceleration, a_{eq}, is estimated by the following equation

$$a_{eq} = 0.7 \cdot \frac{\tau_{max}}{\sigma'_v} \cdot g \tag{T4.26}$$

where τ_{max} = maximum shear stress in the layer obtained from site response analysis, σ'_v = effective overburden pressure calculated, this time, with respect to the ground surface elevation at the time of earthquake.

c) Liquefaction potential assessment for clean sands

For clean sands (i.e. soils with fines content F_c less than 5%), liquefaction potential is determined by using the chart in Fig. T4.34. The zone to which a soil layer belongs is determined from the equivalent N value, N_{65}, and the equivalent acceleration, a_{eq}. Liquefaction potential is assessed as follows:

– Zone I has a very high possibility of liquefaction. Decide that liquefaction will occur. Cyclic triaxial tests are not required.
– Zone II has a high possibility of liquefaction. Decide either to determine that liquefaction will occur, or to conduct further evaluation based on cyclic triaxial tests.

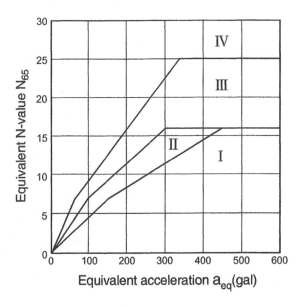

Fig. T4.34. Liquefaction potential assessment based on equivalent acceleration and equivalent N-values (Ministry of Transport, Japan, 1999).

– Zone III has a low possibility of liquefaction. Decide either to determine that that liquefaction will *not* occur, or to conduct further evaluation based on cyclic triaxial tests. When it is necessary to allow a significant safety margin for a structure, decide either to determine that liquefaction *will* occur, or to conduct further evaluation based on cyclic triaxial tests.
– Zone IV has a very low possibility of liquefaction. Decide that liquefaction will not occur. Cyclic triaxial tests are not required.

d) Correction for silty or plastic soils

For a soil with fines content higher than 5%, the SPT N value is corrected according to the following three cases:

Case 1: The plasticity index $I_P < 10$ or the fines content $5 < F_c < 15\%$.
An equivalent N-value obtained from Eqn. (T4.25) is corrected through the following equation:

$$N_{65}^* = N_{65}/C_{FC} \tag{T4.27}$$

where: N_{65}^* = a corrected equivalent N-value, C_{FC} = a correction factor based on the fines content (Fig. T4.35). Using the corrected equivalent N-value, the liquefaction potential is evaluated in the same manner as c) by referring to Fig. T4.34.

Case 2: The plasticity index $10 < I_P < 20$, and the fines content $F_c > 15\%$.
Two corrected equivalent N-values are calculated as follows:

$$N_{65}^* = N_{65}/0.5 \tag{T4.28}$$

$$N_{65}^{**} = N + \Delta N \tag{T4.29}$$

$$\Delta N = 8 + 0.45(I_P - 10) \tag{T4.30}$$

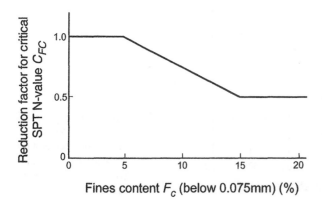

Fig. T4.35. Reduction factor for critical SPT N-value based on the fines content (Ministry of Transport, Japan, 1999).

The two corrected equivalent N-values are plotted in Fig. T4.34 against the equivalent acceleration, and the zone to which a soil layer belongs is determined as follows:

> When N_{65}^{**} is in Zone I, liquefaction potential is evaluated by Zone I.
> When N_{65}^{**} is in Zone II, liquefaction potential is evaluated by Zone II.
> When N_{65}^{**} is in Zone III or IV, and when N_{65}^{*} is in Zone I, II, or III liquefaction potential is evaluated by Zone III.
> When N_{65}^{**} is in Zone III or IV, and when N_{65}^{*} is in Zone IV, liquefaction potential is evaluated by Zone IV.

Case 3: The plasticity index $I_P > 20$, and the fines content $F_c > 15\%$.
A corrected equivalent N-value is calculated using Eqns. (T4.29) and (T4.30), and then liquefaction potential is evaluated using Fig. T4.34.

(2) Cyclic triaxial test (2_{nd} step)
When the liquefaction potential cannot be determined from the grain size distribution and SPT N-value (i.e. for zones II and III), it is evaluated based on undrained cyclic triaxial tests using undisturbed soil samples. Based on the relationship between the cyclic stress ratio and number of cycles to cause liquefaction in the laboratory tests, the cyclic resistance $(\tau_l/\sigma_c')_{N_1 = 20}$ for use in the liquefaction potential assessment is obtained by reading off the cyclic stress ratio at 20 cycles of loading (Fig. T4.36). Using this cyclic resistance, the in-situ liquefaction resistance R_{max} is obtained using the following equation:

$$R_{max} = \frac{0.9}{C_k} \frac{(1+2K_0)}{3} \left(\frac{\tau_l}{\sigma_c'} \right)_{N_1 = 20}$$

(T4.31)

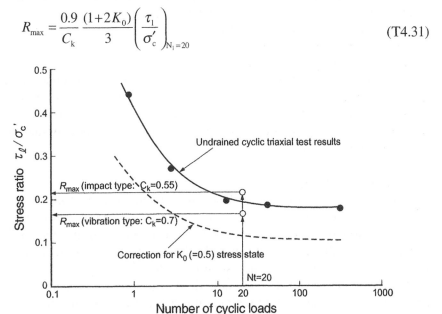

Fig. T4.36. Correction of laboratory liquefaction resistance for in-situ condition (Ministry of Transport, Japan, 1999).

In this equation, the following corrections are applied
- Stress correction: The stress correction for in-situ (K_0) conditions from the triaxial test condition (isotropic consolidation with mean confining pressure of σ'_c).
- Input motion correction: The correction for irregularity of earthquake motion (e.g. impact type/vibration type) from the harmonic motion in the cyclic triaxial test. The correction factors are given as
      Impact type input motion $C_k = 0.55$
      Vibration type input motion $C_k = 0.7$

The seismic stress ratio $L_{max} = \tau_{max}/\sigma'_c$ is calculated through site response analysis. The liquefaction potential (safety factor) F_L is given as

$$F_L = \frac{R_{max}}{L_{max}} \qquad\qquad (T4.32)$$

If $F_L < 1.0$, the soil layer is expected to liquefy.

Laboratory data on pore-water pressure increase can be also obtained and used for design of remediation methods against liquefaction (see Technical Commentary 6).

T4.5   Geotechnical Characterisation Strategy with Reference to the Analysis Requirements

In seismic analysis of port structures, appropriate geotechnical characterisation should be made with respect to the following:
- soil profile and ground water table;
- initial effective stress state;
- physical and mechanical properties of soils.

In particular, a reasonable strategy should be established for adequate geotechnical characterisation with respect to the mechanical properties of soils required for seismic response analysis. According to the hierarchy of analysis procedures addressed in Section 5.1 in the Main Text, the following strategy may be suggested (see Table T4.10):

(1) Simplified analysis

In simplified analysis, local site effects are evaluated based on the thickness of the deposits and the average stiffness to a specified depth (generally few tens of meters), or over the entire deposit above the bedrock. This information is then used to establish the site classification, leading to the use of specified site amplification factors or site dependent response spectra.

The subsoil characterisation required for the simplified analysis may be achieved based on the existing regional information on site category or by field correlation using standard penetration tests (SPT) or cone penetration tests (CPT) data. Liquefaction potential of sandy soils may also be evaluated based on SPT/CPT through empirical liquefaction criteria. At this level, index properties and conventional static tests may effectively support the geotechnical character-isation and/or the site classification.

(2) Simplified dynamic analysis

In simplified dynamic analysis, local site effects are typically evaluated using equivalent linear modelling in total stresses. The subsoil is idealised into hori-zontal layers of infinite lateral extent and one-dimensional (1D) wave propaga-tion analysis is used. The pre-failure, strain-dependent parameters (G and D) can be adequately characterised by means of geophysical tests (Section T4.2.2), cyclic/dynamic laboratory tests (Section T4.2.3), and combining the field and laboratory measurements through the methods suggested in Section T4.2.4. In the absence of direct measurements by suitable field and laboratory tests, it is common practice to use empirical correlations with penetration test data and index properties (Section T4.3). These correlations are typically shown in design guidelines (e.g. EPRI, 1993; Port and Harbour Research Institute, 1997). The applicability of these correlations should be confirmed with regional data before using them in practice. This is especially important when special or regional soils are encountered in design, including highly organic soils, cemented soils, soils with fissures, weathered soils, or soils of volcanic origin.

In simplified dynamic analysis, liquefaction potential can be evaluated based on a comparison of computed shear stresses during the design earthquake and the results of cyclic laboratory tests, and/or based on SPT/CPT data. The liquefac-tion potential evaluation based on the shear-wave velocity V_S may also be advan-tageous. The advantages of using V_S may be distinctive for gravelly soils because their penetration resistance may be misleading unless measured by the Becker Penetration Test (BPT) (Youd and Idriss, 1997). However, caution must be exert-ed with the straightforward application of shear-wave velocity (valid for small-strains) to liquefaction analysis (which is a large strain mechanism). For instance, the liquefaction potential of weakly cemented soils may be misinterpreted by the method using V_S, since these soils typically show unusually higher V_S relative to their likely liquefaction resistance. It is preferable to apply two or more field-based procedures to achieve consistent and reliable evaluation of liquefaction resistance.

(3) Dynamic analysis

Dynamic analyses of soil-structure interaction (SSI) attempt to account for the combined, or coupled, response of the structure and the foundation soils. Unlike the simplified procedures wherein the response of the structure is evaluated using the free-field soil response as input, SSI analyses incorporate the behaviour of both soil and structure in a single model. Numerical finite element (FEM) or finite difference (FDM) methods are commonly used for advanced SSI analy-ses. In this type of analysis, local site effects and liquefaction potential are often not evaluated independently but are evaluated as a part of the soil-structure

Table T4.10. Geotechnical investigation methods with reference to analysis requirements.

| Type of analysis | Aspects | Analysis method | Input seismic action | Soil parameters | Geotechnical investigation methods | | | | | | |
					Gathering existing data	Field PT	Field GT	Laboratory Index Prop.	Laboratory Static	Cyclic/Dynamic DT	Cyclic/Dynamic LT
Simplified SRA		Site category	a_{max} at bedrock	Soil profile and classification	■	■		▨			
	LIQ	SPT/CPT lique-faction charts	a_{max} at surface	N/q_c	■	■		▨			
	SSI Quay walls/Breakwaters	Pseudo-static/Empirical	Output of SRA	c, ϕ, μ_B, δ	■	■		▨	▨		
	Piled wharf	Response spectrum	Output of SRA	Output of $p:y$ curves	■	■					
Simplified dynamic	SRA	1D total stress analysis	$a(t)$ at bedrock	$V_s(z)$ profile $G:\gamma, D:\gamma$ curves	■	■	▨	▨		▨	
	LIQ Field based approach	SPT/CPT/V_s liquefaction charts	a_{max} at surface	$N/q_c/V_s$	■	■	▨	▨			
	Lab. based approach	1D total stress analysis	$\tau(t)$ at surface	Cyclic strength curves	■			▨			■
	SSI Quay walls/Breakwaters	Newmark/Parametric	Output of SRA	c, ϕ, μ_B, δ	■	■		▨			▨
	Piled wharf	Pushover/Response spectrum	Output of SRA	$p:y$ curves	■	■		▨	▨	▨	▨

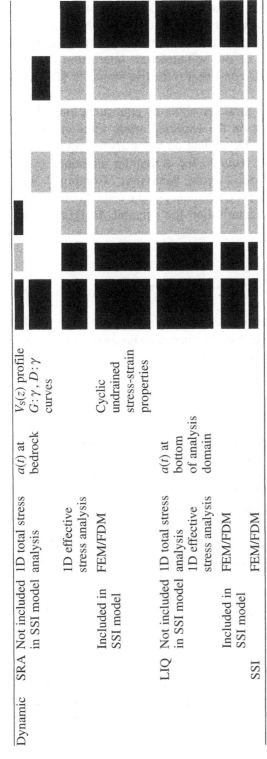

Key: SRA = Seismic Response Analysis; LIQ = Liquefaction; SSI = Soil Structure Interaction. PT = Penetration Tests, GT = Geophysical Tests, DT = Pre-failure/Deformation Tests, LT = Failure/Liquefaction Tests.

■ used as a standard

▨ use depends on design conditions

interaction analysis of port structures. If the bottom boundary of the analysis domain for soil-structure interaction analysis differs from the bedrock (i.e. when the bedrock is too deep for soil-structure interaction analysis), the local site effects below the bottom boundary of the soil-structure interaction analysis domain may be evaluated based on 1D non-linear (effective stress) or 1D equivalent linear (total stress) analysis.

In this category of analyses, soil is idealized either by equivalent linear or by an effective stress model, depending on the expected strain level in the soil deposit during the design earthquake. Depending on the analysis requirements, laboratory cyclic tests may often be required for evaluating undrained cyclic properties in terms of effective stress-strain behaviour. Combined with field investigations, including SPT/CPT/V_S, laboratory cyclic tests should achieve the most adequate geotechnical characterisation for dynamic analysis of port structures.

APPENDIX: INTERPRETATION OF GEOPHYSICAL TESTS

The interpretation of borehole geophysical tests can be performed by the analysis of the recorded signals in the time domain (time domain analysis) or in the frequency domain (frequency domain analysis).

A.1   Time domain analysis

Figure A1 shows a typical recording of the source signal (an impact) and receiver signals (generated waveforms) from a CH test in a soft clay deposit (Simonelli and Mancuso, 1999). Visual analysis allows the determination of the travel times of the waves between the source and the receivers, by picking the trigger time (T) at the source, and the first arrivals of the waveforms at the receivers (P for compression waves and S1 for shear-waves). Moreover, the travel times between the receivers, located along a straight array, can be determined by comparing the first arrivals S1, or even by picking the times of characteristic points of the waveforms as peaks (see open circles in the figure) or zero-crossing points.

In the conventional procedure, the shear-wave velocity V_S is given by the ratio between the travel path of the waveform L_w and the travel time T_w

$$V_S = \frac{L_w}{T_w} \tag{A.1}$$

Direct wave velocities (between the source and the receivers) and interval wave velocities (between the two receivers) can be determined; their values can be different, due to the 'spreading' of the waves during propagation (Hoar, 1982).

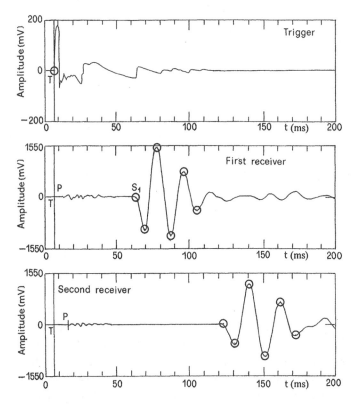

Fig. A1. Example of time domain analysis of a CH test with two receivers.

The waveform velocity between two receivers located along a straight line can be determined by the Cross Correlation function, provided that the signals are acquired in digital form. The Cross Correlation $[CC_{xy}(\tau)]$ of two signals $x(t)$ and $y(t)$ is defined as

$$CC_{xy}(\tau) = \int_{-\infty}^{+\infty} x(t)y(t+\tau)\mathrm{d}t \qquad (A.2)$$

where τ is the variable time delay between signals $x(t)$ and $y(t)$.

Hence, a Cross Correlation value is given by the sum of the products of the two signal amplitudes, after having shifted the second record, $y(t)$, with respect to the first, $x(t)$, by a time interval τ. The Cross Correlation function yields different values by varying the time delay τ, and reaches its maximum CC_{max} when the degree of correlation between the signals $x(t)$ and $y(t+\tau)$ is maximum (i.e. when the waveforms are practically superimposed). The time delay τ_{max} corresponding to CC_{max} is the best estimate of the travel time of the whole waveform between the two receivers.

The Cross Correlation shown in Fig. A2 is obtained using two receiver signals (S waves) of Fig. A1. The Cross Correlation values have been normalised with

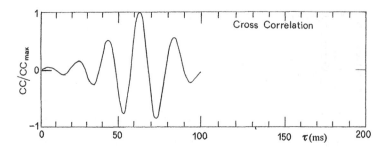

Fig. A2. Cross-correlation function of the receiver signals in Fig. A1.

respect to the maximum CC_{max}, hence the function varies between -1 and 1. The travel time is approximately 62 ms.

A.2   Frequency domain analysis

By means of the Fourier analysis, a generic signal $x(t)$ can be decomposed into a sum of a series of sinusoidal functions, each one characterised by its amplitude (A), phase (Φ) and frequency (f) (see Fig. A3). The sine functions are the frequency components of the signal $x(t)$, and their complex representation in the frequency domain is defined as the Linear Spectrum (L_x) of the signal $x(t)$.

The transformation of a signal in the correspondent Linear Spectrum can be performed by means of the algorithm Direct Fourier Transform (DFT); inversely, it is possible to convert a Linear Spectrum into the correspondent time record by means of the algorithm Inverse Fourier Transform (IFT). The frequency content of a signal is effectively represented by the amplitude versus frequency plot of the Linear Spectrum; in Fig. A4, such diagrams are plotted for the receiver (R1 and R2) signals in Fig. A1.

Each sine function of a Linear Spectrum can be effectively represented by a vector (with length proportional to the amplitude, and direction equal to the phase of the sine function), and therefore, by a complex number. The rules and tools of complex algebra are applicable to signal analysis in the frequency domain. In the following paragraphs, the most useful functions for frequency domain analysis are reviewed.

The Cross Power Spectrum CPS_{xy} of two signals $x(t)$ and $y(t)$ is the product of the correspondent Linear Spectra L_x and L_y

$$CPS_{yx}(f) = L_y(f)\ L_x^*(f) \tag{A.3}$$

Each frequency component (f) of the CPS has an amplitude [$A_{CPS}(f)$] equal to the product of the amplitudes of the same frequency components of the linear spectra, and phase [$\Phi_{CPS}(f)$] equal to their phase difference. The $A_{CPS}(f)$ allows

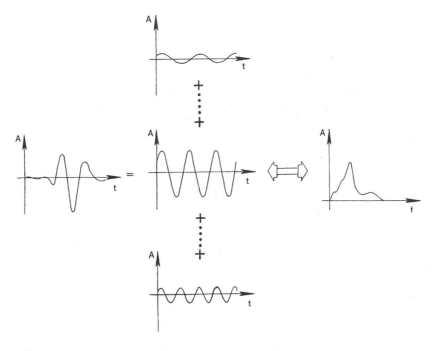

Fig. A3. Transformation of an amplitude-time signal to an amplitude-frequency signal by means of the DFT.

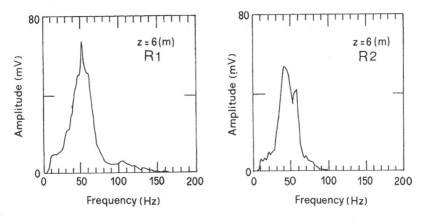

Fig. A4. Linear Spectra of two signals propagating between two receivers in a Cross-Hole (CH) test.

one to determine the main frequency components of the propagating waveform, while the $\Phi_{\text{CPS}}(f)$ allows one to determine the delay of phase and then the travel time of each component.

Generally, the *CPS* function is repeatedly applied to several couples of signals, relative to the same propagation phenomenon. In fact, repeating the signals and averaging them drastically reduces the noise which usually affects signals during propagation.

The *CPS* obtained from the signals similar to those of Fig. A1 is shown in Fig. A5. In the upper part, the $A_{CPS}(f)$ shows that the main frequency components of the propagating waveform lie in the range from 25 to 75 Hz. In the lower part, the values of the $\Phi_{CPS}(f)$ are conventionally plotted between $\pm$ 180°; actually the real diagram should be obtained by translating the points A, B, C etc., respectively, in A′, B′, C′ etc.; the $\Phi_{CPS}(f)$ values start from zero and gradually decrease as frequency increases.

The Coherence function $[CO_{xy}(f)]$ between $x(t)$ and $y(t)$ signals is an algorithm which allows one to determine the level of noise affecting the frequency components of the propagating waveform.

The CO_{xy} function, as the *CPS*, is applied to repeated couple of signals, relative to the same propagation phenomenon. The coherence and the corresponding signal to noise ratio $[S/R]_{xy}$ are determined as:

$$CO_{xy}(f) = \frac{\overline{CPS_{yx}(f)} \cdot \overline{CPS}^{*}_{yx}(f)}{\overline{APS_{yy}(f)} \cdot \overline{APS_{xx}(f)}} \qquad [S/R]_{xy}(f) = \frac{CO_{xy}(f)}{1 - CO_{xy}(f)} \qquad (A.4)$$

where APS_{xx} and APS_{yy} are the Auto Power Spectrum of each signal, given by the square of the linear spectrum of the signal (e.g. $APS_{xx} = L_x(f)\, L^{*}_x(f)$). The bar over the symbols *CPS* and *APS* indicate the averaged value of repeated signals.

The CO_{xy} values range from 0 to 1, as shown in Fig. A6, for a series of two signals from CH tests. The diagram clearly shows that the effect of noise is very low in the range from 25 to 75 Hz.

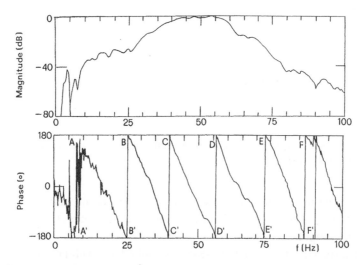

Fig. A5. Cross Power Spectrum of two signals from a CH test.

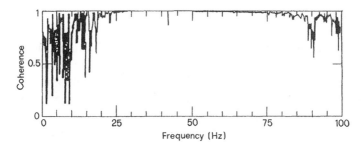

Fig. A6. Coherence of two signals from a CH test.

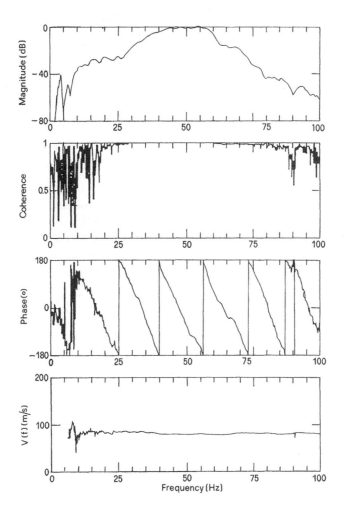

Fig. A7. Phase velocity diagram from a CH test.

A.3   Spectral analysis and propagation velocity

The Cross Correlation function (previously defined in the time domain analysis paragraph), and hence the travel time and the correspondent waveform velocity, can be immediately determined utilising the frequency domain analysis. The Cross Correlation of two signal $x(t)$ and $y(t)$ is given by the Inverse Fourier Transform of the Cross Power Spectrum of the two signals

$$CC_{xy}(\tau) = \int_{-\infty}^{+\infty} x(t)y(t+\tau)\mathrm{d}t = IFT[CPS_{yx}(f)] \qquad (A.5)$$

The Cross Power Spectrum allows for the determination of the velocity of each of the frequency components of the propagating waveform, by means of its phase $[\Phi_{CPS}(f)]$. Using the delay of phase $\Delta\phi(f)$ for a given frequency component (f), wave propagation between two points at known distance is represented by the corresponding number of wavelengths N_w given by

$$N_w(f) = -\frac{\Delta\phi(f)}{360°} \qquad (A.6)$$

If r_{12} is the distance between the two points, the wavelength $\lambda(f)$ and the velocity $V_{ph}(f)$, defined as phase velocity, of the component (f) are given by

$$\lambda(f) = \frac{r_{12}}{N_c(f)} = -\frac{r_{12}}{\Delta\phi(f)/360°}; \quad V(f) = \lambda f = -\frac{r_{12} \cdot f}{\Delta\phi(f)/360°} \qquad (A.7)$$

The phase velocity diagram $V_{ph}(f)$, calculated for a series of signals $x(t)$ and $y(t)$ from a CH test is shown in Fig. A7, together with the relative $A_{CPS}(f)$, $CO_{xy}(f)$ and $\Phi_{CPS}(f)$. It is clear that, for frequencies greater than 25 Hz, the $V_{ph}(f)$ diagram is practically horizontal, slightly varying around the value of about 78 m/s, which can be assumed as the group velocity of the whole waveform.

TC5: Structural Design Aspects of Pile-Deck Systems

T5.1 Design Performance and Earthquake Levels
    T5.1.1 Performance goal
    T5.1.2 Design earthquake levels
    T5.1.3 Limit states
    T5.1.4 Seismic load combinations
T5.2 Modelling Aspects
    T5.2.1 Soil-structure interaction
    T5.2.2 Movement joints
T5.3 Methods of Analysis for Seismic Response
    T5.3.1 Method A: Equivalent single mode analysis
    T5.3.2 Method B: Multi-mode spectral analysis
    T5.3.3 Method C: Pushover analysis
    T5.3.4 Method D: Inelastic time-history analysis
    T5.3.5 Method E: Gross foundation deformation analysis
T5.4 Structural Criteria for Pier and Wharf Piles
    T5.4.1 Deformation capacity of pile plastic hinges
    T5.4.2 Implication of limit states
    T5.4.3 Moment-curvature characteristics of piles
    T5.4.4 Primary structural parameters and response of piles
    T5.4.5 Plastic hinges
    T5.4.6 Shear strength of piles
    T5.4.7 Design strength for deck members
    T5.4.8 Batter and timber piles
T5.5 Pile/Deck Connection Details
    T5.5.1 Steel-shell piles
    T5.5.2 Prestressed piles
    T5.5.3 Practical connection considerations
    T5.5.4 Capacity of existing substandard connection details
T5.6 Existing Construction
    T5.6.1 Structural criteria for existing construction
    T5.6.2 Strengthening of an existing structure
    T5.6.3 Deterioration of waterfront structures

TECHNICAL COMMENTARY 5

Structural Design Aspects of Pile-Deck Systems

This technical commentary will further illustrate the general design methodologies developed in the Main Text and present some numerical guidelines for the performance of pile-deck structures. Piles are a common and significant element of waterfront construction. This technical commentary will provide guidance in pile design.

The following guidance is directly based on Technical Report TR 2103-SHR 'Seismic Criteria For California Marine Oil Terminals' by J. Ferritto, S. Dickenson, N. Priestley, S. Werner and C. Taylor, July 1999. This report was sponsored by the California State Lands Commission, Marine Facilities Division. The structural criteria defined herein are based largely on the work of Nigel Priestley. These criteria are developed from a compilation of current practice in North America, with state-of-the-art technology for evaluating seismic damage potential. Although the guidance addresses structural design of concrete, steel and wooden piles, the focus of attention is directed to concrete piles, reflecting the practice in North America. Design criteria adopted in other regions, where steel piles have been typically used, notably in Japan, might be different from this guidance as briefly discussed in Technical Commentary 7.

T5.1 DESIGN PERFORMANCE AND EARTHQUAKE LEVELS

T5.1.1 Performance goal

The general performance goals were discussed in Chapter 3 in the Main Text. This technical commentary focuses on pile-deck structures, specifically piers and wharves. The reader is referred to Tables 3.1 to 3.3 in the Main Text. In this technical commentary, piers and wharves are considered as Grade A structures and are expected to perform as follows:

(1) To resist earthquakes of moderate size that can be expected to occur one or more times during the life of the structure (Level 1) without structural damage of significance (Degree I: Serviceable).

(2) To resist major earthquakes that are considered as infrequent events (Level 2), and to maintain environmental protection and life safety, precluding total collapse but allowing a measure of controlled inelastic behaviour that will require repair (Degree II: Repairable).

T5.1.2   Design earthquake levels

This section utilizes the following earthquake levels presented in Chapter 3 in the Main Text as defined events.

Level 1 – (L1) Ground motions with a 50% probability of exceedance in 50 years exposure. This event has a return time of 75 years and is considered a moderate event, likely to occur one or more times during the life of the facility. Such an event is considered a strength event.

Level 2 – (L2) Ground motions with a 10% probability of exceedance in 50 years exposure. This event has a return time of 475 years and is considered a major event. Such an event is considered a strength and ductility event.

T5.1.3   Limit states

Serviceability Limit State (Degree I: Serviceable) All structures and their foundations should be capable of resisting the Level 1 earthquake without sustaining damage requiring post-earthquake remedial action.

Damage Control Limit State (Degree II: Repairable) Structures and their foundations should be capable of resisting a Level 2 earthquake, without collapse and with repairable damage, while maintaining life safety. Repairable damage to structures and/or foundations, and limited permanent deformation are expected under this level of earthquake.

T5.1.4   Seismic load combinations

Wharves and piers may be checked for the following seismic load combinations, applicable to both Level 1 and Level 2 earthquakes:

$$(1 + k_v)(D \oplus rL) \oplus E \tag{T5.1}$$

$$(1 - k_v)D \oplus E \tag{T5.2}$$

where D = Dead Load
   L = Design Live Load
   r = Live Load reduction factor (depends on expected L present in actual case; typically 0.2 but could be higher)

E = Level 1 or Level 2 earthquake load, as appropriate

k_v = Vertical seismic coefficient (typically one half of the effective peak horizontal ground acceleration in g)

$\oplus$ = Load combination symbol for vector summation

Note: seismic mass for E should include an allowance for rL, but need not include an allowance for the mass of flexible crane structures.

Effects of simultaneous seismic excitation in orthogonal horizontal directions, indicated by x and y in Fig. T5.1, may be considered in design and assessment of wharves and piers. For this purpose, it will be sufficient to consider two characteristic cases:

$$100\% \; E_x \oplus 30\% \; E_y \qquad\qquad\qquad\qquad\qquad (T5.3)$$

$$30\% \; E_x \oplus 100\% \; E_y \qquad\qquad\qquad\qquad\qquad (T5.4)$$

where E_x and E_y are the earthquake loads (E) in the principal directions x and y, respectively.

Where inelastic time history analyses in accordance with the requirements of Method D in Section T5.3 are carried out, the above loading combination may be replaced by analyses under the simultaneous action of x and y direction components of ground motion. Such motions should recognize the direction-dependency of fault-normal and fault-parallel motions with respect to the structure principal axes, where appropriate.

T5.2   MODELLING ASPECTS

This section further illustrates the modelling aspects of pile-deck structures summarized in Technical Commentary 7.

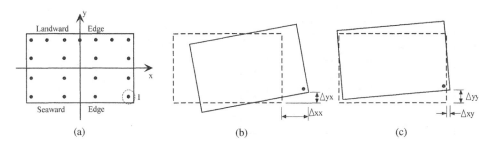

(a)   (b)   (c)

Fig. T5.1. Plan view of wharf segment under x and y seismic excitations.
(a) Plan view.
(b) Displaced shape, x excitation.
(c) Displaced shape, y excitation.

T5.2.1   Soil-structure interaction

Figure T5.2(a) represents a typical transverse section of a wharf supported on a soil foundation comprised of different materials, including alluvial sand and gravel, perhaps with clay lenses, rip-rap, and other foundation improvement materials. The most precise modelling of this situation would involve inelastic finite element modelling of the foundation material to a sufficient depth below, and to a sufficient distance on each side, such that strains in the foundation material at the boundaries would not be influenced by the response of the wharf structure. The foundation would be connected to the piles by inelastic Winkler springs at sufficiently close spacing so that adequate representation of the pile deformation relative to the foundation material, and precise definition of the in-ground plastic hinging, would be provided. Close to the ground surface, where soil spring stiffness has the greatest influence on structural response, the springs should have different stiffnesses and strengths in the seaward, landward, and longitudinal (parallel to shore) directions, as a consequence of the dike slope.

The pile elements would be represented by inelastic properties based on moment-curvature analyses. Since the key properties of piles (namely strength and stiffness) depend on the axial load level which varies during seismic response, sophisticated interaction modelling would be necessary.

It is apparent that structural modelling of the accuracy and detail suggested above, though possible, is at the upper limit of engineering practice, and may be incompatible with the considerable uncertainty associated with the seismic input. In many cases, it is thus appropriate to adopt simplified modelling techniques. Two possibilities are illustrated in Fig. T5.2. In Fig. T5.2(a), the complexity may be reduced by assuming that the piles are fixed at their bases

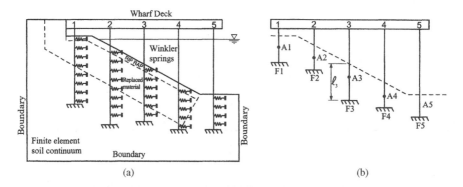

Fig. T5.2. Modelling soil-structure interaction for a wharf.
(a) Soil-structure continuum model.
(b) Equivalent depth-to-fixed model.

with the seismic input applied simultaneously and coherently to each pile base. The soil springs also connect the piles to the rigid boundary. Thus the assumption is made that the deformations within the soil are small compared with those of the wharf or pier.

In Fig. T5.2(b), the complexity is further reduced by replacing the soil spring systems by shortening the piles to 'equivalent fixity' piles, where the soil is not explicitly modelled, and the piles are considered to be fixed at a depth which results in the correct overall stiffness and displacement for the wharf or pier. Where different soil stiffnesses are appropriate for opposite directions of response, average values will be used, or two analyses carried out, based on the two different stiffnesses, respectively. It should be noted that, though this modelling can correctly predict stiffness, displacements and elastic periods, it will over-predict maximum pile in-ground moments. The model will also require adjustment for inelastic analyses, since in-ground hinges will form higher in the pile than the depth of equivalent fixity for displacements.

Geotechnical guidance will be needed to determine the appropriate depth to fixity (e.g. l_3 in Fig. T5.2(b)), but an approximate value may be determined from the dimensionless charts of Fig. T5.3. A typical value of $l_3 = 5D_p$ where D_p is the pile diameter, may be used as a starting point for design.

T5.2.2   Movement joints

Long wharf structures, and less commonly piers, may be divided into segments by movement joints to facilitate thermal, creep and shrinkage movements. Typically, the joints allow free longitudinal opening, but restrains transverse displacement by the incorporation of shear keys. Modelling these poses problems, particularly when elastic analysis methods are used. Under relative longitudinal displacement, the segments can open freely, but under closing displacements, high axial stiffness between the segments develops after the initial gap is closed. Relative transverse rotations are initially unrestrained, but, after rotations, are sufficient to close one side of the joint. Further rotation implies relative axial displacement, resulting in axial compression between the segments. The behaviour can be modelled with a central shear spring allowing longitudinal displacement, but, restraining transverse displacements as per the two axial springs, one at each end of the joint, as suggested in Fig. T5.4. These two springs have zero stiffness under relative opening displacement, but have a very high compression stiffness after the initial gap Δ_g is closed.

Clearly, it is not feasible to represent this nonlinear response in an elastic modal analysis. Even in an inelastic analysis, accurate modelling is difficult because of the need to consider different possible initial gaps, and difficulty in modelling energy dissipation by plastic impact as the joint closes.

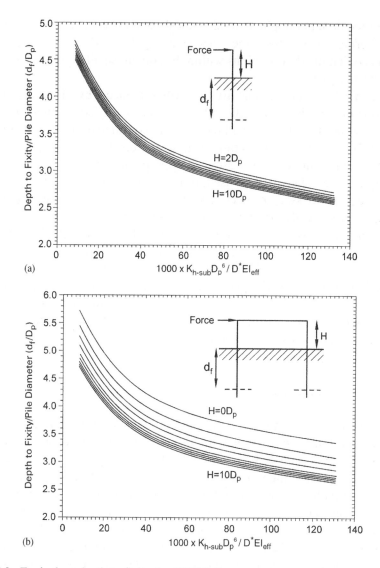

Fig. T5.3. Equivalent depth to fixity for CIDH piles.
(a) Free-head pile.
(b) Fixed-head pile.

T5.3   METHODS OF ANALYSIS FOR SEISMIC RESPONSE

This section further illustrates the methods of analysis for pile-deck structures discussed in Chapter 5 in the Main Text and in Technical Commentary 7. According to the three analysis types discussed in the Main Text, Methods A through E discussed below may be classified as follows:

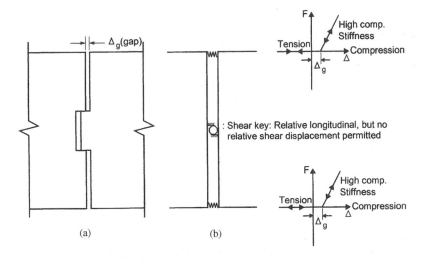

Fig. T5.4. Modelling of movement joints.
  (a) Joint details.
  (b) Joint modelling.

Simplified Analysis: Method A
Simplified Dynamic Analysis: Method C combined with Methods A and/or B
Dynamic Analysis: Method D
Method E is a gross deformation soil analysis and can be used with all the methods. It may take the form of a simple Newmark type analysis as in Simplified Dynamic Analysis or a complex nonlinear effective stress analysis as in Dynamic Analysis.

The use of Method A for Simplified Analysis is justified for preliminary analysis or Grade C structures, provided that the parameters used in Method A have been derived based on a comprehensive set of Method C type parameter studies for typical pile-deck systems proposed for design (see Section T7.1.3 in Technical Commentary 7). A pushover analysis can be used in Method A to establish structure stiffness and resistance.

  In comparison with many other structures such as multistory buildings or multispan bridges, wharves and piers are frequently structurally rather simple. As such, analyses to determine seismic response can often utilize comparatively simple analysis procedures, as discussed in Sections T5.3.1 through T5.3.4, without significant loss of accuracy. Complexity tends to come more from the high significance of soil-structure interaction, from significant torsional response resulting from the typical increase in effective pile lengths from landward to seaward sides (or ends) of the wharf (or pier), and from interaction between adjacent wharf or pier segments separated by movement joints with shear keys, rather than from structural configuration.

It has been well documented that the peak response computed for pile-deck structures will vary depending on the analytical method employed (e.g., Methods A, B, or D in the following text). It should be noted, however, that the modeling of member stiffness and soil-structure interaction effects has been found to yield a greater influence on the computed peak response of the structure than the variations in computed response due to the application of various analytical techniques. Indeed, there is no point in carrying out a sophisticated time-history analysis unless detailed and accurate simulation of member stiffness and strength has preceded the analysis.

T5.3.1   Method A: Equivalent single mode analysis

Long wharves on regular ground tend to behave as simple one-degree-of-freedom structures under transverse response. The main complexity arises from torsional displacements under longitudinal response, and from interaction across shear keys between adjacent segments as discussed above. An individual segment will respond with a significant torsional component when excited by motion parallel to the shore or along the wharf. Dynamic analyses by Priestley (1999) indicate that the torsional response is reduced when segments are connected by shear keys, and with multiple-segment wharves, the torsional response of the inner segments is negligible. It is thus conservative to estimate the displacement response based on the behaviour of a single wharf segment between movement joints except when calculating shear key forces.

Extensive inelastic time-history analyses of single and multi-segment wharves have indicated that an approximate upper bound on displacement response of the piles could be established by multiplying the displacement response calculated under pure transverse (perpendicular to the shore) excitation by a factor, taking into account the orthogonal load combinations of Eqns. (T5.3) and (T5.4), including torsional components of response. For the critical shorter landward line of piles

$$\Delta_{max} = \Delta_t \sqrt{1 + \left(0.3 \left(1 + 20 \frac{e_{ex}}{L_L} \right) \right)^2} \qquad \text{(T5.5)}$$

where

Δ_t = displacement under pure translational response
e_{ex} = eccentricity between centre of mass and centre of rigidity
L_L = longitudinal length of the wharf segment

Note that the corner piles of the line of piles at the sea edge of the wharf will be subjected to slightly higher load/displacement than given by Eqn. (T5.5), but these piles will not be critical because of their greatly increased flexibility and hence increased displacement capacity compared with the landward piles.

The shear force across shear keys connecting adjacent wharf segments cannot be directly estimated from a Method A analysis. However, time-history analysis of multiple wharf segments indicates that the highest shear forces will occur across movement joints in two-segment wharves, and that these forces decrease, relative to segment weight, in inverse proportion to wharf segment aspect ratio L_L/L_T, where L_L and L_T denote the longitudinal and transverse plan dimensions of the wharf segment pile group. The following expression may be used as an approximate upper bound to the shear key force V_{sk}:

$$V_{sk} = 1.5(e_{ex}/L_L)V_{\Delta T} \tag{T5.6}$$

where $V_{\Delta T}$ is the total segment lateral force found from a pushover analysis at the level of displacement ΔT calculated for pure translational response at the appropriate limit state.

Method A is of adequate accuracy for design of many simple structures when supplemented by a Method C pushover analysis, and is particularly useful for the preliminary stages of design, even of relatively important and complex structures. Final design verification may be made by one of the more sophisticated analysis methods.

The approximate lateral stiffness of a wharf structure analyzed using Method A should be determined from a pushover analysis, as discussed in relation to Method C.

T5.3.2   Method B: Multi-mode spectral analysis

Spectral modal analysis is the most common method used for estimating maximum displacement levels, particularly when deck flexibility is significant. However, as noted above, when multiple segments of long wharves are connected by movement joints with shear keys, it is difficult to model the interactions occurring at the movement joints adequately because of their non-linear nature.

It is thus doubtful if it is worth modelling the joint, particularly if conditions are relatively uniform along the wharf. It should also be recognized that it is unlikely that there will be coherency of input motion for different segments of a long wharf. Longitudinal motion of the wharf is likely to be reduced as a consequence of impact across joints and restriction of resonant build-up.

It is thus reasonable for regular wharf conditions to consider an analysis of single wharf segments between movement joints as 'stand-alone' elements. It will also be reasonable in most cases to lump several piles together along a given line parallel to the shore and to provide an analytical 'super pile' with the composite properties of the tributary piles in order to reduce the number of structural elements, which can be excessive, in a multi-mode analysis. Since the modal analysis will be used in conjunction with a pushover analysis, it will be worth considering a further reduction in the number of structural elements by representing the composite stiffness of a transverse line of piles, found from the pushover analysis, by a single pile, and using a damping level appropriate to the expected displacement as discussed in more detail in the following section.

In most stand-alone analyses of single wharf or pier segments, there will be only three highly significant modes: two translational and one torsional. These will generally be closely coupled. As a consequence, it will not generally be difficult to satisfy the 95% participating mass requirement, which is larger than commonly specified in design codes. Note that, to correctly model the torsional response, it is essential to model the torsional inertia of the deck mass. This can either be done by distributing the deck mass to a sufficiently large number of uniformly distributed mass locations (with a minimum of 4 located at the radius of gyration of the deck area), or by use of a single mass point with rotational inertia directly specified. Deck mass should include a contribution for the mass of the piles. Typically, adding 33% of the pile mass, from deck level to the point of equivalent foundation fixity, is appropriate.

Because of the typical close coupling of the key modes of vibration, the modal responses should be combined using the Complete Quadratic Combination (CQC) rule (Wilson et al., 1981).

T5.3.3   Method C: Pushover analysis

Typically, as a result of large variations in effective depth to fixity of different piles in a given wharf or pier, the onset of inelastic response will occur at greatly differing displacements for different piles. This is illustrated in Fig. T5.5, representing the response of a 6 m segment of a typical wharf loaded in the transverse (perpendicular to shoreline) direction. There are a number of important consequences to this sequential hinging:

- It is difficult to adequately represent response by an elasto-plastic approximation, which is the basis of the common force-reduction factor approach to design.
- The appropriate elastic stiffness to be used in analysis is not obvious.
- Different piles will have greatly different levels of ductility demand, with the shortest piles being the most critical for design or assessment.
- The centre of stiffness will move from a position close to the landward line of piles to a position closer to the centre of mass as inelasticity starts to develop first in the landward piles.

As a consequence, it is not directly feasible to carry out an elastic analysis based on the typically assumed 5% equivalent viscous damping, and then determine member forces by reducing elastic force levels by a force-reduction factor. Among other failings, this will grossly underestimate the design forces for the longer piles. It is thus strongly recommended that a key aspect of the design or analysis process be a series of inelastic pushover analyses in both transverse and

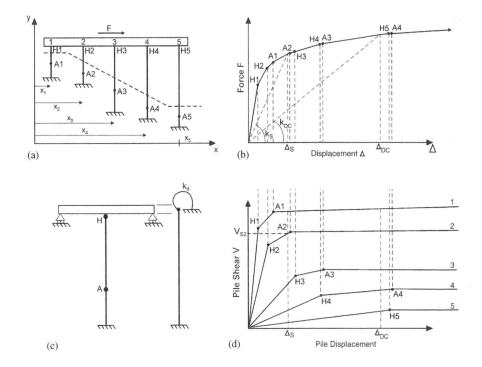

Fig. T5.5. Pushover analysis of a wharf segment.
    (a) Transverse section for analysis.
    (b) Force-displacement response and hinge sequence for transverse excitation.
    (c) Pile model for longitudinal analysis.
    (d) Individual pile longitudinal force-displacement response.

longitudinal directions, where two dimensional (2D) sections of the wharf or pier are subjected to incremental increases in displacement allowing the inelastic force-displacement response (e.g. Fig. T5.5(b)), the sequence of hinge formation, and the magnitude of inelastic rotation Θ_p developing in each hinge to be determined. The results of such analyses can be used to:

• determine an appropriate stiffness for a Method A or B analysis,
• determine appropriate damping levels for elastic analyses, and
• determine peak plastic rotations of critical hinges at the maximum displacements predicted by the Method A or B analyses.

These aspects are discussed in more detail in the following sections (1) through (4).

(1) Elastic stiffness from pushover analysis

It is recommended that elastic analyses used for Method A or B analysis be based on the substitute structure analysis approach (Gulkan and Sozen, 1974), where the elastic characteristics are based on the effective stiffness to maximum displacement response anticipated for the given limit state, and a corresponding level of equivalent viscous damping based on the hysteretic characteristics of the force displacement response (e.g. Fig. T5.5(b)). Thus, for the transverse response, the stiffness for a 6 m tributary length of wharf would be k_S or k_{DC} for the serviceability and damage control limit states, respectively, based on the expected limit states displacements Δ_S and Δ_{DC}. Since these displacements will not be known prior to a Method A or B analysis, some iteration will be required to determine the appropriate stiffnesses.

For a Method A analysis, the transverse period can then be directly calculated from the stiffnesses and tributary mass in the usual manner. For a Method B analysis, using a reduced number of piles as suggested above, longitudinal pushover analyses will be required on characteristic sections of a wharf or pile. As suggested in Fig. T5.5(c), the characteristic element for a wharf may be taken as a pile plus deck section extending midway to adjacent piles on either side. A set of pushover analyses for each of the characteristic piles, A through E, can then be carried out and plotted as shown in Fig. T5.5(d). Again, based on the expected limit states displacements, the total stiffness of the tributary length of wharf considered can be calculated as

$$K_L = \sum n_i V_i / \Delta_{ls} \tag{T5.7}$$

where n_i is the number of piles in row i in the tributary width considered, and V_i is the pile shear force at the limit state deflection Δ_{ls}.

The centre of rigidity measured relative to the arbitrary datum shown in Fig. T5.5(a) is given by

$$x_r = \frac{\sum (n_i V_i x_i)}{\sum (n_i V_i)} \tag{T5.8}$$

Thus, for a Method B analysis, the complete longitudinal and transverse stiffness of a given length of wharf may be represented by a single 'super pile' located at the longitudinal centre of the length modelled by the pile, and located transversely at the position defined by Eqn. (T5.8), with stiffness values as defined above. Note that different stiffnesses and centres of rigidity will normally apply for the serviceability and damage control limit states.

(2) Damping

The substitute structure approach models the inelastic characteristics of structures by elastic stiffness and damping levels appropriate for the maximum response displacements rather than using the initial elastic parameters. The advantage to using this approach is that it provides a more realistic representation of peak response, and eliminates the need to invoke force-reduction factors to bring member force levels down from unrealistically high elastic values to realistic levels. The damping level is found from the shape of the complete

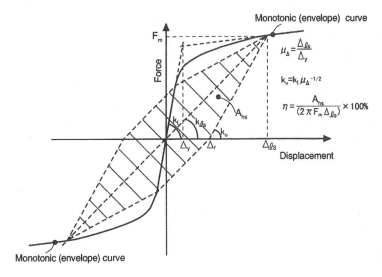

Fig. T5.6. Equivalent viscous damping for substitute-structure analysis.

hysteresis loop for a single cycle of displacement at maximum response, as illustrated in Fig. T5.6. In this, the skeleton force-displacement curve for both directions of response is calculated by a pushover analysis, as illustrated in Fig. T5.5(b). The unloading curve is based on a modified Takeda approach (Takeda et al., 1970), where the unloading stiffness k_u is related to the structure ductility μ_Δ, defined in Fig. T5.6, and the initial stiffness k_i by

$$k_u = k_i \mu_\Delta^{-1/2} \tag{T5.9}$$

The residual displacement Δ_r is thus given by

$$\Delta_r = (\Delta_{ls} - F_m / K_u) \tag{T5.10}$$

The remainder of the stabilized hysteresis loop is constructed, as shown in Fig. T5.6. The equivalent viscous damping, as a percentage of critical damping, is then given by

$$\eta = A_{hs} / (2\pi \cdot F_m \Delta_{ls})\ 100\% \tag{T5.11}$$

where A_{hs} is the area of the stabilized loop, shown by hatching in Fig. T5.6.

   For initial analyses of wharves or piers supported on reinforced or prestressed concrete piles, damping values of 10% and 20% at the serviceability and damage control limit states will generally be appropriate.

(3) Capacity design checks
When the pushover analyses are based on the conservative estimates of material properties for the wharf or pier elements, the member forces resulting from the

analyses represent lower bound estimates of the forces developed at the appropriate limit state. Although this may be appropriate for determining required strength of plastic hinge regions, and for ensuring that calculated limit state displacements are not exceeded, the results will not be appropriate for determining the required strength of members. It is clear that if material strengths exceed the specified strength, then the flexural strength of plastic hinges may significantly exceed the dependable strength, and as a consequence, the shear forces in piles and deck members may be as much as 20–30% higher than predicted by the pushover analysis.

In order to ensure that a conservative upper limit on the possible strength of members or actions, which are to be protected against inelastic response, is achieved, a second set of pushover analyses should be carried out where the material strengths adopted represent probable upper bounds. This approach, based on capacity design principles, is discussed in detail in Priestley et al. (1996).

(4) Iteration considerations

When checking a completed design, iteration will be needed to ensure compatibility between the Method A or B elastic analysis used to determine response displacements, and the stiffness and damping values determined from the pushover analysis. Typically, the necessary adjustments to the substitute structure characteristics take very few cycles to converge with adequate accuracy.

In the design process, the limit state displacements will generally be known before the design strength is established, provided the type and diameter of pile is known. In this case, the iteration required will be centred on determining the correct number and location of piles in order to limit the response displacements to the structural limit state displacements.

T5.3.4   Method D: Inelastic time-history analysis

Inelastic time-history analysis is potentially the most accurate method for estimating the full seismic response of a wharf or pier. It has the capability of determining the maximum displacements and the inelastic rotations in plastic hinges from a single analysis. However, to do this, it is necessary to model each pile individually and to simulate the pile/soil interaction by a series of Winkler springs as discussed in Section T5.3.1. This can result in unacceptable matrix sizes for analysis of wharves or piers with very large numbers of piles. An alternative is to combine the analysis with inelastic pushover analyses and represent groups of piles by equivalent 'super piles' for modal analysis as discussed in Sections T5.3.2 and T5.3.3. Although this somewhat reduces the attraction of the time-history analysis, it makes the analysis more tractable, and still enables several advantages of the method, not available with Method A or B, to be retained.

A disadvantage of the method is that considerable variations in response can be obtained between two different spectrum-compatible acceleration records. As a consequence, it is essential to run an adequately large number of simulations using different acceleration records, and to average the results. A minimum of five spectrum-compatible records is recommended.

The following points need to be considered before undertaking a time-history analysis:

1. A full simulation of the wharf or pier will require use of a computer program capable of modelling three dimensional (3D) response. There are comparatively few 3D inelastic time-history programs available at present (March 2000), and experience with them outside of research applications is rather limited. Frequently, approximations must be made relating to hysteresis rules and strength interactions in orthogonal directions which make the added sophistication of time-history analysis of reduced utility.

2. If the deck has sufficient rigidity to justify its approximation as a rigid element both in-plane and out-of-plane, a two dimensional (2D) plan simulation may provide adequate accuracy. However, special multi-direction spring elements with realistic hysteresis rules are required for such a simulation.

3. Time-history analysis enables different stiffnesses to be used for different directions of response. Thus, provided the necessary hysteresis rules are available, it will generally be possible to model the higher stiffness for movement into the shore than for movement away from the shore, and to have a separate stiffness for longitudinal response.

4. It is comparatively straightforward to model the interactions across shear keys, using inelastic time-history analysis.

5. When modelling reinforced or prestressed concrete members, degrading stiffness models such as the Modified Takeda rule should be adopted. There is little point in carrying out time history analyses if simplified rules such as elasto-plastic or even bilinear stiffness rules are adopted.

6. Care needs to be exercised into how elastic damping is handled. It is common to specify 5% elastic initial stiffness related damping. This can greatly overestimate the damping at high ductilities. In fact, since the hysteretic rules available in the literature have generally been calibrated to experimental results, the justification for adding elastic damping, which should be apparent in the experimental hysteresis loops, is of doubtful validity. There is thus a case for ignoring elastic damping if levels of inelastic response are high. However, at low levels of displacement response, the simplifications inherent in the hysteresis rules generally mean that the damping is underestimated for 'elastic' or near-elastic response.

7. Results from a time-history analysis should always be compared with results from a simplified approach (e.g. a Method A or B analysis) to ensure that reasonable results are being obtained.

T5.3.5   Method E: Gross foundation deformation analysis

As discussed above, the assumption will normally be made that the foundation material is competent when carrying out any of the analyses described above. However, examination of the performance of wharves and piers in past earthquakes reveals that liquefaction and foundation sliding or slumping are common as discussed in Technical Commentary 2. Analysis techniques to estimate the sensitivity of the foundation to such failures, and the extent of deformation to be expected, are dealt with in Technical Commentary 7. Where geotechnical analyses indicate that moderate permanent deformations are to be expected, the structure should be analyzed under the deformed soil profile to estimate the influence on the structure. This will generally require at least a 2D analysis incorporating discrete modelling of piles and soil springs, with foundation deformations applied at the boundary ends of the springs as appropriate.

Structural deformations due to inertial effects alone occur relatively early in the time history as the peak ground motions are experienced. Conversely, structural behavior associated with ground failures, soil settlement and gross foundation deformations are likely to take place much later in the time history, or even well after the strong ground motions have ended. Indeed, liquefaction failure may well occur some minutes after the ground motion has ceased. Consequently, it is not necessary to combine the results from a Method E analysis with the results of a Method A to D analysis. It will be sufficient to confirm that response of each type is individually satisfactory.

T5.4   STRUCTURAL CRITERIA FOR PIER AND WHARF PILES

T5.4.1   Deformation capacity of pile plastic hinges

The ability of wharves and piers to respond inelastically to seismic excitation depends on the displacement capacity of pile plastic hinges. This displacement capacity will depend on types of piles used in the structure, their length, cross-section dimensions, axial load, and material properties. A brief discussion of the deformation capacities of different pile types is included below. This is based on an examination of their moment-curvature characteristics.

Until recently, piles were designed neglecting the confinement effects of the reinforcement in the concrete. Work by Joen and Park (1990) shows the significant increase in moment capacity by considering the effect of spiral confinement on the concrete. This work uses the Mander et al. (1988) concrete model for confined and unconfined concrete. Typically, the ultimate compressive concrete strain of unconfined concrete is about 0.003 for use in computing flexural strength as reported by Priestley et al. (1992). For confined concrete, the following ultimate compressive strain ε_{cu} may be used (Priestley et al., 1992):

$$\varepsilon_{cu} = 0.004 + (1.4\rho_s f_{yh}\varepsilon_{sm})/f'_{cc} \geq 0.005 \tag{T5.12}$$

where ρ_s effective volume ratio of confining steel

$\quad\quad\;\; f_{yh}$ yield stress of confining steel

$\quad\quad\;\; \varepsilon_{sm}$ strain at peak stress of confining reinforcement, 0.15 for grade 40 and 0.12 for grade 60 (see Table T5.1 for designation of reinforcing bars)

$\quad\quad\;\; f'_{cc}$ confined strength of concrete approximated by $1.5 f'_c$

$\quad\quad\;\; f'_c$ ultimate concrete compressive stress at 28 days

T5.4.2 Implication of limit states

Two levels of design earthquake are considered for this study.

Level 1 Earthquake; Serviceability Limit State Level 1 has a high probability of occurrence during the life of the pier or wharf. Under this level of excitation, the structure should satisfy the serviceability criterion of continued functionality immediately after the Level 1 earthquake. Any repairs required should be essentially cosmetic. Structural damage requiring repair is not permitted. Note that this does not imply a requirement for elastic response which would limit concrete strains to about 0.001 and steel strains to yield strain. Concrete structures may be considered serviceable, without any significant decrement to structural integrity, provided concrete strains reached during maximum seismic response to the Level 1 earthquake do not cause incipient spalling of concrete cover and if residual crack widths after the earthquake are small enough to negate the need for special corrosion protection measures such as crack grouting. Note that this latter requirement implies that significantly larger crack widths might momentarily exist during seismic response since these have no effect on corrosion potential.

Table T5.1. Tensile and yield stresses of concrete reinforcing bars in North America (1.0 psi = 6.9 kPa).

Designation	Minimum yield (psi)	Minimum tensile (psi)
Grade 40*	40,000	70,000
Grade 50**	50,000	80,000
Grade 60	60,000	90,000
Grade 70	70,000	80,000
Grade 75	75,000	100,000

*Previously designated as Intermediate.
**Previously designated as Hard.

Thus, the performance of potential plastic hinges should be checked under Level 1 earthquake to ensure that maximum material strains do not exceed the limits defined for structural performance limit states.

Note that this check will normally only be required at the pile/deck interface, since curvatures will be higher there than in any potential in-ground hinges. Also, the cover concrete of the potential in-ground plastic hinges are confined by the lateral pressure developed in the soil, and therefore the concrete strain at which spalling initiates is higher for the in-ground plastic hinges (Budek et al., 1997). This will increase the corresponding serviceability curvatures for the in-ground plastic hinges, particularly for prestressed piles, where steel strains are unlikely to be critical.

Level 2 Earthquake; Damage-control Limit State   Level 2 earthquake represents an earthquake with a low probability of occurring during the design life of the structure. Repairable damage to structure and/or foundation, and limited permanent deformation are acceptable under this level of earthquake

Since the design philosophy is to restrict inelastic action to carefully defined and detailed plastic hinges in piles, strain limits may again be used to define acceptable response. Distinction needs to be made between the pile/deck hinge and the in-ground hinge locations, since access to the latter will frequently be impractical after an earthquake. As a consequence, the in-ground hinge should satisfy serviceability criteria even under Level 2 earthquake. Thus, the steel strain should not exceed 0.01. However, as noted above, passive confinement by the soil increases the spalling strain and a concrete strain of 0.008 may be adopted. The pile/deck connection hinge will have allowable concrete strains, dependent on the amount of transverse reinforcement provided. As a consequence of these actions, the maximum strains calculated under a Level 2 earthquake must not exceed the strain capacities defined in Section T5.4.3.

T5.4.3   Moment-curvature characteristics of piles

The key tool for investigating the deformation characteristics of piles is moment-curvature analysis. The analysis adopted must be capable of modelling the full stress strain curves of reinforcement and concrete realistically. For the concrete, this entails discrimination between the behaviour of unconfined cover concrete, if present, and that of the confined core, which will have enhanced strength, and particularly enhanced deformation capacity. Elastic, yield plateau, and strain-hardening sections of the steel stress-strain curve should be modelled separately; a simple elasto-plastic or bi-linear representation will be inadequate for mild steel reinforcement, but may be adequate for prestressing steel. The computer code used should provide output of moment and curvature at regular intervals of a sequential analysis to failure, and should identify the peak values of extreme fibre concrete strain, and maximum reinforcement and/or prestressing strain at each increment of the analysis, so that curvatures corresponding to the serviceability and damage-control limit state can be identified.

Curvatures at the serviceability limit state are based on strain limits sufficiently low so that spalling of cover concrete will not occur under the Level 1 earthquake, and any residual cracks will be fine enough so that remedial grouting will not be needed. These are taken to be

Concrete extreme fibre compression strain:	0.004	
Reinforcing steel tension strain:	0.010	Serviceability
Prestressing strand incremental strain:	0.005	Limit Strains
Structural steel (pile and concrete-filled pipe):	0.008	Level 1
Hollow steel pipe pile:	0.008	

The specified incremental strain for prestressing strand is less than for reinforcing steel to ensure that no significant loss of effective prestress force occurs at the serviceability limit state. For steel shell piles, either hollow or concrete filled, it is recommended that the maximum tension strain in the shell be also limited to 0.01, to ensure that residual displacements are negligible. Note that, if the steel shell pile is connected to the deck by a dowel reinforcing bar detail, the connection is essentially a reinforced concrete connection confined by the steel shell. As such, the limit strains for reinforced concrete apply.

A sample moment-curvature response appropriate for a reinforcing steel dowel connection between a prestressed pile and a reinforced concrete deck is shown in Fig. T5.7. For structural analysis, it will often be adequate to represent the response by a bi-linear approximation as shown in the figure. The initial elastic portion passes through the first yield point (defined by mild steel yield strain, or

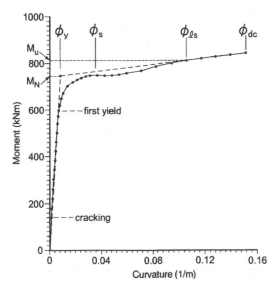

Fig. T5.7. Pile-cap moment curvature.

an extreme fibre concrete compression strain of 0.002, whichever occurs first), and is extrapolated to the nominal flexural strength. This defines the nominal yield curvature, ϕ_y. The second slope of the bi-linear approximation joins the yield curvature and the point on the curve corresponding to the damage-control limit-state strains. These are taken to be

Concrete extreme fibre compression strain:
    Pile/deck hinge: Value given by Eqn. (T5.12), but <0.025
    In-ground hinge: Value given by Eqn. (T5.12), but <0.008
Pile/deck reinforcing steel tension strain: 0.05
In-ground reinforcing steel tension strain: 0.01
Structural steel (pile and concrete filled pipe):   0.035
Prestressing strand:
    Pile/deck hinge: 0.04
    In-ground hinge: 0.015
Hollow steel pipe pile: 0.025

> Damage Control Limit Strains Level 2

Equation (T5.12) defines a safe lower bound estimate of the ultimate compression strain of concrete confined by hoops or spirals (Priestley et al., 1996). Actual compression failure, initiated by fracture of spiral or hoop confinement, will typically not occur until strains are on average 50% larger than the value given by Eqn. (T5.12), providing an adequate margin for uncertainty of input.

Figure T5.8 shows examples of theoretical moment-curvature curves for different types of piles, each with an outside diameter of 61 cm, and each subjected to different levels of axial load. Comparison of the curves for the two reinforced concrete doweled connections of Figs. T5.8(a) and T5.8(b) show the significant influence of the concrete cover on the moment-curvature characteristics. Flexural strength is significantly higher with the reduced concrete cover of Fig. T5.8(b). Sections with high cover (10 cm) show significant reductions in moment capacity when spalling initiates, for all levels of axial load, while piles with 6.4 cm cover exhibit much more satisfactory response. Similar behaviour is apparent for circular or octagonal prestressed piles, as shown for two different values for concrete cover in Figs. T5.8(c) and T5.8(d). Moment-curvature curves for hollow and concrete-filled steel shell piles are shown in Figs. T5.8(e) and T5.8(f). For the hollow steel shell pile, the influence of axial load is rather limited with axial compression tending to reduce the flexural strength. The influence of internal concrete is to increase the moment capacity significantly, particularly when axial load is high. Another advantage of the concrete infill is that it reduces the sensitivity of the steel shell to local buckling.

Rectangular section reinforced or prestressed concrete piles confined by square spirals and a rectangular distribution of longitudinal reinforcing bars or prestressing strand should not be used for new construction, since the confinement provided to the core concrete by the rectangular spirals is of very low

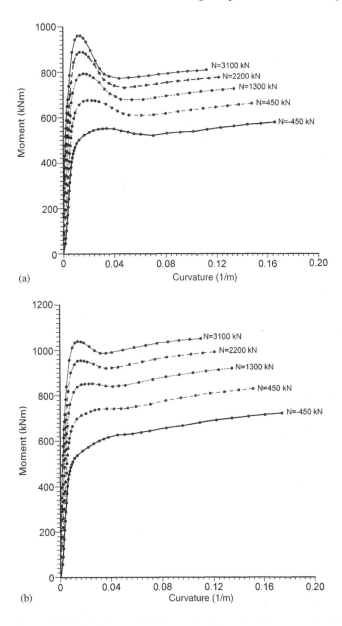

Fig. T5.8. Moment-curvature curves for 610 mm diameter piles.
  (a) Cover = 102 mm (4 in.) RC dowel connection 8#10 bars.
  (b) Cover = 64 mm (2.5 in.) RC dowel connection 8#10 bars.
  (c) Prestressed pile, F/A = 6.4 MPa, cover = 76 mm.
  (d) Prestressed pile, F/A = 6.4 MPa, cover = 25 mm.
  (e) Steel shell pile, t_p = 12 mm.
  (f) Concrete filled steel shell pile, t_p = 12 mm.

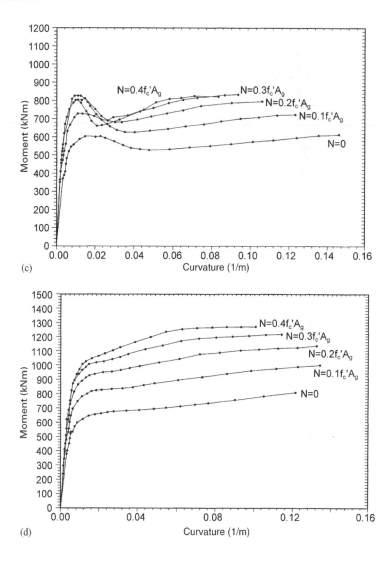

Fig. T5.8. Continued.

efficiency. For assessment of existing structures, the concrete core maximum strain corresponding to the damage-control limit state should not be taken larger than 0.007 at the pile/deck hinge. Higher maximum concrete strains, as given by Eqn. (T5.12), are appropriate if the core of the rectangular pile is confined by a circular spiral and the longitudinal reinforcement or prestressing is also circularly disposed. However, such sections typically have low ratios of concrete core area to gross section area, resulting in flexural strength of the confined core being less than that of the unconfined gross section.

Hollow prestressed piles are sometimes used for marine structures. These, however, have a tendency to implode when longitudinal compression strains at

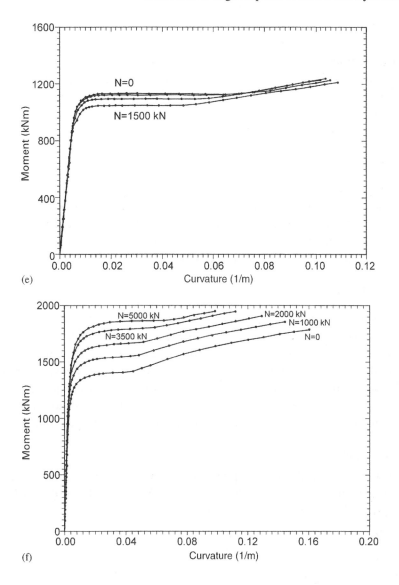

Fig. T5.8. Continued.

the inside surface exceed 0.005. Consequently, the damage control limit state should have an additional requirement to the strain limits defined earlier that inside surface compression strains must not exceed 0.005. Note that to check for this condition, the moment-curvature analysis must be able to model spalling of the outside cover concrete when strains exceed about 0.004 or 0.005. The outside spalling can cause a sudden shift of the neutral axis towards the centre of the section resulting in strains on the inside surface reaching the critical level soon after initiation of outside surface spalling. More information on piles is available in Priestley and Seible (1997).

T5.4.4   Primary structural parameters and response of piles

(1) Elastic stiffness
The effective elastic stiffness may be calculated from the slope of the 'elastic' portion of the bi-linear approximation to the moment-curvature curve (e.g. Fig. T5.7), as

$$EI_{\text{eff}} = M_{\text{N}}/\phi_{\text{y}}$$
(T5.13)

where I_{eff} effective moment of inertia
         M_{N} nominal flexural strength
   The value of the yield curvature, ϕ_{y}, for a reinforced concrete pile or pile/deck connection is rather insensitive to axial load level or longitudinal reinforcement ratio. Results of analyses of a large number of cases indicate that Eqn. (T5.13) may be approximately expressed as a fraction of the gross section stiffness as

$$EI_{\text{eff}}/EI_{\text{gross}} = 0.3 + N/(f_{\text{c}}'A_{\text{gross}})$$
(T5.14)

where N is the axial load level, and A_{gross} is the uncracked section area.
   For prestressed piles, the effective stiffness is higher than for reinforced concrete piles, and values in the range $0.6 < EI_{\text{eff}}/EI_{\text{gross}} < 0.75$ are appropriate. For prestressed piles with reinforced dowel connections to the deck, the effective stiffness should be an average of that for a reinforced and a prestressed connection, or a short length, approximately $2D_{\text{p}}$ long of reduced stiffness appropriate for reinforced pile, should be located at the top of the prestressed pile.

(2) Plastic rotation
The plastic rotation capacity of a plastic hinge at a given limit state depends on the yield curvature, ϕ_{y}, the limit-state curvature, ϕ_{s} for Level 1 or ϕ_{LS} for Level 2, and the plastic hinge length L_{p}, and is given by

$$\Theta_{\text{p}} = \phi_{\text{p}}L_{\text{p}} = ((\phi_{\text{s}} \text{ or } \phi_{\text{LS}}) - \phi_{\text{y}})L_{\text{p}}$$
(T5.15)

(3) Plastic hinge length
The plastic hinge length for piles depends on whether the hinge is located at the pile/deck interface or is an in-ground hinge. Because of the reduced moment gradient in the vicinity of the in-ground hinge, the plastic hinge length is significantly longer there. For pile/deck hinge locations with reinforced concrete details, the plastic hinge length can be approximated by

SI units $L_{\text{p}} = 0.08L + 0.022f_{\text{y}}d_{\text{b}} > 0.044f_{\text{y}}d_{\text{b}}$ (MPa, mm)   (T5.16a)

US units $L_{\text{p}} = 0.08L + 0.15f_{\text{y}}d_{\text{b}} > 0.30f_{\text{y}}d_{\text{b}}$ (ksi, in.)   (T5.16b)

where f_{y} is the yield strength of the dowel reinforcement, of diameter d_{b}, and L is the distance from the pile/deck intersection to the point of contraflexure in the pile.

For prestressed piles where the solid pile is embedded in the deck (an unusual detail in the USA), the plastic hinge length at the pile/deck interface can be taken as

$$L_p = 0.5D_p \tag{T5.17}$$

For in-ground hinges, the plastic hinge length depends on the relative stiffness of the pile and the foundation material. The curves of Fig. T5.9 relate the plastic hinge length of the in-ground hinge to the pile diameter, D_p, a reference diameter $D^* = 1.82$ m, and the soil lateral subgrade coefficient, $k_{h\text{-sub}}$ (N/m^3).

For structural steel sections and for hollow or concrete-filled steel pipe piles, the plastic hinge length depends on the section shape and the slope of the moment diagram in the vicinity of the plastic hinge, and should be calibrated by integration of the section moment-curvature curve. For plastic hinges forming in steel piles at the deck/pile interface, and where the hinge forms in the steel section rather than in a special connection detail (such as a reinforced concrete dowel connection), allowance should be made for strain penetration into the pile cap. In the absence of experimental data, the increase in plastic hinge length due to strain penetration may be taken as $0.25\,D_p$, where D_p is the pile diameter (with a circular cross section) or pile depth (with a non ciruclar cross section) in the direction of the applied shear force.

(4) In-ground hinge location
The location of the in-ground plastic hinge for a pile may be found directly from an analyses where the pile is modelled as a series of inelastic beam elements, and

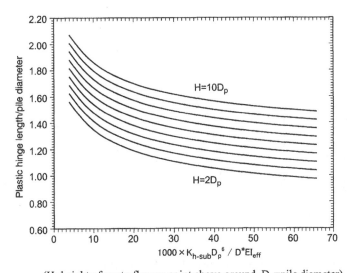

(H=height of contraflexure point above ground, D_p=pile diameter)

Fig. T5.9. In-ground plastic hinge length.

the soil is modelled by inelastic Winkler springs. When the pile/soil interaction is modelled by equivalent-depth-to-fixity piles, the location of the in-ground hinge is significantly higher than the depth to effective fixity, as illustrated in Fig. T5.10 by the difference between points F, at the effective fixity location, and B, the location of maximum moment. Note that when significant inelastic rotation is expected at the in-ground hinge, the location of B tends to migrate upwards to a point somewhat higher than predicted by a purely elastic analysis. It is thus important that its location, which is typically about 1–2 pile diameters below ground surface (such as dike surface beneath wharf), should be determined by inelastic analysis.

(5) Pile force-displacement response

The information provided in the subsections (1) through (4) enables an inelastic force displacement response to be developed individually for each pile. This may be directly carried out on a full 2D section through the wharf, involving many piles, as part of a pushover analysis, or it may be on a pile-by-pile basis, with the pushover analysis assembled from the combined response of the individual piles. With respect to the equivalent-depth-to-fixity model of Fig. T5.10, the pile is initially represented by an elastic member, length L, with stiffness EI_{eff} given by Eqn. (T5.13) or (T5.14), as appropriate, and the deck stiffness represented by a spring k_d, as shown in the same figure. Often, it will be sufficiently accurate to assume the deck to be flexurally rigid, particularly with longer piles.

The deflection and force corresponding to development of nominal strength M_N at the pile/deck hinge can then be calculated. Note that, for elastic deformation calculations, the interface between the deck and pile should not be

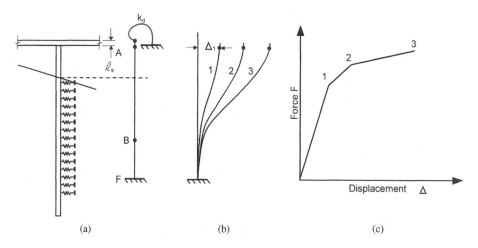

(a)   (b)   (c)

Fig. T5.10. Force-displacement response of an isolated pile.
      (a) Pile equivalent-depth to fixity model.
      (b) Displacement profiles.
      (c) Force-displacement response.

considered rigid. The effective top of the pile should be located at a distance $0.022 f_y d_b$ into the deck to account for strain penetration. This is particularly important for short piles. This additional length applies only to displacements. Maximum moment should still be considered to develop at the soffit of the deck.

The elastic calculations above result in a pile displacement profile marked 1 in Fig. T5.10(b), and the corresponding point on the force displacement curve of Fig. T5.10(c). For the next step in the pile pushover analysis, an additional spring k_{pt} must be added at A, the deck/pile interface (i.e. the deck soffit), to represent the inelastic stiffness of the top plastic hinge. This stiffness can be determined from Fig. T5.7 as

$$k_{pt} = \frac{(M_u - M_N)}{(\phi_{LS} - \phi_y)L_P} \qquad (T5.18)$$

Essentially, this spring is in series with the deck spring. Additional force can be applied to the modified structure until the incremental moment at B is sufficient to develop the nominal moment capacity at the in-ground hinge. The deflection profiles and force-displacement points marked 2 in Figs. T5.10(b) and (c) refer to the status at the end of this increment. Finally, the modified structure, with plastic hinges at the deck/pile interface and B, is subjected to additional displacement until the limit state curvature ϕ_{LS} is developed at the critical hinge, which will normally be the deck/pile hinge. Note that the inelastic spring stiffnesses at the two hinge locations will normally be different, due to differences in the structural details between the in-ground and hinge locations, and due to the different plastic hinge lengths.

The procedure outlined above is sufficiently simple to be carried out by spreadsheet operations, and has the advantage that the post yield moment-rotation stiffness of the hinges can be accurately modelled. This function is not available in all pushover codes.

T5.4.5   Plastic hinges

(1) Material properties for plastic hinges
For both design of new structures and the assessment of existing structures, it is recommended that the moment-curvature characteristics of piles be determined based on probable lower bound estimates of constituent material strengths. This is because strength is less important to successful seismic resistance than is displacement capacity. For the same reason, there is little value in incorporating flexural strength reduction factors in the estimation of the strength of plastic hinges. Flexural strength reduction factors were developed for gravity load design where it is essential to maintain a margin of strength over loads to avoid catastrophic failure. Inelastic seismic response, however, requires that the flexural strength be developed in the design level earthquake. Incorporation of a

strength reduction factor will not change this and at best may slightly reduce the ductility demand. It has been shown (Priestley, 1997) that the incorporation of strength reduction factors in the design of plastic hinges may reduce the ductility capacity slightly, thereby negating any benefits that may have been derived from additional strength.

On the other hand, the use of nominal material strengths and strength reduction factors in design or assessment will place a demand for corresponding increases in the required strength of capacity protected actions, such as shear strength. This is because the maximum feasible flexural strength of plastic hinges, which dictates the required dependable shear strength, will be increased when design is based on conservatively low estimates of material strength. This will have an adverse economic impact on the design of new structures and may result in an unwarranted negative assessment of existing structures.

As a consequence, the following recommendations are made for the determination of moment-curvature characteristics of piles (Priestley et al., 1996):

Concrete compression strength $f'_{ce} = 1.3 f'_c$
Reinforcement yield strength $f_{ye} = 1.1 f_y$
Prestress strand ultimate strength $f_{pue} = 1.0 f_{pu}$

where f'_c is the specified 28 day compression strength for the concrete, and f_y and f_{pu} are the nominal yield and ultimate strength of the mild steel reinforcement and prestressing strand, respectively. For assessment of existing structures, higher concrete compression strength will often be appropriate due to natural aging, but should be confirmed by in-situ testing.

In both new design and assessment of existing structures, the flexural strength reduction factor for pile plastic hinges should be taken as unity.

(2) Confinement of pile plastic hinges
Research by Budek et al. (1997) has shown that lateral soil pressure at the in-ground plastic hinge location helps to confine both core and cover concrete. This research found that, as a consequence of this confinement, for both reinforced and prestressed circular piles, the plastic rotation capacity of the in-ground hinge was essentially independent of the volumetric ratio of confinement provided. As a consequence of this, and also as a result of the low material strains permitted in the in-ground hinge at the damage-control limit state, much lower confinement ratios are possible than those frequently provided for piles. It is recommended that, unless higher confinement ratios are required for pile-driving, the confinement ratio for the in-ground portion of the pile need not exceed

$$\rho_s = \frac{4A_{sp}}{(D_p - 2c_o)s} = 0.008 \qquad \text{(T5.19)}$$

where A_{sp} = area of the spiral or hoop bar

D_p = pile diameter

c_o = concrete cover to centre of hoop or spiral bar

s = spacing of spiral or hoop along the pile axis.

In the vicinity of the potential plastic hinge at the top of the pile, the amount of spiral or hoop reinforcement can be adjusted to ensure that the ultimate compression strain given by Eqn. (T5.12), is adequate to provide the required displacements at the damage-control limit state. Thus, the designer has some ability to optimize the design of pile confinement, dependent on the displacement requirement predicted under the lateral response analysis. The calculated value should be supplied for at least $2D_p$ from the critical section. Because of uncertainties associated with the final position of the tip of a driven pile prior to driving, a longer region of pile should have the increased confinement determined from the above approach. In many cases, the full pile length will conservatively be confined with the volumetric ratio required at the deck/pile hinge location.

As an alternative to the approach outlined above, a prescriptive requirement, modified from bridge design, and defined by Eqn. (T5.20) may be adopted

$$\rho_s = 0.16 \frac{f'_{ce}}{f_{ye}} \left(0.5 + \frac{1.25(N_u + F_p)}{f'_{ce} A_{gross}} \right) + 0.13(\rho_1 - 0.01) \qquad \text{(T5.20)}$$

where N_u = axial compression load on pile, including seismic load

F_p = axial prestress force in pile

ρ_1 = longitudinal reinforcement ratio, including prestressing steel.

The volumetric ratio of transverse reinforcement given by Eqn. (T5.20) may be used as a starting point, but the adequacy of the amount provided must always be checked by comparing displacement demand with capacity.

The pitch of spiral reinforcement provided for confinement should not exceed $6d_b$ nor $D_p/5$, where d_b is the diameter of the dowel reinforcement. For in-ground hinges in prestressed piles, the pitch should not exceed $3.5d_p$ where d_p is the nominal diameter of the prestressing strand.

(3) Addition of mild steel reinforcement to prestressed piles

The use of mild steel dowels to provide moment-resisting connections between piles and decks is common. It is also common to provide additional mild steel reinforcement throughout the length of the pile in the belief that this will enhance the performance of the in-ground hinge. Tests by Falconer and Park (1983) have shown this additional reinforcement to be unnecessary. Provided adequate confinement is provided at a pitch not greater than 3.5 times the prestress strand nominal diameter, dependable ductile response can be assured. Figure T5.11 compares lateral load-displacement hysteresis loops of prestressed piles with and without additional longitudinal mild steel reinforcement, and subjected to high axial load levels. Both piles were able to sustain displacement ductility levels of 10 without failure. The pile with additional mild steel reinforcement was, as expected, stronger than the pile without additional reinforcement, and the loops indicated somewhat enhanced energy dissipation. However, the in-ground hinges will normally only be subjected to moderate levels of ductility demand, for which the added damping provided by the mild steel reinforcement will be of only minor benefit.

It is thus recommended that additional longitudinal mild steel reinforcement be provided in piles only when there is a need to increase the flexural strength.

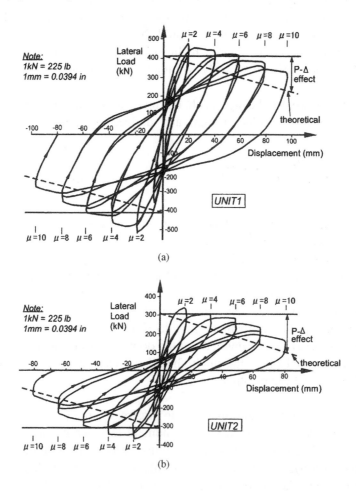

Fig. T5.11. Load-displacement hysteresis loops for prestressed piles (after Falconer and Park, 1983).
  (a) Fully confined pile with additional longitudinal mild steel reinforcement.
  (b) Fully confined pile without additional longitudinal mild steel re-inforcement.

(4) Capacity protection of elastic actions and members
The essence of modern seismic design, is the precise determination of where and how inelastic actions may occur (i.e. by inelastic flexural rotations in specified plastic hinge locations), and the protection of other locations (e.g. the deck) and other actions (e.g. shear) to ensure these remain elastic. This is termed 'capacity design' (Priestley et al., 1996) and is done by ensuring that the dependable strengths of the protected locations and actions exceeds the maximum feasible demand based on high estimates of the flexural strength of plastic hinges. Since development of flexural plastic hinge strength is certain at the design seismic input, the consequence of material strengths significantly exceeding design values will be that corresponding increases will develop in the forces of capacity protected members.

The most consistent method for determination of the required strength of capacity protected actions and members is to carry out a second series of pushover analyses, or dynamic time-history analyses, where the moment curvature characteristics of the pile plastic hinges are based on realistic upper bound estimate of material strengths. The following values are recommended:

Concrete compression strength $\quad f'_{cm} = 1.7 f'_c$
Reinforcement yield strength $\quad f_{ym} = 1.3 f_y$
Prestress strand ultimate strength $\quad f_{pum} = 1.1 f_{pu}$

The design required strength for the capacity protected members and actions should then be determined from the pushover analyses at displacements corresponding to the damage control limit state. Since these force levels will be higher than those corresponding to the serviceability limit state, there is no need to check capacity protection at the serviceability limit state.

A simpler, conservative approach to the use of a second 'upper bound' pushover analyses is to multiply the force levels determined for capacity protected actions from the initial design analysis by a constant factor, representing the maximum feasible ratio of required strength based on upper bound and lower bound material strengths in plastic hinges. This ratio should be taken as 1.4 (Priestley et al., 1996).

T5.4.6   Shear strength of piles

(1) Shear strength of concrete piles
The requirement for capacity protection is that the dependable strength exceeds the maximum feasible demand. Hence, shear strength should be based on nominal material strengths, and shear strength reduction factors should be employed.

Most existing code equations for shear strength of compression members, including the Americal Concrete Institute's (ACI's) 318 equations, which are widely used in the USA, tend to be unreasonably conservative, but do not adequately represent the influence of reduction of the strength of concrete shear resisting mechanisms. An alternative approach (Kowalski et al., 1998), which has been widely calibrated against experimental data and shown to provide good agreement over a wide range of parameter variations, is recommended for assessing the strength of piles. This approach is based on a three parameter model, with separate contributions to shear strength from concrete (V_c), transverse reinforcement (V_t), and axial load (V_p)

$$V_N = V_c + V_t + V_p \qquad \text{(T5.21)}$$

A shear strength reduction factor of 0.85 should be applied to Eqn. (T5.21) to determine the dependable shear strength.

Concrete Mechanism Strength: The strength of the concrete shear resisting mechanisms, which include the effects of compression shear transfer, aggregate interlock, and dowel action, is given by

$$V_c = k_c \sqrt{f_c'} \cdot A_e \qquad \text{(T5.22)}$$

where k_c = factor dependent on the curvature ductility within the plastic hinge region, given by Fig. T5.12

f_c' = concrete compression strength in MPa

$A_e = 0.8 A_{\text{gross}}$ is the effective shear area.

The reduction in k_c with increasing curvature ductility $\mu_\phi = \phi/\phi_y$ occurs as a result of reduced aggregate interlock effectiveness as wide cracks develop in the plastic hinge region. For regions further than $2D_p$ from the plastic hinge location, the flexural cracks will be small, and the strength can be based on $\mu_\phi = 1.0$.

Different values of k_c are provided in Fig. T5.12 for new design and for assessment. It is appropriate to be more conservative for new design than for assessment, since the economic consequences of extra conservatism are insignificant when new designs are considered, but can be substantial when assessment of existing structures are concerned. Different values are also given, depending on whether the pile is likely to be subjected to inelastic action in two orthogonal directions (biaxial ductility) or just in one direction (uniaxial ductility).

Transverse Reinforcement (truss) Mechanism: The strength of the truss mechanism involving transverse spirals or hoops is given by:
Circular spirals or hoops

$$V_t = \frac{\pi}{2} A_{\text{sp}} f_{\text{yh}} (D_P - c_{\text{ex}} - c_o)(\cot \alpha_t)/s \qquad \text{(T5.23)}$$

where A_{sp} = spiral or hoop cross section area

f_{yh} = yield strength of transverse spiral or hoop reinforcement

D_p = pile diameter or gross depth (in the case of a rectangular pile with Spiral confinement)

c_{ex} = depth from extreme compression fibre to neutral axis at flexural strength (see Fig. T5.13)

c_o = concrete cover to centre of hoop or spiral (see Fig. T5.13)

α_t = angle of critical crack to the pile axis (see Fig. T5.13) taken as 30° for assessment, and 35° for new design.

s = spacing of hoops or spiral along the pile axis.

Rectangular hoops or spirals

$$V_t = A_h f_{yh}(D_p - c_{ex} - c_o)\,(\cot \alpha_t)/s \qquad\qquad (T5.24)$$

where A_h = total area of transverse reinforcement, parallel to direction of applied shear cut by an inclined shear crack.

Axial Load Mechanism: The presence of axial compression enhances the shear strength by development of an internal compression strut in the pile between the compression zones of plastic hinges, whose horizontal component V_p opposes the applied shear force. Axial prestress also acts in similar fashion to enhance shear strength. Thus, with reference to Fig. T5.14, the shear strength provided by axial compression is

$$V_p = \Phi_p(N_u + F_p)\tan \alpha_{ax} \qquad\qquad (T5.25)$$

where N_u = external axial compression on pile including load due to earthquake action

F_p = prestress compression force in pile

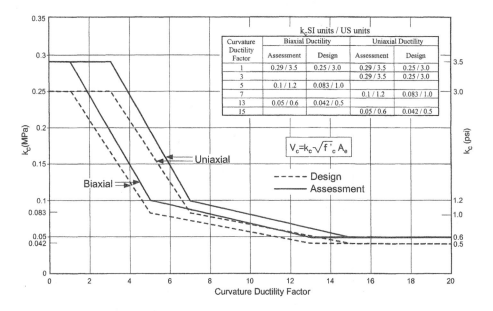

Fig. T5.12. Concrete shear mechanism.

α_{ax} = angle between line joining the centres of flexural compression in the deck/pile and in-ground hinges, and the pile axis (see Fig. T5.14)

Φ_p = 1.0 for assessment, and 0.85 for new design.

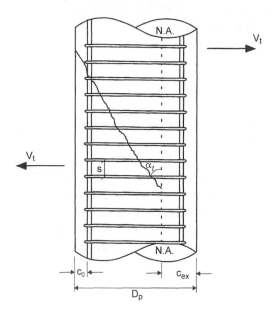

Fig. T5.13. Transverse shear mechanism.

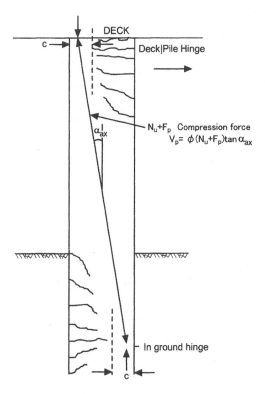

Fig. T5.14. Axial force shear mechanism.

(2) Shear strength of concrete-filled steel shell piles

The flexural and shear strength of concrete-filled steel shell piles can be determined based on normal reinforced concrete theory, assuming full composite action between the shell and the infill concrete. Although some slip between the shell and concrete may occur, it does not appear to significantly influence flexural strength or displacement capacity.

Shear strength can be determined using the equations above, and considering the steel shell as additional transverse hoop reinforcement, with area equal to the shell thickness, and spacing along the pile axis of $s = 1.0$. The contribution of the shell to the shear strength is thus

$$V_{shell} = (\pi/2)t_p f_{yh}(D_p - c_{ex} - c_o) \cot \alpha_t \qquad \text{(T5.26)}$$

where t_p is the thickness of the steel shell.

T5.4.7   Design strength for deck members

The required strength for deck members should be determined by adding the actions resulting from the 'upper bound' pushover response, at the displacement levels corresponding to the Level 2 earthquake, to those resulting from gravity action. Particularly in assessment of existing structures, some redistribution of design actions is appropriate, provided overall equilibrium between internal actions and external forces is maintained.

Dependable strength for flexure and shear actions can be determined in accordance with ACI 318 principles.

T5.4.8   Batter and timber piles

(1) Assessment of wharves and piers with batter piles

Batter piles primarily respond to earthquakes by developing large axial compression or tension forces. Bending moments are generally of secondary importance. The strength in compression may be dictated by material compression failure, by buckling, or more commonly by failure of the deck/pile connection, or by excessive local shear in deck members adjacent to the batter pile. Strength in tension may be dictated by connection strength or by pile pull out. In assessing the seismic performance of wharves and piers with batter piles, the following additional items should be considered

- Pile pullout is ductile and has the potential to dissipate a considerable amount of energy. Displacement capacity is essentially unlimited.
- In compression, displacement capacity should consider the effect of reduction in pile modulus of elasticity at high axial load levels, and the increase in effective length for friction piles, resulting from local slip between pile and soil.

Typical calculations indicate that displacement ductility of two or more are possible.

- Where the prime concern is the prevention of total collapse, it should be recognized that failure of the batter piles does not necessarily constitute failure of the wharf or pier, nor the initiation of collapse. It is possible that, after failure of the batter piles, the wharf or pier may be capable of sustaining higher levels of seismic attack. Although the strength will be diminished as a consequence of the batter pile failure, the displacement capacity of the wharf or pier will generally be greatly increased before the secondary failure stage, involving the vertical piles, develops. Consequently this system, involving only the vertical piles, should be checked independently of the batter pile system.

(2) Assessment of wharves and piers with timber piles

There is little reported evidence of damage to timber piles in earthquakes, despite their wide spread usage for wharf and pier supports for the past 150 years on the West Coast of the USA. This is due to their large displacement capacity, and the typically low mass of the supported wharf, if also constructed of timber, though there are cases of timber piles supporting concrete decks.

Extensive testing of timber piles, both new and used, in both dry and saturated conditions, has been carried out by British Columbia Hydro (B.C. Hydro, 1992). No significant difference between results of new and old, wet and dry, piles was found. The tests were carried out on nominal 12-in. (30 cm) diameter peeled Douglas fir piles, though actual dimensions were as large as 14 in. (35 cm) at the butt end, and as small as 9 in. (22.5 cm) at the tip end. Three types of tests were carried out

1. A simple bending test using a non-central lateral load on a pile simply supported over a length of 27.2 feet (8.28 m).
2. Cantilever bending tests on piles embedded in concrete pile caps with a free length of about 4 feet (1.2 m).
3. In-situ testing of piles embedded in a firm silty-sand foundation with a free length of about 16.4 feet (5 m).

In all cases, the piles were subjected to cyclic loading. Only in the first series of tests were the displacements equal in the opposite loading directions. All piles were subjected to axial loads of about 20 kips (89 kN) throughout the testing. The results indicated that the piles typically exhibited ductile behaviour. Failure generally initiated by compression wrinkling at the critical section, followed by a period of essentially plastic deformation at approximately constant lateral load, and terminated in a tension fracture. The first series of tests gave the lowest results in terms of displacement capacity. This may have been due to the steel loading collar, which applied lateral force to the pile at the location of maximum moment, causing local distress because of the sharp edges of the collar. In fact, failure was not achieved in the other two test series before actuator travel capacity was reached, despite displacement of 1.0 m (40 in.) for the in-situ pile tests.

Back-analyses from the more conservative, (and more extensive) first series (simply supported beams) indicated that the following data might conservatively be used for checking the capacity of existing Douglas fir piles:

Modulus of Elasticity:	10 GPa (1.5×10^6 psi)
Modulus of Rupture:	35 GPa (5000 psi)
Serviceability limit strain:	0.004
Ultimate limit strain:	0.008
Damping at ultimate limit:	2%

The fracture strain has been back calculated from the maximum displacements assuming a linear curvature distribution along the pile. This is because there is insufficient data to define a true ultimate strain, and an effective plastic hinge length. Thus displacement capacity should be based on the same assumptions. Note that this will result in the ultimate displacement capacity of a 12-in. (30 cm) pile with an effective length of 25 feet being about 40 in. (100 cm).

Since the effectiveness of lateral bracing in timber wharves will generally be at best suspect, it is recommended that as a first approximation it conservatively be ignored when assessing seismic performance. This assumes that the bracing connections, or members, will fail or soften to the extent that they will be in-effective after the initial stages of lateral response. Similarly, it may be sufficient to assume that the connections between batter piles and the wharf fail, and to check the 'worst case' condition without the batter piles. Generally, calculations based on these conservative assumptions will still produce displacement demands that are significantly less than displacement capacity.

T5.5   PILE/DECK CONNECTION DETAILS

Connection details between the pile and the deck depend on the type of pile used to support the wharf or pier. Although connection details for buildings and bridges are often detailed to act as pinned connections, piles for wharves and piers are almost always designed for moment resistance at the pile/deck connection. Consequently, only moment-resisting connections will be considered in this technical commentary.

T5.5.1   Steel-shell piles

Steel shell piles will normally be connected to the deck via reinforcing bars and a concrete plug, even when the concrete infill is not continuous down the height of the pile. If the concrete plug is only placed in the vicinity of the connection, care is needed to ensure that shear transfer exists between the concrete and the steel shell. Although this may often be adequately provided by natural roughness of the inside surface of the steel shell, some more positive methods of transfer should be considered. One possibility is the use of weld-metal laid on the inside surface of the steel shell in a continuous spiral in the connection region prior to placing the concrete plug (Park et al., 1983). Dependable flexural strength and extremely large ductility capacity could be achieved for concrete filled steel shell piles when the steel shell was embedded 5 cm inside the deck concrete, and the connection was made by dowel reinforcement properly anchored in the deck. An example of the force-displacement hysteresis response is shown in Fig. T5.15. It will be observed that the flexural strength considerably exceeds the nominal strength, denoted H_{ACI}, when P–Δ effects are included. This is partly a result of the enhanced concrete strength resulting from very effective concrete confinement by the steel shell, and partly due to the steel shell acting as compression reinforcement by bearing against the deck concrete. The remarkable ductility capacity of the pile apparent in Fig. T5.15 has been duplicated in other tests with discontinuous casings at the interface. This pile, subsequent to two cycles at a displacement ductility of $\mu_\Delta = 18$, was subjected to 81 dynamic cycles at $\mu_\Delta = 20$, and then a static half cycle to $\mu_\Delta = 40$ without significant strength degradation.

T5.5.2   Prestressed piles

In the USA, prestressed piles are normally connected to the deck with mild-steel dowel reinforcement, as discussed in the next section. However, other details are possible, and have been successfully tested under simulated seismic conditions.

Pam et al. (1988) tested a number of connection details appropriate for prestressed solid piles. The details were based on 40 cm octagonal piles with

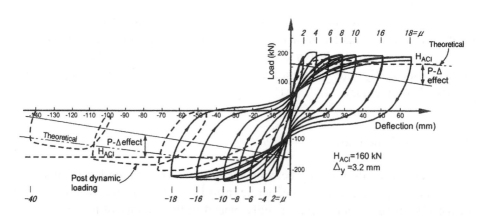

Fig. T5.15. Force-displacement response of concrete-filled steel shell pile connected to deck by dowels.

10–12.5 mm tendons, but the results can be safely extrapolated to the 60 cm piles more commonly used in wharf structures. Two units were tested with the solid end of the pile embedded 80 cm into the deck, with a 10 mm diameter spiral at 15 cm pitch placed around the embedded length of the pile. Two further piles were tested with the strand exposed, enclosed in a 12 mm spiral at 47 mm pitch, and embedded straight in the deck for a distance of 60 cm. In each of the pairs of piles, one pile contained only strand, while the other included 10–20 mm diameter mild steel dowels, thus increasing the flexural strength of the piles. All pile units were subjected to an axial load ratio of $N_u/f_c'A_{gross} = 0.2$, in addition to the axial compression resulting from prestress.

Results for the force-displacement response of these four units are shown in Fig. T5.16. In each case, the piles were capable of developing the theoretical moment capacity of the connection, and to maintain it to high ductility levels. The apparent strength degradation in Fig. T5.16 is a result of P–Δ effects from the moderately high axial load.

The first sign of crushing of the concrete in the plastic hinge region was noted at displacement ductility factors of about 2.0 for the embedded pile and about 2.5 for the embedded strand. The difference was due to the higher total compression force

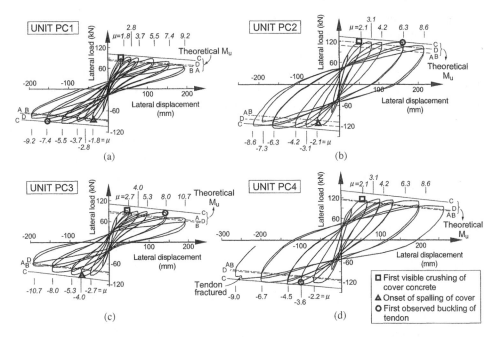

Fig. T5.16. Lateral force-displacement response for four pile-pile cap connection details
($D_p = 400$ mm) (after Pam et al., 1988).
(a) Solid pile embedded 800 mm, no dowels.
(b) Solid pile with standard dowels embedded 800 mm.
(c) Exposed strand embedded 600 mm, no dowels.
(d) Exposed strand and dowels embedded 600 mm.

at the critical section (due to the pile prestress force) when the full pile section was embedded in the pile cap. Since the piles were very slender, with an aspect ratio of 7, these values might be considered as lower bound estimates for a serviceability criterion for the piles. However, this would be compensated by increased yield displacements if the flexibility of the deck had been included in the tests.

In all cases, the strand was fully developed, despite the rather short development length of $48d_b$ in the exposed strand tests, where d_b denotes the bar diameter of the prestressing strand. These results indicate that strand is sufficient to develop moment capacity at the deck/pile interface, and that mild steel dowels are not necessary.

The test units by Pam et al. (1990b) had rather light joint reinforcement surrounding the pile reinforcement in the joint region. However, it must be realized that the deck in these tests was rigidly connected to a reaction frame, which reduced stress levels in the joint region. Practical details should thus not be based on the joint rebar in Pam's tests.

T5.5.3   Practical connection considerations

The most common connection detail consists of dowel reinforcing bars, typically of size #8, #9, or #10, inserted in ducts in the top of the pile after the pile has been driven to the required level, and the top cut to the correct final elevation. Typically the dowel bars in the past have been bent outwards, with the top of the bend below the level of the top mat of the deck reinforcement, as suggested in Fig. T5.17(a). In the past, the top of the bend has often been much lower than shown in this figure.

If hooks are provided on dowels, they must have the tails bent inwards, rather than outwards. The reason for this is that if significant tension force is transferred up to the hook, which is bent outwards, it adds tension stress within the joint region which is already subjected to high tension stress as a result of joint shear forces. There is then a tendency for the diagonal joint shear crack to propagate and bend horizontally outside the hook, particularly if the hooks are below the top layer of deck reinforcement, as illustrated in Fig. T5.17(a). The problem is compounded if the top of the dowel hooks is lower in the deck than shown in Fig. T5.17(a). Proper force transfer between the pile and deck becomes increasingly doubtful, and spalling of the soffit concrete may occur, as has been observed in response to moderate earthquakes in Southern California.

If the hooks are high, and are bent inwards as shown in Fig. T5.17(b), the bend results in anchorage forces directed back towards the compression corner of the deck/pile connection, resulting in a stable force-transfer mechanism.

Although satisfactory force transfer is expected with the detail of Fig. T5.17(b), it can be difficult to construct. This is because tolerances in the final position in plan of a driven pile may be as high as +/−15 cm. In such cases,

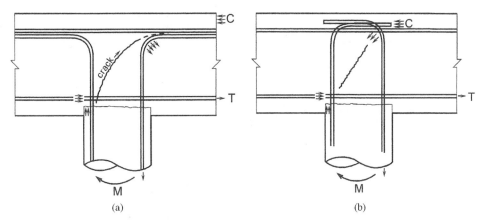

Fig. T5.17. Anchorage with hooked dowels.
      (a) Hooks low, bent out.
      (b) Hooks high, bent in.

interference between the bent dowels and the deck reinforcement can cause excessive placing difficulties, which may result in one or more dowel bars being omitted. Design details need to be developed that are simple and insensitive to pile position. This means that bent dowels should not generally be used.

Two alternative details, developed for the Port of Los Angeles (Priestley, 1998) are shown for typical 60 cm prestressed piles in Fig. T5.18. The first uses straight bar development up as high as possible into the deck, but with bars not terminating more than 10 cm below the top surface. This is combined with spiral reinforcement to control the potential joint shear cracking. The second uses reduced embedment length of the dowels, with partial anchorage provided by enlarged end upstands (end bulbs), with bars lapped with headed vertical bond bars. The lap and joint are again confined by a spiral.

Straight-bar Embedment Detail: The straight bar embedment detail of Fig. T5.18(a) is directly analogous to details used for moment-resisting column/cap-beam connections for seismic response of bridges (Priestley et al., 1996). The required embedment length, l_e, is thus given by

$$\text{SI units} \quad l_e = 0.3 d_b \frac{f_{ye}}{\sqrt{f_c'}} \quad \text{(MPa, mm)} \quad \text{(T5.27a)}$$

$$\text{US units} \quad l_e = 0.3 d_b \frac{f_{ye}}{\sqrt{f_c'}} \quad \text{(psi, in.)} \quad \text{(T5.27b)}$$

where the dowel bar diameter d_b is in mm or inches, and the material strengths are in MPa or psi.

For typical concrete strengths of about $f_c' = 30$ MPa and dowel strength of $f_{ye} = 450$ MPa, Eqn. (T5.27) indicates that it is feasible to anchor #8 to #10

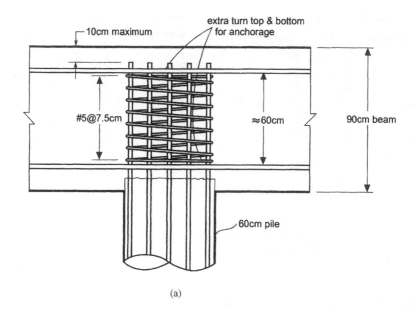

(a)

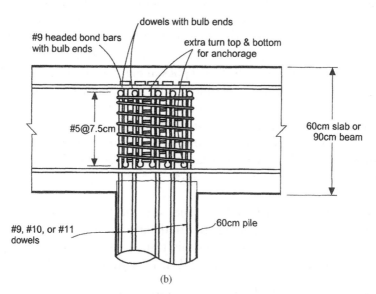

(b)

(Note: Take dowels up to as high as possible: 7.5cm from C.G. top or bottom mat okay)
#9 headed bond bars; 75cm long for 90cm beam; 45cm long for 60cm beam

Fig. T5.18. Anchorage details for dowels.
        (a) Anchorage detail for dowels in 90 cm beam.
        (b) Anchorage detail for dowels with upstands.

dowels by straight embedment in a 90 cm deep deck beam, allowing 10 cm from the top of the dowel to the deck surface. Anchorage of a #11 bar in a 90 cm deck would require a somewhat higher deck compression strength. Note that the use of the specified 28 day concrete compression strength in Eqn. (T5.27) is very conservative, given expected conservatism in concrete batch design, and expected strength gain before occurrence of the design earthquake, and it would be more realistic, particularly for assessment of existing structures to use a more characteristic strength, say $1.2 f_c'$. This would still be considerably less than the probable strength at the age of the concrete when subjected to seismic loading.

Priestley (1998) requires that, if additional external joint reinforcement is not provided, the anchored bars must be enclosed in spiral or hoop confinement with the effective volume ratio of the confining steel, given by

$$\rho_s = \frac{0.46 A_{sc}}{D' l_a} \left[\frac{f_{yc}^o}{f_s} \right]$$

(T5.28)

where $f_{yc}^o = 1.4 f_y$ is the dowel overstrength bar capacity
$\quad\quad f_s = 0.0015 E_d$
$\quad\quad l_a$ = actual embedment length provided
$\quad\quad A_{sc}$ = total area of dowel bars in the connection
$\quad\quad E_d$ = dowel modulus of elasticity
$\quad\quad D'$ = diameter of the connection core, measured to the centreline of the spiral confinement.

The detail shown in Fig. T5.18(a) is adequate for embedment of 8 #11 dowels in a 90 cm deck beam.

Headed Rebar Detail: This detail, shown in Fig. T5.18(b), is designed to allow the dowels to terminate below the top layer of beam reinforcement. Anchorage is improved by using an upstand on the end of the dowel. In this design, 50% of the dowel anchorage force is transferred by the upstand, and 50% by bond, using special headed reinforcement bond bars. This allows the minimum embedment length to be reduced to 50% of that given by Eqn. (T5.27). The top of the dowel should be as high as practicable, but in this case, may be below the top mat of reinforcement. With this detail, all bars sized up to #11 can safely be anchored in a 90 cm deck beam, and all sizes up to #10 can be anchored in a 60 cm deep deck beam.

The total required area of the headed bond bars is $A_t = 0.65 A_{sc}$. Spiral confinement around the connection must be provided in accordance with Eqn. (T5.28).

Details similar to those shown in Fig. T5.18(a) have been tested for bridge designs (Sritharan and Priestley, 1998) and found to perform well, with displacement ductilities typically exceeding $\mu_\Delta = 6$. The detail of Fig. T5.18(b) was recently tested for the Port of Los Angeles, and found to have excellent response (see Fig. T5.19), with failure finally occurring outside the connection at high displacement ductility. Only minor cracking developed in the joint region.

Steel H-Section Piles: Connection of steel H-Piles to pile caps or decks is normally by partial embedment in the deck. Full moment-resisting connections are rarely attempted with this detail. If moment-resistance is required of the connection, sufficient transverse reinforcement must be placed around the pile head to enable the pile moment to be transferred by bearing in opposite directions at the upper and lower regions of the pile embedment.

T5.5.4   Capacity of existing substandard connection details

Many existing piers and wharves will have connection details similar to that shown in Fig. T5.17(a), where the dowels bend outwards from the pile centreline, and the top of the 90° bend, h_d above the soffit, is well below the deck surface. Typically, the dowels will not be enclosed in a spiral within the joint region, and there is thus a probability of joint shear failure.

There are no known test data related to the detail of Fig. T5.17, but poor performance of such details in the recent moderate Northridge earthquake, and similarity to bridge T-joints, which have been extensively tested, leads to a need for conservative assessment.

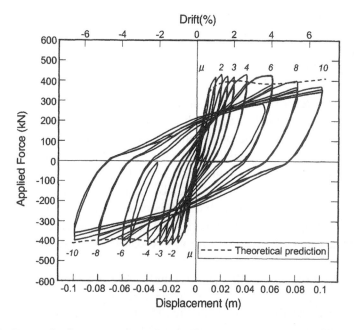

Fig. T5.19. Force-displacement response of pile connected to deck with headed rebar (detail of Fig. T5.18).

A relevant procedure is available in Priestley et al. (1996), and is adapted in the following for wharves and piers with reinforced concrete decks.

1. Determine the nominal shear stress in the joint region corresponding to the pile plastic moment capacity.

$$v_j = \frac{M^\circ}{\sqrt{2}h_d D_p^2} \qquad (T5.29)$$

where M° is the overstrength moment of the plastic hinge, and h_d is the depth of the pile cap. The overstrength moment is the maximum possible moment that could occur in the pile. It is an upper bound assessment of capacity to ensure adequate shear and is usually based on multiplying the ultimate moment by some factor (usually 1.4 times the nominal moment capacity).

2. Determine the nominal principal tension stress in the joint region

$$p_t = \frac{-f_a}{2} + \sqrt{\left(\frac{f_a}{2}\right)^2 + v_j} \qquad (T5.30)$$

where

$$f_a = \frac{N}{(D_p + h_d)^2} \qquad (T5.31)$$

is the average compression stress at the joint centre, caused by the pile axial compression force N. Note, if the pile is subjected to axial tension under the seismic load case considered, then the value of N, and thus f_a, will be negative.

3. If p_t calculated from Eqn. (T5.30) exceeds $0.42\sqrt{f_c'}$ MPa, joint failure will occur at a lower moment than the column plastic moment capacity M°. In this case, the maximum moment that can be developed at the pile/deck interface will be limited by the joint principal tension stress capacity, which will continue to degrade as the joint rotation increases, in accordance with Fig. T5.20. The moment capacity of the connection at a given joint rotation can be found from the following steps:

   a) From Fig. T5.20, determine the principal tension ratio $p_t/\sqrt{f_c'}$ corresponding to the given joint rotational shear strain, referring to the T-joint curve.

   b) Determine the corresponding joint shear stress force from

   $$v_j/\sqrt{p_t(p_t - f_a)} \qquad (T5.32)$$

   where f_a is determined from Eqn. (T5.31).

   c) The moment at the pile/deck interface can be approximated by

   $$M_c = \sqrt{2}v_j h_d D_p^2 \le M^\circ \qquad (T5.33)$$

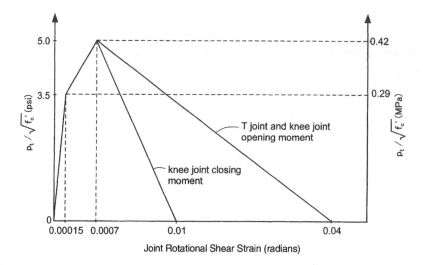

Fig. T5.20. Degradation of effective principal tension strength with joint shear strain (after Priestley et al., 1996).

This will result in a reduced strength and effective stiffness for the pile in a pushover analysis. The maximum displacement capacity of the pile should be based on a drift angle of 0.04.

T5.6   EXISTING CONSTRUCTION

The criteria presented above were developed from a compilation of current practice by many agencies combined with state-of-the-art technology for estimation of seismic damage potential. It is not a revolutionary step forward but rather an evolution of design. The developed integrated criteria specify:
1. The required structure performance under expected loads.
2. Specification of the expected loads.
3. Specification of strain limits to ensure structural response limits to achieve performance requirements.

The overall effect on the design, selection of pile sizes and cost of the pier is not expected to be great; however, the assurance in meeting performance goals is thought to be substantially enhanced.

In the application of these criteria to existing construction, it is thought that the objective of a uniform set of performance goals should be maintained across the waterfront. Where adequate capacity is lacking in the existing system, it is thought better to strive for the performance goal, develop several candidate upgrade alternatives, and then perform an economic/risk analysis to determine what is the most cost effective solution, considering the potential for a damaging earthquake and the existing lateral force system. This approach varies from alternative procedures outlined in current codes, wherein retrofit and restrengthening measures are initially designed using criteria for new construction, then the computed response (e.g., loads, moments, drift) is reduced by the application of semi-empirical reduction factors for existing structures. The adherence to performance standards for design is recommended over the application of simplified code-based reduction factors, as any single reduction coefficient is not optimal over a range of structures.

T5.6.1   Structural criteria for existing construction

The discussion of existing structures is of major importance, since there are many existing facilities in use and relatively few new facilities being constructed. Many of the existing structures were built during periods when seismic standards were not well defined. In general, existing-structure performance criteria may be related to new-structure performance criteria in terms of the degree of hazard, the amount of strength required, the extent of ductility demand allowed, or the level of design ground motion. The structure once built does not 'know' that it is expected to perform to a 'new' or an 'existing' structure criteria; it responds according the principles of structural dynamics.

The approach taken herein is to recognize that the goals for both new and existing facilities should be the same. The structural parameters that are used to control the response should be the same. What is of significance is that existing structures have been in place for a period of time and have a shorter remaining life than new facilities. Thus, existing facilities face a shorter exposure to natural hazards. This major factor suggests that the design ground motions be allowed to differ based on the differing remaining life spans of the structures. A prudent course must be charted to select reasonable goals for existing structures to the adverse impact on the economics of port operations while ensuring public mandates for preservation of the environment.

The approach taken in this criterion is to utilize a factor, α_y, to relate the existing-structure exposure time to the new-construction exposure time taken as 50 years. The value of α_y is equal to or less than 1.0. The value of α_y is used to define the exposure time for use in the Level 1 and Level 2 earthquake return times, as shown in the following sections. Determination of α_y establishes the seismic loading. The performance goals and response limits for existing construction remain the same as for new construction; only the loading is reduced. This applies to all elements including the structure, the foundation and all associated lifelines.

Seismic reviews of existing waterfront construction directed by requirements of regulatory agencies can utilize the above criteria for new construction as the target goal requirement for performance. In general, the existing structure must satisfy the new structure performance limit states under earthquake peak ground acceleration levels corresponding to reduced exposure times as follows:

Existing-Structure Exposure Time = α_y(New-Construction Exposure Time)

where New-Construction Exposure Time is 50 years.

The requirement for evaluation of the seismic resistance and possible upgrade is triggered when the load on the structure changes, such as when the operation of the structure is changed, or when the structure requires major repairs, or modifications to meet operational needs, or when recertification is required.

Determining α_y When it is shown to be impossible or uneconomical to achieve new construction levels of performance, ($\alpha_y = 1$), an economic/risk analysis using procedures shown by Ferritto et al. (1999) can be performed to determine the most appropriate facility exposure time and level of seismic design upgrade. Various alternative upgrade levels are considered ranging from the existing condition to the maximum achievable. Each alternative should be examined to determine the cost of the upgrade, the cost of the expected earthquake damage over the life of the structure, the impact of the damage on life and the environment, and the structure's operational requirements. The choices of upgrade levels are made by the design team based on a strategy consistent with requirements of life safety, operational needs, cost effectiveness, and protection of the environment. In no case should α_y be less than 0.6 of the 50-year exposure time for new construction (30 years).

As a minimum, analysis should be conducted and data developed for the cases of α_y equal to 1.0 and 0.6. Additional cases are encouraged:

Level 1 – An earthquake with a 50% probability of exceedance in $\alpha_y \times$ (50 years) exposure.

Level 2 – An earthquake with a 10% probability of exceedance in $\alpha_y \times$ (50 years) exposure.

Table T5.2 shows the shortest allowable return times.

Peer review: Review of the results of the analysis by an independent peer review panel of experts in structural engineering, geoscience, earthquake engineering, seismic risk analysis, economics, and environmental engineering/science may be required. In addition, the findings from the analysis can be reviewed by the public and other stakeholders.

Allowance for Deterioration
In evaluating existing construction, it is most important to:
- Evaluate the actual physical conditions of all structural members to determine the actual sizes and condition of existing members.
- Provide an allowance for corrosion and deterioration.

Table T5.2. Shortest allowable return periods for existing structures.

Probability of non-exceedance (%)	Exposure time (years)	Return time (years)	Nominal return time (years)
50	30	43	60
10	30	285	300

- Evaluate the properties of the construction materials, considering age effects in computing yield strengths. Average actual material properties should be used in the evaluation.
- Evaluate the existing structure details and connections, since this is often the weakest link and source of failure.
- Determine the design methodology used by the original designers at the time the structure was designed and constructed.
- Evaluate displacement demands and capacities. Previous code requirements did not emphasize the need for ductility and the failure to include shear and containment reinforcing is most common in existing construction. This has led to numerous structure failures especially when batter piles have been used.

T5.6.2   Strengthening of an existing structure

Various methods of strengthening existing structures are possible. They include:
Plating. Where the top of flange is not accessible for adding cover plates, reinforcement can be added to the web plate. The beam should be relieved of load before the reinforcement is added. When cover plates are added, the flange to web connection and the web plate stresses at the toe of the flange should be investigated.
Composite Action. Beam section properties can be materially increased by causing the concrete slab to act as a composite with the beam. The slab serves as a top cover plate.
Prestressing. Jacks can be used effectively to reduce stresses in existing flanges. Cover plates should be welded before removing jacks.
Shear Reinforcement. Vertical stirrups serve as hangers that support the beam from the uncracked portion of concrete near the column.
Flexural Reinforcement. Longitudinal reinforcement can be added effectively, if positive means for preventing separation and for transferring horizontal shear are used. Composite materials can be glued to the underside of pier decks to increase section capacity. Composite rods can be inserted and epoxied in groves cut in the topside of pier decks to add reinforcement.
Pile/Column Reinforcement. Pile/Column sections may be strengthened by adding concrete with longitudinal and lateral reinforcement, or by adding unreinforced concrete restrained by hoop bars. Wrapping by composite materials has

been used very effectively for bridge columns. Mandrels have been applied to piles and filled with concrete.

Other design considerations for existing structures include:

Compatibility. The design details should encompass any inherent incompatibility of old and new materials. Provision should be made to resist separation forces. New concrete should have a different modulus of elasticity, coefficient of thermal expansion, and shrinkage than old concrete. Consider differing expansion effects due to differing absorption of moisture. Provide resistance against 'curling' due to thermal gradients. Compatibility of connectors must be considered. For example, rivets or bolts are not compatible with welds. Friction bolts are not compatible with rivets. Creep is an important factor.

Dead Load Versus Live Load Stresses. Unless the load on a structure is relieved (for example, by removal or by jacking), the existing framing will continue to carry:

a) the full dead load of the construction,

b) any part of the live load which is in place when the new framing is connected, and

c) a proportionate share of the live load subsequently added.

The new framing will carry only a part of the live load. As a result, under the final loading condition, the stresses in the new and existing material of the same or similar members will be different, often radically so. For example, assuming a 1:1 ratio of dead to live load and of new to existing material in the cross section of a given member and disregarding plastic deformation, the stress in the existing material would be three times the stress in the new. As a result, the new material cannot be stressed up to allowable values without simultaneously overstressing the existing sections. It is necessary either to provide an excess of new material, or to relieve the load on the structure before strengthening. This may not apply, if plastic deformation of the structure (and its associated, increased deflection) can be permitted.

T5.6.3   Deterioration of waterfront structures

(1) Causes of deterioration

The more common causes of deterioration associated with steel, concrete, and timber waterfront structures are as follows.

Steel Structures. For steel structures deterioration is caused by:

a) corrosion,

b) abrasion,

c) impact.

Concrete Structures. In concrete structures deterioration is caused by:

a) corrosion of reinforcement,

b) chemical reactions,

 c) weathering,

 d) swelling of concrete, and

 e) impact.

Timber Structures. In timber structures deterioration is caused by:

 a) corrosion and abrasion of hardware,

 b) borer attack,

 c) decay, and

 d) impact.

(2) Preventive measures in design and construction

Steel Structures All parts that will be subject to corrosion, should be accessible for inspection and repair. If not, then they should be encased with concrete or some other long-life, high-resistance type of coating. Shapes should be selected that have a minimum of exposed surface. Detailing should be designed so that accumulations of dirt and debris will be avoided. Avoid narrow crevices that cannot be painted or sealed. Draw faying surfaces into tight contact by use of closely spaced stitch rivets, bolts, or welds. Prime faying surfaces before assembly. In general, the ability of framing to shed water is the single most important factor in inhibiting corrosion and deterioration of coatings. If the potential for ponding is unavoidable, provide drain holes. Drain holes should be a minimum of 47 in. (10 cm) in diameter to prevent clogging. The use of sacrificial metal should be avoided in favour of using protective coatings.

The thickness of metal and section properties should be determined by considering the loss of the section as established in MIL-HDBK-1002/3, Steel Structures (U.S. Navy, 1987), unless corrosion protection is provided. Typical average corrosion rates for bare carbon steel are shown in Table T5.3.

The minimum thickness should not be less than 0.40 in. (1 cm). When the required minimum thickness is excessive, corrosion protection using approved products or cathodic protection should be used. When coatings are used, care must be exercised in driving the piles to preclude damage to the coatings. Additionally, consideration must be given to abrasion of the piles by contact with a fender. Tips of all steel H piles, having a thickness of metal less than 0.5 in. (1.3 cm) and driven to end bearing on sound rock by an impact hammer, should be reinforced.

Concrete Structures Good quality is the important factor in obtaining a dense concrete. This, in turn, is the most important factor in preventing penetration of moisture, which is the primary cause of deterioration of concrete. Do not use poorly graded aggregate, or a water-cement ratio greater than 6 gal (22.71 l)/sack of cement (94 pounds = 418 kN), reduce to 5 gal (18.92 l)/sack of cement for thinner sections such as slabs and wherever clear cover over reinforcement is 2 in. (5 cm) or less. Watertight concrete can be obtained by using air entrainment (maximum 6% by volume) and a water-cement ratio not greater than 5 gal/sack of cement.

Table T5.3. Typical corrosion rates for bare carbon steel (1.0 ft = 0.3048 m).

Zone (elevation from pile tip)	Average corrosion rate (mm/year)
Imbedded Zone (0 to 15 ft)	2
Erosion Zone (15 to 21 ft)	10
Immersed Zone (21 to 29 ft)	8
Atmospheric Zone (29 to 35 ft)	12

Types of cement (ACI Committee 116, 2000) affect the deterioration of concrete. Type III (high early strength) cement is excessively susceptible to sulphate attack, and should not be used. In general, avoid the use of Type I cement in a saltwater environment. Type IX (sulphate-resistant) cement should be used. The use of Type V (high resistant) cement is seldom required.

Provision should be made for an adequate number of expansion joints. Use types of expansion joints such as double bents with movement taken up by bending of the piles or electrometric pads with some form of joint sealer.

In tropical climates and in areas subject to salt spray, consider the use of galvanized or plastic coated reinforcing bars. If plastic coated bars are used, attention should be given to bond stresses. Excessively rich mixes, over 6 bags per yd^3 (0.764 m^3), should be avoided, as excess cement tends to enhance the potential chemical reaction with seawater. For most aggregates, alkali-aggregate reaction can be prevented by specifying maximum alkali content of the cement (percent Na_2O, plus 0.658 times percent K_2O) not to exceed 0.6%. In a surf zone, the concrete cover and streamline sections should be increased to prevent abrasion. Calcium chloride (as an accelerator) should not be used in prestressed concrete and concrete exposed to seawater. The use of calcium nitrite or other chemicals as a deterrent to corrosion of embedded reinforcing steel, is not an adequate substitute for good quality concrete and adequate cover. This is not to say that these additives do not have merit; however, use of coated reinforcing bars may be required. Timber jackets for concrete piles and stone facing for concrete seawalls work extremely well to prevent deterioration due to corrosion of reinforcement, weathering, and chemical attack. They tend to isolate the concrete from chemical constituents in the environment, insulate against freezing, and keep the reinforcing bars free of oxygen.

Reinforced concrete has been used as one of the major construction materials at the waterfront. Since concrete is much weaker in tension, cracking would be expected to occur when the tensile stresses in the concrete were exceeded, typically at a numerical level equal to about 10% of the maximum compressive stress. Cracking is a normal occurrence in concrete members under flexural load. When the concrete cracks, the section moment of inertia is reduced; generally the cracked moment of inertia is about 30 to 60% of the gross moment of inertia, depending on the axial load level and reinforcement content.

In a marine environment, it is desirous to control the cracking to prevent corrosion of the reinforcing steel. Confining steel is used to increase concrete strength, ductility and shear strength. An initial prestress force is used in piles as a mechanism for improving concrete performance by keeping the cracks closed. It has been noted that crack widths of 0.18 to 0.22 mm are sufficiently small to preclude deterioration of the reinforcement so that an allowable crack width may be approximated at about 0.25 mm. It is not possible to directly equate the crack width to an allowable tensile strain since crack spacing is not known; however, corrosion has not been a problem when reinforcing stress has been restricted to a tension of 117 kPa or less under service loads. At concrete compressive strains below 0.0021, the compression concrete does not evidence damage and crack widths under cyclic load should be acceptable. Occasional larger loads may be sustained without deterioration, as long as a permanent offset does not occur and the prestress forces can close the cracks. Reinforcement deterioration is most pronounced in the presence of oxygen, such as in a pier pile, where the pile is freestanding out of water or in the splash zone. At deep-water depths or in soil, the oxygen content is reduced such that pile reinforcement deterioration is less. Large loads causing loss of the concrete cover result in loss of pile capacity and facilitate deterioration; such conditions can be repaired if accessible by jackets around the pile. Loss of concrete cover may begin at displacement ductility of about 1.5 to 2.0.

Timber Structures. Timber structures should conform to the following criteria:
a) Design detail should minimize cutting, especially that which must be done after treatment.
b) Design detail should provide for ventilation around timbers. Avoid multiple layers of timbers, as decay is enhanced by moist conditions at facing surfaces. Curb logs should be set up on blocks. Walers (i.e. edge pieces of timber such as timbers used as a fender parallel to the pier) should be blocked out from face of pier. Thin spacers between chocks and wales (i.e. essentially the same as walers), and gaps between deck and tread planks should be provided.

TC6: Remediation of Liquefiable Soils

TECHNICAL COMMENTARY 6

Remediation of Liquefiable Soils

Remediation of liquefiable soils can be an effective technique to improve the performance of port structures subject to earthquake loading. This commentary outlines current techniques of seismic remediation, based on Japanese practice, and proposes design approaches for areas of soil improvement and techniques to assess the influence of remediated zones on existing structures. The approach discussed here draws upon the methods presented in the Handbook on Liquefaction Remediation of Reclaimed Land by the Port and Harbour Research Institute, Japan (1997).

T6.1   OVERVIEW

T6.1.1   Soil improvement and structural solutions

Remediation solutions against liquefaction can be divided into two broad categories:
  1) soil improvement to reduce the probability of liquefaction;
  2) structural design to minimize damage in the event of liquefaction.
Both approaches can be further subdivided. Soil improvement techniques are aimed at:
  a) improving the performance of the soil skeleton under earthquake shaking;
  b) increasing the rate of dissipation of excess pore water pressure; or
  c) a combination of a) and b).
Similarly, structural solutions can be divided into i) those that reinforce the structure in terms of strength and stiffness, and ii) those that effectively reduce the consequences of liquefaction in neighbouring or underlying ground through the use of flexible joints or other structural modifications. Again, in practice a combination of these two measures is often adopted.
  The basic strategies for liquefaction remediation are summarized in Fig. T6.1 and these follow the outline laid down above. The following sections describe the different approaches currently in use. Before discussing these in detail, however,

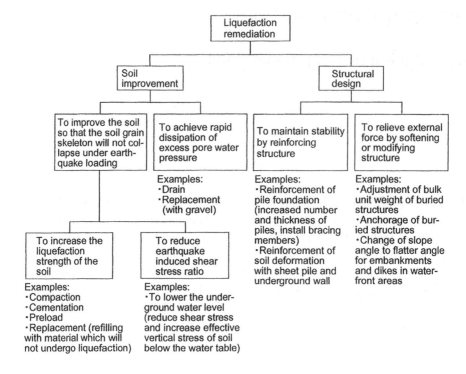

Fig. T6.1. Basic strategy for liquefaction remediation.

it is important to summarize the standard design process or procedure for liquefaction remediation.

T6.1.2 Standard design procedure for liquefaction remediation

A flowchart illustrating a standard design procedure for liquefaction remediation is shown in Fig. T6.2. Once the strategy has been determined, then it is common practice to select a method first, and then to compare the advantages and disadvantages of different solutions for the particular project. Typically, a number of different solutions will need to be assessed before a final decision can be reached. Sometimes, solutions will be combined to achieve the optimum design. The procedure illustrated in Fig. T6.2 is intended to be used as a guide only; it may often be more efficient to screen out remediation methods that are clearly unsuitable at an early stage based on a rough assessment of the likely area to be treated and any project specific constraints.

A combination of two or more remediation methods is often very effective. A typical example might be the combination of a low noise/low vibration method, such as the use of drains, combined with compaction around the improved area, to constrain or confine the overall site. At present, however, formal procedures for

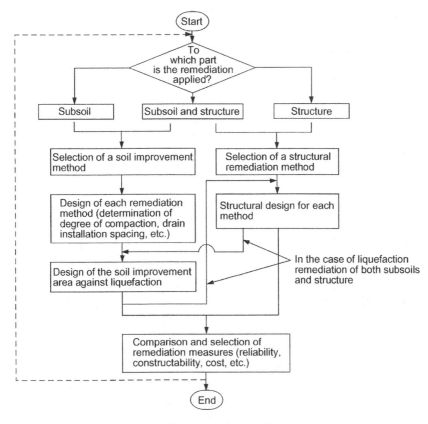

Fig. T6.2. Standard procedure for liquefaction remediation.

selecting the most efficient combination of methods for a specific site have not been developed and the engineer must use judgement and experience to reach the optimum solution.

T6.2   OUTLINE OF REMEDIAL MEASURES AGAINST LIQUEFACTION
(Japanese Society of Soil Mechanics and Foundation Engineering, 1988; 1991)

Remediation methods can be grouped into different categories, depending on the objective of the treatment. In this section, eight different categories are proposed, and these are discussed below. Table T6.1 provides a summary of a wide range of current techniques based on these categories.

(1) Compaction
Methods based on compaction are used to reduce the susceptibility of the soil to liquefaction by densifying sandy soil using vibration and impact. The

Table T6.1.   Remedial measures against liquefaction.

Principle	Methods Designation	Depth	Summary	Influence on the surrounding areas	Remarks
Compaction	Sand compaction pile method	About GL–35 m	Insert steel pipe casing underground Install sand compaction pile by forcing out sand during extraction of casing, and compact the sand pile and the natural soil in a horizontal direction simultaneously	This method produces high levels of noise and vibration The extent of noise and vibration differs according to the type of construction equipment used	Compaction efficiency is high for soil with fines content of less than 25 to 30%. SPT N-value increases to about 25 to 30
	Vibro-rod method	About GL–20 m	Compaction of soil is conducted using vibratory penetration of a rod and filling with additional sand from the ground surface	This method produces noise and vibration that are slightly less than those produced by the sand compaction pile method	Compaction efficiency is high with respect to soil having fines content of less than 15 to 20%. SPT N-value increases to about 15 to 20
	Vibroflotation method	About GL–20 m	The surrounding soil is compacted with water sprayed from nozzle tip end and with horizontal vibration from vibrator with a built-in eccentric load Additional sands are filled into the voids in the subsoil profile	This method produces little noise and vibration compared to other methods based on compaction	Compaction efficiency is high with respect to soil having fines content of less than 15 to 20%. SPT N-value increases to about 15 to 20

	Dynamic compaction method	About GL–10 m	The soil is compacted with an impact load by dropping a weight of 10 to 30 tf from a height of about 25 to 40 m above the ground	This method produces high levels of vibration and large impact	Compaction is difficult when the fines content is high
Pore water pressure dissipation	Gravel drain method	About GL–20 m	The gravel pile is installed by inserting gravel into a casing placed at prescribed position and then extracting the casing. Excess pore water pressure is dissipated through the gravel pile during earthquakes	Influence on the surrounding areas is minimal	This method is often used when compaction is difficult. If fines content of the subsoil is high and permeability is low, application of this method is difficult
	Attachment of drainage device for steel pipes or sheet piles	—	Piles with a drainage device are inserted into the subsoil as liquefaction remedial measures	Noise and vibration depend on the installation method used	The pile prevents uplift and settlement of upper-structures.
Cementation and solidification	Deep mix method	About GL–30 m	Stabilizing material such as cement is mixed and solidified in the soil. There are two approaches: one is total improvement which solidifies all the soil, the other is partial improvement which makes a solidified wall in the subsoil profile	Influence on the surrounding areas is minimal	Liquefaction remediation is possible even when there is a possibility of liquefaction in the soil under the existing structure, such as for an embankment, by installing an improvement wall at the periphery

Table T6.1. Continued.

Principle	Methods Designation	Depth	Summary	Influence on the surrounding areas	Remarks
	Premix method	—	Landfill is deposited by adding and mixing a stabilizing material such as cement in advance in the soil. Liquefaction remediation following the landfill becomes unnecessary	Water quality control is necessary since sand mixed with a stabilizing material is dropped into water	This method can reduce the construction period because liquefaction remediation can be accomplished simultaneously with landfilling, and treatment can be performed in a large volume
Replacement	Replacement method	About GL-5 m	The liquefiable soil is replaced with a non-liquefiable material	Handling of the excavated soil is necessary	The construction is reliable since the liquefiable soil is replaced with non-liquefiable material
Lowering of ground water level	Deep well method	Water level lowering of about 15–20 m	The perimeter of the liquefiable soil area is surrounded with a cut off retaining wall constructed with sheet piles, and the ground water level is lowered using deep wells	Caution should be exercised with regard to consolidation and settlement of the soil and influence on the surrounding area from the lowering of the ground water level	This method can be applied to the subsoils below existing structures. The long-term operating cost should be considered

Shear strain restraint	Underground wall method	—	Liquefaction remediation can be achieved by surrounding the liquefiable soil using underground walls made of grouting or sheet piles, and restraining the shear deformation of subsoil during earthquakes	Influence on the surrounding areas is minimal	This method can also prevent the propagation of excess pore water pressure from the surrounding liquefied soil
Preload	Soil overlay	Depending on embankment width	Preload is applied by overlaying soils in order to overconsolidate the subsoil	Influence on the surrounding areas is minimal	The stress propagation in the soil should be considered since there is load dispersion
	Deep well method	Water level lowering of about 15–20 m	The ground water level of liquefiable soil is lowered temporarily and thereafter restored	Caution should be exercised with regard to consolidation and settlement of the soil and influence on the surrounding area from the lowering of the ground water level	This method can be applied to the subsoils below existing structures
Structural strengthening	Pile foundation	—	The strength of the pile increases resistance even if liquefaction occurs	Influence on the surrounding areas is minimal	—
	Sheet pile	—	Shear deformation after liquefaction is restrained by surrounding the soil with sheet piles	There is noise and vibration during sheet pile installation	This method can also prevent propagation of excess pore water pressure from the surrounding liquefied soil
	Top-shaped foundation	—	Used for spread foundation by laying top-shaped concrete blocks on the soil	Influence on the surrounding areas is minimal	This method is applied to small buildings such as residential housing

effectiveness of compaction depends on the soil gradation; when the fines content of the soil is large, compaction is difficult. Compaction is used extensively, and there are many case histories. Examples of compaction methods include sand compaction piles (Mizuno et al., 1987), vibrating rod (Sato et al., 1988), vibroflotation (Shimizu and Takano, 1979) and dynamic compaction (Tanaka and Sasaki, 1989). These methods usually cause vibration and noise during the compaction process and result in increased lateral earth pressures; hence the influence of such methods on the surrounding ground and adjacent structures needs to be carefully considered. Recent developments of the vibroflotation method are aimed at reducing these environmental impacts.

(2) Pore water pressure dissipation

A second approach is to control the development of excess pore water pressure in sandy soils by increasing the rate of dissipation through the installation of permeable stone or gravel columns or drains. The gravel drain and plastic drain methods (Iai and Koizumi, 1986) are typical examples. The installation of drains usually causes less vibration and noise than the compaction method and hence has the advantage that the influence on the surrounding area is reduced. This approach requires complex calculations and the measurement of many parameters to establish the design pile installation spacing. In some cases, for example in soils with low permeability and a high fines content, the drainage method may be impractical.

The installation of drains does not generally densify the natural ground and the method relies on drainage of excess pore water pressures to reduce the likelihood of liquefaction. In the event that the excess pore water pressures generated are greater than anticipated and approach 100% of the initial vertical effective stress, then Yoshimi and Tokimatsu (1991) have warned that large deformations may suddenly be induced, and suggested that a site improved in this way may lack 'ductility'. Drainage solutions are therefore often used when the application of other methods (such as compaction) is considered too difficult.

Variations of the drainage method have also been developed. Noda et al. (1988), for example, proposed a water drainage device attached to a pile that combines the benefit of increased pore water pressure dissipation in the ground with the pile itself.

(3) Cementation and solidification

Liquefaction remedial measures based on solidification include the deep mix method (Suzuki et al., 1989) and premix method (Zen, 1990). The deep mix method improves the ground by mixing a stabilizing material into sandy soil to cement the soil skeleton. The method is versatile and can be used in different ways, such as directly below existing structures, or to create a continuous underground 'wall' around a structure or site, see (6) and (8) below.

The premix method works in a similar manner to deep mix methods by mixing stabilizing materials with natural soil. Premix is particularly relevant to the improvement of soil where substantial landfill is taking place, such as filling to create reclaimed land. The stabilizing material is added to the landfill soil or sand in advance and then the treated soil is placed underwater using a tremie chute or bottom dump barge. The use of pre-mixing can reduce construction periods by removing the need to treat the ground after fill placement.

Aspects that require particular attention under this approach are quality control of the treated subsoil profile, mix control of the stabilizing material, and control of water quality in the surrounding area.

(4) Replacement (Japanese Society of Soil Mechanics and Foundation Engineering, 1982)

The replacement method replaces the original liquefiable soil with non-liquefiable material. This may comprise gravel, or soil mixed with cement. Consideration must be given to the sourcing of suitable replacement material and to the disposal of the excavated soil (spoil). The stability of the excavation itself requires careful assessment.

(5) Lowering of ground water level (Omori, 1988)

Lowering the ground water level within a soil profile provides two benefits. Firstly, the effective confining stress is increased at depths below the original ground water table, and secondly, the upper soil layers can be effectively de-watered.

The ground water level can be reduced by active or passive means. In most circumstances, an active system (such as pumping from wells) is necessary to maintain a lowered ground water level at or below the design requirement. Active systems require to be operational at all times to provide security. In some circumstances, passive drainage can be used. Both approaches require maintenance and will therefore impose operating costs; these need to be assessed.

(6) Shear strain restraint (Yamanouchi et al., 1990)

A continuous underground wall can be installed around a liquefiable site with the intent of reducing soil shear strain levels to acceptable levels under earthquake loading. The method has advantages where access to the liquefiable soils or foundation is difficult, such as under an existing building or structure. The principal disadvantage of this approach is the complexity of the analysis and design process and justification of the proposed solution.

(7) Preload (Matsuoka, 1985)

Preload methods overconsolidate the soil, thus increasing the liquefaction resistance. This is typically achieved using a surcharge on the ground surface, but can also be achieved by temporarily lowering the ground water level (e.g. beneath existing structures). The method is applicable to soil with a large fines content. Detailed evaluation of the underlying soil strata is necessary, and it may not be practical to achieve large overconsolidation ratios where the existing vertical load is large. Settlement of existing structures needs to be evaluated. Case histories of this approach are limited and design procedures are not yet fully developed.

(8) Structural remedial measures

An alternative to remediating or constraining the liquefiable soil deposits is to strengthen the foundation system. Piles or sheet piles can be used to found the structure on denser deposits at depth, reducing the consequences for the structure of liquefaction in the soil beneath (Nasu, 1984). Installation of piles can also densify looser soils through increased lateral confinement, giving a further benefit. The design of the strengthened foundation system needs to take into account the changing dynamic response of the structure as liquefaction takes place and the consequences of lateral or downslope ground movements and settlements. The effects of liquefaction on services need to be assessed.

The methods outlined above at (1)–(8) and summarized in Table 6.1 provide a general overview of remediation solutions. In the following sections, four of these approaches are discussed in detail: compaction, drainage, premix and preload.

T6.3   COMPACTION METHOD: DESIGN AND INSTALLATION

Compaction of ground or fill as a remedial measure against liquefaction can be achieved by a range of techniques, such as vibrating a steel rod pushed into the ground or by forcing sands or gravel into the liquefiable subsoils using a steel casing.

T6.3.1   Determination of the SPT N-value to be achieved by compaction

The SPT N-value of the subsoils to be achieved following compaction can be determined using the flow chart shown in Fig. T6.3. This illustrates how the effect of compaction affects the earthquake response characteristics of the subsoil profile.

The minimum area of soil improvement necessary for a specific structure, may often be limited to the area under the structure itself (Section T6.7). In such cases, because of the proximity of the unimproved soil to the foundation, there may be some interaction between the liquefied and non-liquefied areas, leading to further degradation of the foundation. In these cases, it is particularly important that the degree of compaction of the improved soil is adequate and that liquefaction of the

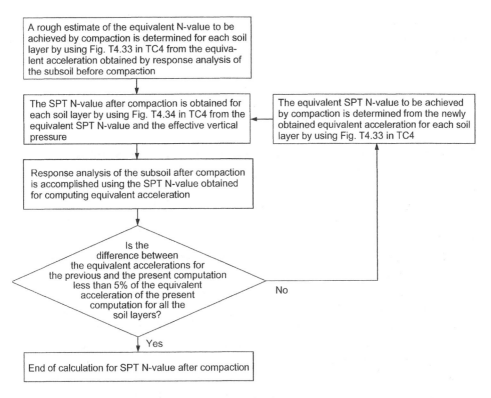

Fig. T6.3. Determination procedure of SPT N-value to be achieved by compaction.

foundation is fully prevented. The equivalent N-value to be achieved in order to satisfy this condition is 16 or higher, which corresponds to a relative density of 80% or higher at an effective vertical pressure at a depth less than about 20 m. Such a target value for the SPT blowcount is near the upper limit of what may be achieved by the compaction method.

In other situations, an impermeable underground wall may be constructed around the perimeter of a structure, greatly reducing the possibility of liquefaction in the improved soil in the interior. In this case, where liquefaction of the outlying areas will have only minimal impact on the improved soil underneath, then the equivalent N-value to be achieved by compaction may be determined following Section T4.4.2 (Technical Commentary 4). In particular, the target blowcount can be determined from the boundary of areas II and III in Fig. T4.34. For important structures requiring a considerable margin of safety, the equivalent N-value to be achieved by compaction is determined from the boundary of zones III and IV in Fig. T4.34.

There may be cases where the SPT N-value assessed following the guidelines above is unlikely to be achievable using the compaction method. One approach is to make a more detailed assessment of the liquefaction resistance of the soil by cyclic triaxial testing, Section T4.4.2 (Technical Commentary 4), with a view to developing a revised SPT N-value. Alternatively, the use of other remediation methods such as preload, or cementation and solidification methods may be considered.

T6.3.2   Designing compaction

The degree of compaction attainable by compaction methods is affected by many factors, including those listed below (Mitchell, 1981).
i)   Characteristics of the soil (grain size distribution and fines content, i.e. the percentage by weight of grains smaller than 0.075 mm)
ii)   Degree of saturation and level of the ground water table
iii)   Relative density of the soil before compaction
iv)   Initial stresses in the soil layer before compaction (effective vertical stress)
v)   Skeleton structure and cementation of soil before compaction
vi)   Distance from a source of vibration during compaction
vii)   Properties of sand fill
viii)   Characteristics of compaction methods (types of installation equipment, vibration capacity of the equipment, installation method, expertise of the operating engineers)
    Because of the range of parameters affecting the degree of compaction that can be achieved, the design of a compaction method should be based on case histories and recent field testing wherever possible. Field tests should be performed at a test site where the characteristics of the soil are demonstrably similar, using comparable compaction plant and techniques. Field tests of the proposed compaction method considerably enhance the reliability of the design.

T6.4   DRAINAGE METHOD: DESIGN AND INSTALLATION

The objective of the drainage method is to enhance the overall drainage capacity of the subsoil profile by installing drains made of very permeable materials below the ground surface. An individual drain is often similar to a pile in shape, formed using gravel and crushed stone and installed in a distributed layout similar to a pile configuration (Seed and Booker, 1977). Other types of drains, however, have been proposed such as drains shaped like a wall, drains surrounding a structure (Sakai and Taniguchi, 1982) and drainage devices that are attached to the back of steel piles or sheet piles (Noda et al., 1990). Rubble or gravel, used as backfill behind quay walls may also be considered as providing drainage if a permeable sheet is used to prevent the migration of sand. In recent years, pipes made of synthetic resin (i.e. plastic) with many small holes or perforations have also been used.

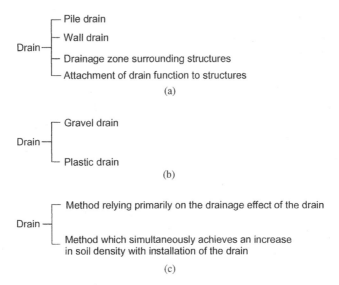

Fig. T6.4. Types of drainage methods used as liquefaction remedial measures.
(a) Types of drainage methods according to shape.
(b) Types of drainage methods according to materials used.
(c) Types of drainage methods according to installation methods.

The remainder of this section focuses on the design and installation of a pile drain using gravel and crushed stone.

There are a number of variations of the method of forming pile shaped drains. One variant is the gravel compaction method (Kishida, 1983) which combines drainage and compaction. In recent years, another method has been developed which increases soil density by pushing gravel into the soil with a steel rod. Care must be taken with these methods to avoid liquefaction of the surrounding soils and clogging of the drain due to strong vibrations during installation. Various drainage methods are summarized in Fig. T6.4.

Vibroflotation also uses gravel and crushed stone as fill material to form stone columns. This is similar in concept to the gravel column used in the drainage method. With vibroflotation, however, there is a high possibility of sand migration into the stone columns during installation. Vibroflotation, therefore, is not used as a method to improve drainage but only to compact the surrounding ground.

In the drainage method, slight settlement accompanying water drainage and a slight increase in excess pore water pressures are inevitable. Compared to the compaction method, however, the drainage method is a relatively recent development; there are few documented case histories of their performance in earthquakes and design approaches are still being debated.

Yoshimi and Tokimatsu (1991) have also questioned the deformation characteristics of loose sandy soil improved by the drainage method. They suggest that the 'ductility' of the ground may be limited, implying that as excess pore water pressures increase under stronger ground shaking, there may be a sudden sharp increase in the level of deformation in the soil. A slight change in the level of shaking caused by earthquakes may lead to a significantly increased level of deformation.

For new constructions where there are minimal restrictions on the choice of method, or for the remediation of existing structures for which a large safety factor is considered appropriate, other approaches such as replacement with gravel, or stone backfill are considered preferable.

Pile drains are designed using the procedure shown in Fig. T6.5. As shown in this figure, the sequence of activities to be followed is (1) decide on the drainage material, (2) conduct soil surveys and in-situ tests to establish the natural subsoil profile, and finally (3) establish the installation spacing.

The recommended design approach is based on establishing a balance between the rate of excess pore water pressure increase due to earthquake shaking and the rate of dissipation of pore water pressure due to drainage. In soils of high permeability, drainage of excess pore water pressure is coupled with the generation of excess pore water pressure. The rate of increase of excess pore water pressure is increased by high amplitudes of earthquake motion, sudden and intense shaking,

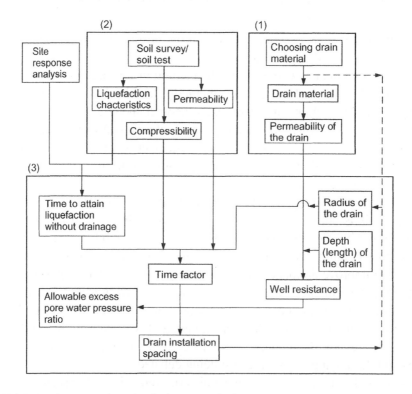

Fig. T6.5. Design procedure for drainage method.

and in loose saturated soils. The rate of dissipation of pore water pressure increases if the coefficient of volume compressibility of the natural soil is small and the coefficient of permeability is high. The object of the drainage method is to increase the natural rate of dissipation by installing closely spaced drains with drainage material of higher permeability than the surrounding ground.

Conditions under which the use of drains is likely to be less effective may be summarised as follows:

a) low natural permeability of the subsoil profile;
b) high coefficient of volume compressibility of the natural soil;
c) low liquefaction strength (cyclic resistance) of the natural soil;
d) high amplitude earthquake motion;
e) sharp or impact-type earthquake motions;
f) deep liquefiable soil layers.

T6.5   PREMIX METHOD: DESIGN AND INSTALLATION

The premix method uses a small amount of stabilizing material and a chemical additive, mixed with landfill, backfill, or fill inside cellular structures to create a new fill material that is less susceptible to the generation of excess pore water pressure and hence more resistant to liquefaction. After processing, the new fill is transported and placed at the designated location to create a stabilized reclaimed land area (Zen, 1990).

Unlike other liquefaction remediation measures, which are only relevant to existing subsoil profiles, the premix method provides a remediation option for new landfill, where the proposed filling method is anticipated to result in material which would otherwise be susceptible to liquefaction.

T6.5.1   Principle and application of the method

The premix method improves the soil matrix by partially cementing the points of contact of the individual soil particles. This effectively reduces the capacity of the soil to generate excess pore water pressure by increasing its stiffness and simultaneously provides some cohesion or cementation of the landfill, such that it can sustain higher levels of excess pore water pressure without exhibiting large deformations.

The method is applicable to construction sites, where it is being proposed to use imported sandy fill material. Examples of applications are shown in Fig. T6.6, which includes hypothetical construction cases. Careful consideration must be given to the specific conditions at each site or project since the installation method and characteristics of the treated sand often differs widely. It is recommended that

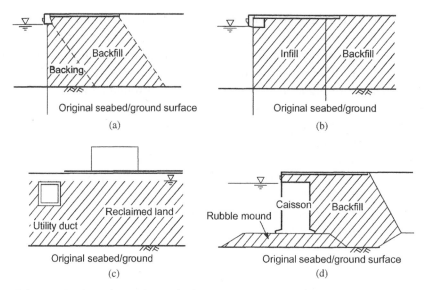

Fig. T6.6. Application of premix method.
(a) Sheet pile.
(b) Cell structure.
(c) Utility duct, reclaimed land.
(d) Caisson.

the premix method should be used only after performing laboratory and/or field tests to confirm the suitability of the approach.

There are a number of advantages to the premix method:
1 remedial treatment after the landfill is completed become unnecessary, and therefore time on site can be shortened and site costs may be reduced,
2 the strength of the treated soil can be reasonably predicted,
3 dredged sand can be exploited as fill material,
4 large, deep-water installations can be completed readily (using a conveyor, chute or bottom dump barge),
5 commonly available plant can be used throughout the process,
6 noise and vibration during fill placement are greatly reduced compared to the conventional vibro-rod and sand compaction methods,
7 the effects on adjacent structures caused by the placement and compaction are reduced compared to other methods,
8 lateral earth pressures on adjacent structures in the long term are also likely to be reduced.

There are, however, four disadvantages:
1 there may be substantial variation in the strength of the treated subsoil profiles, depending on the type of sand and placement procedure,
2 the stabilizing material or chemical additives can be costly,

3   the environmental impact on water quality may need to be considered,
4   the method is limited to sites where the subsoil is to be excavated or replaced
     or to newly reclaimed land.

T6.5.2   Design method

Figure T6.7 outlines the design procedure. The premix method differs from other
remedial treatments against liquefaction in that the treatment of the fill occurs
prior to placement, rather than afterwards. Quality control procedures can be used
to test the fill prior to placement. The method can then reduce the length of time
required on site for reclamation projects, resulting in ground with the required
strength being achieved soon after the completion of filling. Determining the
required strength of the treated fill is the first consideration in the design process,

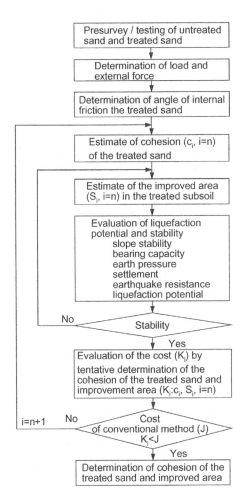

Fig. T6.7. Design procedure of premix method.

and this will depend on the type of structure and facilities to be constructed. Laboratory and/or field tests may be necessary to confirm design assumptions concerning the strength of the fill that can be achieved using the proposed additives. Lateral earth pressures on adjacent structures, bearing capacity and liquefaction susceptibility of the treated fill must be determined. The required quantity of cement or chemical additive can then be calculated and the installation cost estimated. However, soil strength and the area and volume of fill to be treated will vary depending on the proportion of additive used. It is recommended, that the effects on cost and construction programme of using a range of different percentages of additive are considered before a final design is reached.

T6.6   PRELOAD METHOD: DESIGN AND USE

The preload method for liquefaction remediation is based on laboratory findings that overconsolidation of sand increases liquefaction resistance (Yamazaki et al., 1992; Ishihara and Takatsu, 1979; Matsuoka, 1985). In the field, overconsolidation can be achieved either by placing a surcharge or by lowering the ground water level by pumping.

The preload method is recommended for liquefaction remediation in the following four situations:

1   when the subsoil profile consists of sandy soil with a high fines content and it is difficult to apply compaction and/or drainage methods,
2   when the subsoil profile consists of combined layers of sand and clay and it is necessary to improve the clay layer,
3   when it is necessary to lower the ground water level to facilitate excavation, and
4   when it is required to improve the ground under an existing structure.

As with the premix method, it is recommended that in-situ field tests be used to verify the preload method in advance of construction.

T6.6.1   Liquefaction characteristics of overconsolidated soil

(1) Increase in liquefaction resistance by overconsolidation (Yamazaki et al., 1992)

The liquefaction resistance of sandy soil increases with overconsolidation. Figure T6.8 shows cyclic triaxial test results for Niigata Eastern Harbour sand. The overconsolidation ratio (OCR) is defined by the ratio of σ_p' over σ_c', where σ_p' is the maximum effective confining pressure that the sample has been subjected to previously and σ_c' is the current effective confining pressure as shown in Eqn. (T6.1).

$$OCR = \frac{\sigma'_p}{\sigma'_c} \qquad (T6.1)$$

In laboratory element testing, the cyclic stress ratio (CSR or liquefaction strength), τ_l/σ'_c is defined as the stress ratio to cause liquefaction (+/– 5% strain) in 20 load cycles. Figure T6.8 shows how the CSR varies as a function of number of load cycles. The overconsolidation ratio also has an influence on the CSR, as shown in Fig. T6.9. Ishihara and Takatsu (1979) proposed that the ratio of the cyclic stress ratio to reach liquefaction of overconsolidated sands compared to normally consolidated sands was in proportion to the square root of the overconsolidation ratio

$$\left(\frac{\tau_l}{\sigma'_c}\right)_{OC} = \left(\frac{\tau_l}{\sigma'_c}\right)_{NC} \times \sqrt{OCR} \qquad (T6.2)$$

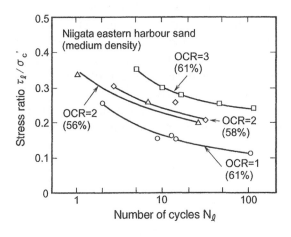

Fig. T6.8. Cyclic shear stress ratio and number of load cycles.

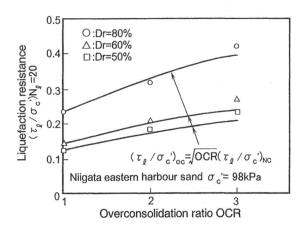

Fig. T6.9. Overconsolidation ratio and liquefaction strength.

where

$(\tau_l/\sigma_c')_{NC}$: Liquefaction strength for normally consolidated sands

$(\tau_l/\sigma_c')_{OC}$: Liquefaction strength for overconsolidated sands

This function is represented by the solid line in Fig. T6.9. For an OCR = 2, for example, the liquefaction strength is increased by 40%; if the OCR = 4, then the liquefaction strength is doubled.

(2) Consolidation or stress history

The influence of consolidation history (cycles of pre-consolidation) in comparison to the overconsolidation effect noted above appears to be minimal. The ◊ mark in Fig. T6.8 indicates the result of an experiment when the consolidation history ($\sigma_c' = 100 \sim 200$ MPa, OCR = 2) was repeated 5 times. It is apparent by comparing this result with the data point for a single pre-consolidation cycle that the repetition of cycles has had little effect.

(3) Influence of fines content

The effect of consolidation history on the mechanical behaviour of soil is generally greater in clay than sand. If sandy soil contains a high fines content, F_c, of clay and silt, then the overconsolidation effect may be expected to be higher. Figure T6.10 shows results from the work of Ishihara et al. (1976) on the relationship between the fines content and the overconsolidation ratio expressed in terms of the liquefaction strength (here defined as the cyclic shear stress ratio to cause liquefaction in 30 cycles). It is apparent in this figure that the effect of overconsolidation on liquefaction strength increases with increasing fines content.

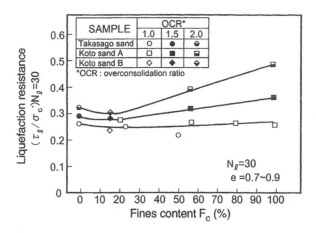

Fig. T6.10. Fines content and liquefaction strength.

T6.6.2   Design and installation procedure

A flowchart illustrating design and installation for the pre-load method is shown in Fig. T6.11. The procedure is summarised as follows:

(1) Estimate the overconsolidation ratio (OCR) of the natural ground

(a) Undisturbed samples are obtained from the field before improvement. Cyclic triaxial tests are performed on normally consolidated specimens to obtain the liquefaction strength of the insitu soils.

(b) Using the cyclic triaxial test results, the liquefaction safety factor F_L is obtained following the procedure described in Section T4.4.2 (Technical Commentary 4).

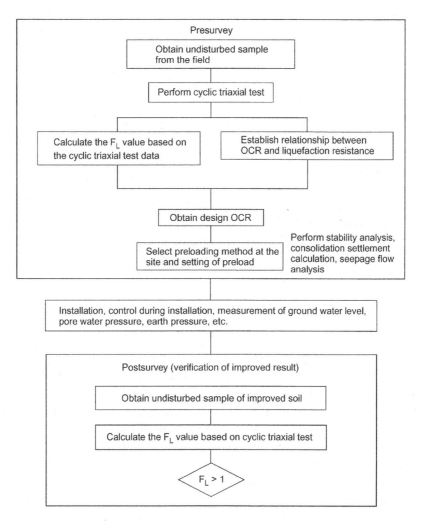

Fig. T6.11. Design/installation flow chart.

(c) The natural deposits are susceptible to liquefaction under earthquake loading for $F_L < 1$. For such deposits, the liquefaction strength must be improved by overconsolidation by a factor of at least $1/F_L$ to achieve a factor of safety greater than 1 against liquefaction.

(d) The relationship between the overconsolidation ratio and liquefaction strength is then obtained from further cyclic triaxial tests on overconsolidated specimens of the field samples.

(e) The overconsolidation ratio for the subsoil profile with respect to the specified liquefaction strength can then be deduced.

(2) Select method of preloading

Two methods are generally adopted to achieve overconsolidation of a subsoil profile; firstly, the application of a surcharge by embankment load, and secondly, the temporary lowering of the ground water level using deep wells (Japanese Society of Soil Mechanics and Foundation Engineering, 1988).

For the surcharge approach, the vertical effective stress increment within the soil is calculated from the prescribed overconsolidation ratio and then the height and area of the embankment are determined. Stability of the embankment and foundation against slip failure should be checked.

Lowering of the ground water level is appropriate for larger vertical effective stress increments. The permeability of the soil must be determined and a seepage flow analysis carried out in advance.

The choice of preload method may depend on other construction aspects of the project. For example, if sand drains are to be installed to assist drainage from in situ clay layers, then the surcharge method may be more appropriate. If the construction works involve dewatering to facilitate excavations, then the dewatering system may also enable overconsolidation of the liquefaction susceptible sand layers to be achieved in parallel.

T6.7   DESIGN OF LIQUEFACTION REMEDIATION

The design requirements for a specific soil improvement area will be based on an assessment of the mechanism of damage likely to be induced in the natural soils by liquefaction. In particular, it is critical to determine whether the damage will be caused primarily by reduction in the shear resistance of the liquefied soil, or by additional external forces due to excessively large displacements. Large displacements of the liquefied subsoil may affect the foundations of existing and new structures, or cause serious damage to lifeline facilities (such as gas, water and telecommunication pipelines and conduits), and special consideration will be necessary in respect of these cases. There are many other cases, however, in which remediation measures against loss of shear resistance will provide benefits against the risk of large displacements; ground compaction being one example.

There is a relatively long experience of the use of remediation measures against loss of shear resistance compared to experience of measures to resist excessively large displacements, and techniques to achieve this are still being studied and developed. For this reason, this section focuses on the design of a soil improvement area to resist loss of shear resistance.

Generally, even if the soil is predicted to undergo liquefaction over a wide area, it may be possible to limit the area requiring soil improvement to the area that controls the stability of the structure. For example, the zone within the subsoil that contributes predominantly to the stability of spread foundation structures is the part directly below and immediately around the structure; ground far away from the structure does not contribute to the same extent. A key issue is therefore to establish how far the soil improvement needs to extend laterally from the structure. This can be determined by following the procedure summarised below (Iai et al., 1987; Iai et al., 1991).

Secondly, it is necessary to determine the extent of the zone of soil improvement in the vertical direction. Standard practice is to improve the soil down to the deepest part of the liquefiable soil layer. If soil improvement is performed only within the top section of a soil layer that then experiences liquefaction below this elevation, the upward seepage flow of pore water generated may still result in loss of strength and stiffness and even liquefaction in the upper, improved sections.

The design procedure for a soil improvement area presented below is based around use of the compaction method. The procedure may, however, be adapted for other methods with appropriate modification by considering the appropriate characteristics of the improved soil (such as permeability and cyclic strength/deformation). (In the following discussion, 'liquefaction' is defined as that state in which the excess pore water pressure ratio of loose and medium sand reaches 1.0, i.e. when the pore water pressure u equals the vertical effective stress σ_v'.)

(1) Propagation of excess pore water pressure into the improved zone

Outside the zone of improved soil, excess pore water pressures will exceed those inside, resulting in a hydraulic gradient driving fluid into the zone of improved soil. This is a complex issue, as the deformation characteristics of dense saturated sand are highly non-linear. As a simplified design procedure, this phenomenon may be addressed as follows:

For excess pore water pressure ratios $u/\sigma_v' < 0.5$, the effect of the excess pore water pressure increase may be ignored, because laboratory test data indicate a very small strain generation below this level. For $u/\sigma_v' > 0.5$, however, it is necessary to take into account the effect of excess pore water pressure increase. Shaking table tests and seepage flow analyses suggests that the pore water pressure ratio is $u/\sigma_v' > 0.5$ within an area defined by the square ABCD in Fig. T6.12. In this area, the soil shear resistance must be reduced for the purposes of the design. The tests also indicate that an area defined by the triangle ACD exhibited unstable characteristics. This area should therefore be assumed in the design to be liquefiable and treated accordingly.

The exception to this recommendation is when a drain or impermeable sheet or zone has been installed at the perimeter of the improved area in order to shut out the inflow of pore water into the improved area. Under these conditions, the area corresponding to the square ABCD need not be included as part of the improvement plan.

(2) Pressure applied by the liquefied sand layer
At the boundary between liquefied and non-liquefied ground, there is a dynamic force as indicated in Fig. T6.13, and a static pressure corresponding to an earth pressure coefficient $K = 1.0$ which acts on the improved ground due to the liquefaction of the surrounding soils. These forces may greatly exceed the forces acting in the opposite direction from the non-liquefied ground. For a retaining structure backfilled with soil and subject to active static and dynamic lateral earth pressures at EF as shown in Fig. T6.14, the area of soil improvement must be large enough that there is no influence of liquefaction in the active failure zone. To accommodate the net outward force, it is essential to check that sufficient shear resistance can be mobilised along the passive failure surface GC.

(3) Loss of shear strength in liquefied sand layer
In the simplified design procedure, unimproved soil of loose or medium relative density should be considered to have negligible shear strength after liquefaction (i.e. the soil is treated as a heavy fluid). Since the shear strength of the improved ground in triangle ACD also cannot be relied upon (see (1) above), then the improvement area should be wide enough to obtain sufficient bearing capacity from the shear resistance along the solid lines EFG and HI in Fig. T6.15.

In practice, lateral pressure from surrounding liquefied sand layers may contribute to the stability of certain structures. Figure T6.15 shows how the dynamic earth pressure may be subtracted from the enhanced static lateral earth pressure

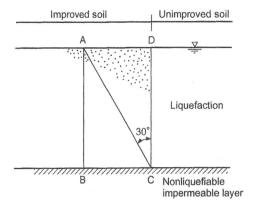

Fig. T6.12. Area of softening due to seepage flow.

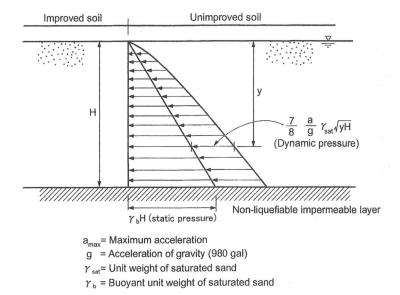

Fig. T6.13. Pressures applied at the boundary of improved soil.

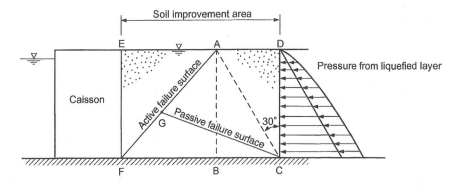

Fig. T6.14. Schematic diagram for investigation of stability with respect to pressures applied from the liquefied sand layer.

(based on an earth pressure coefficient $K = 1.0$) in certain stability calculations. The section on which the pressure from the liquefied sand layer is applied, can be assumed to be along the lines GG' or II'. Although this figure shows only five examples of foundation types, the same principles can be applied to other design arrangements.

(4) Change in dynamic response of the improved soil
The dynamic response of the improved (compacted) soil area will be altered by the influence of liquefaction in the adjacent unimproved soil and by the migration

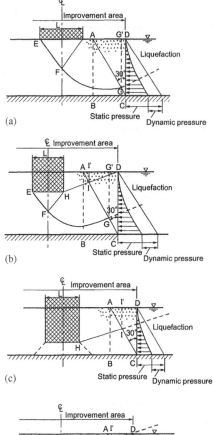

Fig. T6.15. Schematic diagrams for investigation of stability for determining the soil improvement area.
(a) Shallow foundation.
(b) Foundation with shallow
embedment.
(c) Foundation with deep
embedment.
(d) Pile foundation.
(e) Underground structure
(when passive resistance
is necessary in order to
maintain stability of the
structure).

of excess pore water pressures. This can result in a change in the inertia (body)
force applicable to the compacted zone. In order to assess the influence of this
change on the stability of the structure and the soil improvement zone, a separate
detailed evaluation may be necessary.

The nature of the evaluation depends on the extent of the area over which the excess pore water pressure ratio exceeds 0.5 and the general subsoil profile. If this area is small compared to the entire improved area, and if the surface profile of the entire improved area is relatively flat, then the change in the dynamic response of the improved zone is likely to be small (see for example Fig. T6.14). However, in extreme cases, where high excess pore water pressures are predicted close to a structure or with sloping ground or excavations, a two-dimensional earthquake response analysis may be required.

(5) Design of soil improvement area for port and harbour structures

Key parameters for any assessment of the area of soil improvement include the angle of internal friction for sands and the anticipated intensity of earthquake shaking. A stability analysis method will also need to be specified. Standard design procedures adopt a simplified, pseudo-static approach to the prediction of earthquake loads and response. Despite this, experience suggests that with appropriate care and selection of parameters, these methods generally provide satisfactory design solutions.

For example, the seismic coefficient and stability analysis methods used for port and harbour structures in Japan are described in the Technical Standard for Port and Harbour Facilities and Commentaries (Ministry of Transport, Japan, 1989). In this procedure, earth pressures caused by earthquakes loading are computed by the Mononobe–Okabe theory (Mononobe, 1924; Okabe, 1924) and stability analysis is performed by Bishop's method (Bishop, 1955) using the specified angle of internal friction for sands and design seismic coefficients.

The soil improvement area required for a typical gravity quay wall based on this approach, is shown in Figure T6.14. For anchored sheet pile walls, the zone of improvement in the vicinity of the anchor wall must be carefully considered. For flexible anchor walls, soil improvement must extend to sufficient depth below the tie rod level to ensure that the adequate capacity is maintained to resist the tie rod force.

T6.8 INFLUENCE ON EXISTING STRUCTURES DURING SOIL IMPROVEMENT

One further aspect of a remediation design is to consider in detail the influence of the installation method on existing structures. Particular experience has been gained in cases where the compaction method has been used to improve the backfill sand behind flexible (sheet pile) quay walls. Several case histories are discussed below.

T6.8.1   Influence on bending stress in a sheet pile wall by compaction

Figure T6.16 illustrates the increase in bending stress in a steel pipe sheet pile wall as a function of the proximity to the wall of compaction using the vibro-rod method (Toyota et al., 1988). In this example, the main wall comprised a steel pipe sheet pile quay wall (to a depth of −10 m), as shown in Fig. T6.17. Soil improvement was carried out by compacting the ground using the vibro-rod method over the area and depths indicated by the broken line in the figure (with a maximum depth of −20 m). The soil profile is shown in Fig. T6.18 (Toyota et al., 1988; Sato et al., 1988).

The bending stress in the wall shows an increase per rod installation, with the distance from the wall being the distance measured from the driven position of the vibro-rod. Clearly, the bending moments in the main wall will also be affected by rod installation to either side along the quay, and these results are therefore representative of the plane strain condition. The bending stress in the sheet pile wall increases gradually with the number of rod installations.

The results reported by Toyota et al. (1988), do not address the effect of compaction at distances closer than 5 m to the main wall. The Ministry of Transport, Japan, First District Port and Harbour Construction Bureau (1984) has published data for the increase in bending stress caused by vibro-rod and gravel drain installation at a distance of just 2 m from the main wall. These data show an increase of 30 MPa in the bending stress in the steel sheet piles per rod installation using the vibro-rod method and an increase of 25 MPa per gravel pile caused by the

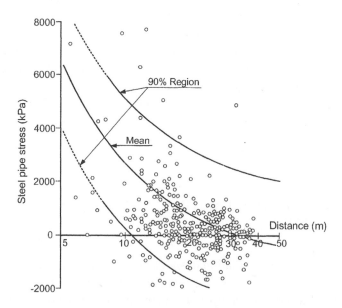

Fig. T6.16. Increase in bending stress in a steel pipe pile wall during installation of the rod compaction method.

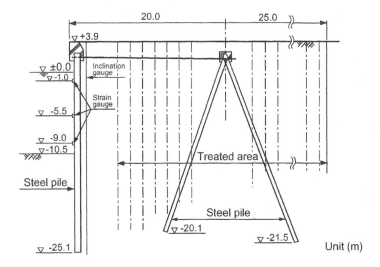

Fig. T6.17. Location of instruments for monitoring the effect of compaction installation for a quay wall.

installation of gravel drains. When compaction was carried out by the vibro-rod method behind the wall following the installation of gravel drains, no significant additional increase in bending stress was found. This study, therefore, suggests that the combined use of gravel drains and compaction may be an effective solution for soil improvement close to existing structures.

Such case histories provide valuable guidance for remediation design. However, the actual stresses and strains generated in an existing structure due to the compaction effect will vary depending on a wide range of factors and it is, therefore, recommended that a test installation is conducted in all cases to quantify any potentially damaging effects from the proposed improvement method.

T6.8.2   Excess pore water pressure and vibration resulting from compaction

Figure T6.19 shows the excess pore water pressures and ground accelerations measured during soil improvement compaction works at a number of ports and harbours in Japan. Though the accelerations shown in the figure are quite large, the frequency of vibration generated by compaction is relatively high, at around 10 Hz; and direct comparison of these accelerations with levels of earthquake shaking used in earthquake resistant design of structures is inappropriate. Nevertheless, the figure shows a rapid or steep increase in the ground acceleration at distances closer than 5 to 10 m from the installation. Clearly, the effects of such vibrations need to be considered carefully in the design of any installation close to existing structures.

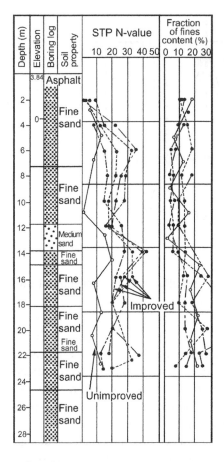

Fig. T6.18. Soil conditions at monitoring site.

Similarly, Fig. T6.20 shows the excess pore water pressure that may be generated due to compaction. These observed levels of excess pore water pressure also suggest that appropriate care must be taken to assess the influence of the installation process on any structures within a distance of around 5 to 10 m. The installation of drains may be used to control the effects of subsequent vibro-rod compaction, but in this case, care must be taken to ensure that the drains do not become clogged by fines driven by the high hydraulic gradients local to the installation.

T6.8.3   Influence of earth pressure increased by compaction

Compaction of soil will enhance its resistance with respect to liquefaction, because both the relative density of the soil and the horizontal effective stress (and therefore the confining pressure) are increased. However, the increase in horizontal stress or earth pressure may also have an effect on existing structures nearby, such as a retaining wall or quay wall, potentially affecting the capacity of the

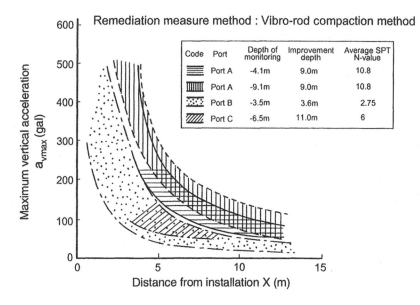

Fig. T6.19. Attenuation of acceleration.

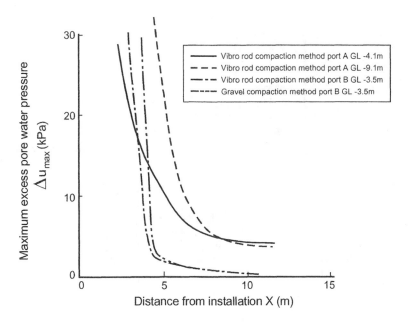

Fig. T6.20. Attenuation of excess pore water pressure.

structure under earthquake loading. It may not be clear whether this effect is positive or negative and therefore, in the remediation design, the consequences of any increase in lateral earth pressure on adjacent structures must be carefully assessed.

Kameoka and Iai (1993) report an analytical study carried out to quantify this effect for a selected configuration of a quay wall. Forced displacements corresponding to the deformation of an initially loose soil (relative density 35%) caused by compaction were generated within a zone behind a wall as shown in Fig. T6.21. This forced displacement resulted in increased earth pressures on the sheet pile wall as shown in Fig. T6.22 (the open circles in the figure indicate the initial earth pressure prior to compaction). An earth pressure ratio was defined to indicate the ratio of the earth pressure induced by compaction over the earth pressure without compaction. The effect of the increase in relative density of the soil on the earth pressure was ignored.

Figure T6.23 shows the results in terms of bending moment and displacement from several analyses of a quay wall with various initial earth pressures. As the initial horizontal earth pressure is increased (by simulating the process of compaction of the fill behind), the initial (pre-earthquake) bending moment in the wall also increases, as expected. The earth pressure after the earthquake, however, decreased. A similar tendency was noted in the displacements of the sheet pile; here it was found that the outward displacements caused by the earthquake

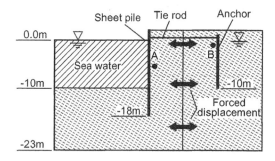

Fig. T6.21. Cross section of quay walls for the analyses.

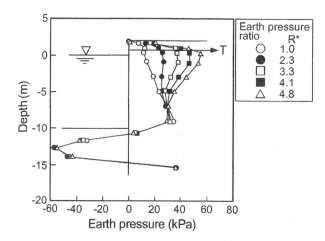

Fig. T6.22. Earth pressure distribution after induced displacement.

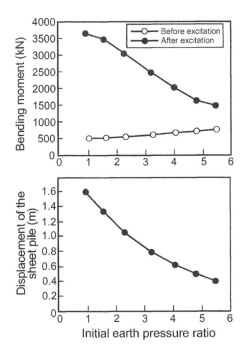

Fig. T6.23. Relationship between the initial soil pressure and earthquake-induced bending moment and displacement.

shaking were reduced compared to the outward displacement without compaction. This effect is similar to the concept of pre-stressing of structures, and was also noted in centrifuge model tests of cantilever retaining walls retaining dry sand, where densification on successive earthquake shaking events increased the in-situ lateral earth pressure (Steedman, 1984). In some cases, the increase in bending stress and displacement caused by compaction may significantly reduce the structural capacity of the wall (or tie backs) or cause unacceptable local damage to services. However, provided this is not the case, then there may be a beneficial effect on the overall resistance of sheet pile walls to earthquakes from this form of soil improvement.

TC7: Analysis Methods

TECHNICAL COMMENTARY 7

Analysis Methods

This technical commentary addresses analysis methods for port structures. These methods are classified into three types according to the level of sophistication as discussed in Chapter 5 in the Main Text. Simplified analysis is appropriate for evaluating approximate threshold limit for displacements and/or elastic response limit and an order-of-magnitude estimate for permanent displacements due to seismic loading. Simplified dynamic analysis is possible to evaluate extent of displacement/stress/ductility/strain based on assumed failure modes. Dynamic analysis is possible to evaluate both failure modes and the extent of the displacement/stress/ductility/strain.

T7.1 SIMPLIFIED ANALYSIS

T7.1.1 Loads on gravity/sheet pile/cellular quay walls

The conventional practice for evaluating seismic stability of retaining walls, including gravity, sheet pile, and cellular walls, is based on pseudo-static approaches. In this procedure, a seismic coefficient, expressed in terms of the acceleration of gravity, is used to compute an equivalent pseudo-static inertia force for use in analysis and design. The actual dynamic behaviour of retaining walls is much more complex than treated in the pseudo-static approach. However, this approach has been the basis for the design of many retaining structures in North America, Japan and other seismically active regions around the world (Whitman and Christian, 1990; Tsuchida, 1990).

(1) Stability assessment

In the pseudo-static approach, the stability of gravity type walls is evaluated with respect to sliding, overturning, and bearing capacity. Sliding is often the critical condition for a wall having a large width to height ratio. Overturning or loss of bearing capacity becomes the critical condition for a wall having a small width to height ratio. In the limit state under strong shaking, the instability with respect to overturning and/or bearing capacity is much more serious than that for sliding because tilting of the wall, if excessive, will lead to collapse. Thus, it is common practice to assign a higher safety factor for overturning and bearing capacity than for sliding (Ministry of Transport, Japan, 1989; Ebeling and Morrison, 1992). In addition to these limit states, estimates of tilting are warranted for evaluating post-earthquake stability of walls (Whitman and Liao, 1984; Prakash et al., 1995). Loss of bearing capacity has been conventionally evaluated based on the circular slip analysis. The design parameters, including the internal friction angle of foundation soil, for use with this simplified analysis can be found in the relevant design codes (e.g. Ministry of Transport, Japan, 1999).

Stability of anchored sheet pile walls is evaluated with respect to gross stability and stresses induced in structural components. In particular, gross stability is evaluated for a sheet pile wall to determine the embedment length of the sheet piles into competent foundation soils. Stability is also considered for an anchor of the sheet pile wall to determine the embedment length and the distance from the wall. Stresses are evaluated for the wall, anchor, and tie-rod. In the ultimate state, the rupture of tie-rods results in catastrophic failure of the wall and, therefore, this mode of failure must be avoided. Thus, it is common practice to assign a large safety factor for tie-rods (Ministry of Transport, Japan, 1989; Ebeling and Morrison, 1992). A less established issue is whether the wall or the anchor should be the first to yield. Excessive displacement of the anchor is undesirable (Gazetas et al., 1990). Balanced movement of the anchor, however, reduces the tension in the tie-rods and the bending moment in the wall. Thus, there are differences in the design practices of anchored sheet pile quay walls (Iai and Finn, 1993). These differences can result in a variation of up to a factor of three in the forces used for designing tie-rods and anchors.

(2) Active earth pressures

In the pseudo-static approach, the earth pressures are usually estimated using the Mononobe–Okabe equation (Mononobe, 1924; Okabe, 1924). This equation is derived by modifying Coulomb's classical earth pressure theory (Coulomb, 1776; Kramer, 1996) to account for inertia forces. In the uniform field of horizontal and (downward) vertical accelerations, $k_h g$ and $k_v g$, the body force vector, originally pointing downward due to gravity, is rotated by the seismic inertia angle, ψ, defined by (see Fig. T7.1)

$$\psi = \tan^{-1}\left[\frac{k_h}{1 - k_v}\right] \tag{T7.1}$$

The Mononobe–Okabe equation is, thus, obtained by rotating the geometry of Coulomb's classical solution through the seismic inertia angle, ψ, and scaling the magnitude of the body force to fit the resultant of the gravity and the inertia forces

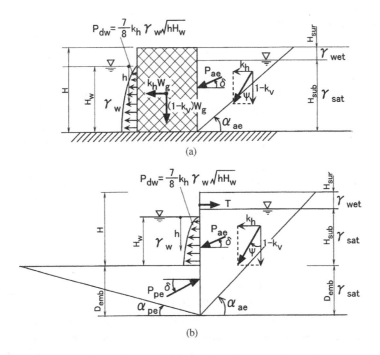

Fig. T7.1. Active and passive earth pressures.
    (a) Gravity quay wall.
    (b) Sheet pile quay wall.

(Mononobe, 1924; Whitman and Christian, 1990). For a vertical wall having a friction angle, δ, between the backfill and the wall, and retaining a horizontal backfill with an angle of internal friction, ϕ, the dynamic active earth pressure coefficient, K_{ae}, is given by

$$K_{ae} = \frac{\cos^2(\phi - \psi)}{\cos\psi \, \cos(\psi + \delta)\left[1 + \sqrt{\dfrac{\sin(\phi + \delta)\sin(\phi - \psi)}{\cos(\delta + \psi)}}\right]^2} \qquad (T7.2)$$

For completely dry soil, the dynamic active earth thrust, which acts at an angle, δ, from the normal to the back of the gravity type wall of height, H, (see Fig. T7.1(a)) is given by

$$P_{ae} = K_{ae} \tfrac{1}{2} \gamma_d (1 - k_v) H^2 \qquad (T7.3)$$

where γ_d is the unit weight of the dry backfill. In case of a uniformly distributed surcharge q_{sur}, γ_d should be substituted with $(\gamma_d + (q_{sur}/H))$. The surcharge q_{sur} used for seismic design is typically half of the surcharge used for static stability design. Other horizontal loads, including a fraction of the horizontal bollard pull, can also be considered in the design.

(3) Passive earth pressures

A similar expression is given for the dynamic passive earth thrust, P_{pe}. The dynamic passive earth pressure coefficient, K_{pe}, for the soil in front of the wall with a level mudline is given by Eqn. (T7.2) with a minus sign in front of the square root symbol. In this case, H in Eqn. (T7.3) is the depth of the embedment of the wall, D_{emb} (see Fig. T7.1(b)). In both active and passive cases, the point of application of the resultant force is typically chosen at a level of $0.40H$ to $0.45H$ (Seed and Whitman, 1970).

Another approach that gives similar results to the Mononobe–Okabe method is that of Seed and Whitman (1970). This approach calculates separately the dynamic component of the active and passive earth pressure coefficients as

$$\Delta K_{ae} = \frac{3}{4}k_h, \quad \Delta K_{pe} = -\frac{17}{8}k_h$$

(T7.4)

This dynamic component acts at a point $0.5H$ to $0.6H$ from the base of the wall.

A complete set of equations may be found in the design codes and manuals (e.g. Ministry of Transport, Japan, 1989; Ebeling and Morrison, 1992). In referring to these codes and manuals, a careful look at the sign convention of the friction angle, δ, is needed. In much of the Japanese literature, the friction angle, δ, is defined as positive whenever the soil drags the wall down. The same is true for American literature for the active earth pressure, but for the passive earth pressure, the friction angle, δ, is defined as positive when the soil drags the wall up, as shown in Fig. T7.1(b). Since this paper is written in English, the sign convention of American literature is followed here.

(4) Fully and partially submerged fills

The Mononobe-Okabe equation was derived for dry yielding backfill retained by a rigid wall. When the backfill is saturated with water, it has been common practice to adopt the assumption that pore water moves with the soil grains. Considering a fully saturated Coulomb wedge, the horizontal inertia force is proportional to the saturated total unit weight, γ_{sat}, and the vertical gravity force is proportional to the buoyant unit weight ($\gamma_b = \gamma_{sat} - \gamma_w$). Thus, the modified horizontal seismic coefficient is given by (Amano et al., 1956)

$$k_h' = \frac{\gamma_{sat}}{\gamma_b}k_h$$

(T7.5)

Using the modified horizontal seismic coefficient in Eqns. (T7.1) through (T7.3) with a unit weight γ_b in Eqn. (T7.3), will give P_{ae} and P_{pe} for a fully submerged backfill. The vertical seismic coefficient k_v will not be modified for the submerged condition.

For a partially submerged soil wedge as shown in Fig. T7.1(a), the equivalent unit weight for the soil wedge may be obtained by computing the weighted average of the unit weights based on the volume of soil in the failure wedge above and below the ground water table. This results in the equivalent unit weight as (e.g. Ebeling and Morrison, 1992)

$$\gamma_e = \gamma_{\text{wet}} \left[1 - \left(\frac{H_{\text{sub}}}{H} \right)^2 \right] + \gamma_b \left(\frac{H_{\text{sub}}}{H} \right)^2 \tag{T7.6}$$

The modified seismic coefficients will be obtained by modifying Eqn. (T7.5) as follows:

$$k_h' = \frac{\frac{1}{2} q_{\text{sur}} H + \frac{1}{2} \gamma_{\text{wet}} H_{\text{sur}}^2 + \gamma_{\text{wet}} H_{\text{sub}} H_{\text{sur}} + \frac{1}{2} \gamma_{\text{sat}} H_{\text{sub}}^2}{\frac{1}{2} q_{\text{sur}} H + \frac{1}{2} \gamma_{\text{wet}} H_{\text{sur}}^2 + \gamma_{\text{wet}} H_{\text{sub}} H_{\text{sur}} + \frac{1}{2} \gamma_b H_{\text{sub}}^2} k_h \tag{T7.7}$$

where

   q_{sur} : a uniformly distributed surcharge
   γ_b :   buoyant unit weight

The active and passive earth thrusts, P_{ae} and P_{pe}, for the partially submerged backfill are obtained through Eqn. (T7.3) by using a unit weight γ_e and k_h'.

An alternative procedure for taking into account a partially submerged soil situation is to idealize the effect of the overlaying soil layer as a surcharge to the soil layer of interest (e.g. Ministry of Transport, Japan, 1999). Since the contribution for surcharge from the layer above the water table to the saturated layer below is equal to $\gamma_{\text{wet}} H_{\text{sur}}$, γ_d in Eqn. (T7.3) should be substituted by $(\gamma_b + (q_{\text{sur}} + \gamma_{\text{wet}} H_{\text{sur}})/H_{\text{sub}})$. This leads to an alternative equation to define the modified seismic coefficient as

$$k_h' = \frac{q_{\text{sur}} + \gamma_{\text{wet}} H_{\text{sur}} + \gamma_{\text{sat}} H_{\text{sub}}}{q_{\text{sur}} + \gamma_{\text{wet}} H_{\text{sur}} + \gamma_b H_{\text{sub}}} k_h \tag{T7.7'}$$

These results can be further generalized for multi-layered partially submerged soils.

(5) Coarse-grained fills

For extremely coarse-grained fill (e.g. gravel and rock fill), the soil skeleton and pore water act independently. The permeability k is often used as a primary parameter to dictate the pore water movement relative to the soil skeleton (Matsuzawa et al., 1985). Values ranging from 10^{-2} cm/s to 10 cm/s have been quoted to define the transition zone between the two states where water and soil move independently, and where the two move as a single body. However, no clear consensus in practice exists for the limiting value of permeability for these conditions. Within the transition zone, a percentage of the pore water remains attached to the soil, and the rest moves independently. The pore water movement, however, is also affected by other factors, including the compressibility co-efficient of the soil skeleton, the rate of cyclic loading, and boundary conditions for pore water flow around the porous fill. These issues need to be thoroughly investigated before making any recommendations in design practice.

It is difficult to develop a completely logical set of assumptions for extremely coarse-grained fill. One procedure is to assume that the total thrust is made up of:

1) a thrust from the soil skeleton, computed using Eqns. (T7.1) through (T7.7) with γ_d in place of γ_{sat} in Eqn. (T7.7) or Eqn. (T7.7'); and

2) the effect of the pore water by assuming Westergaard dynamic water loading (Westergaard, 1933; to be discussed later, see Eqn. (T7.8)).

This procedure, however, is not totally consistent, since the effect of the increased pore pressures due to the dynamic water pressure is ignored in the computation of the thrust from the soil skeleton (Ebeling and Morrison, 1992).

(6) Effects of excess pore water pressure

Another significant, yet poorly defined issue involves the procedures recommended for evaluating the effects of excess pore water pressures generated in saturated backfill on the dynamic earth pressures exerted on the wall. One approach is to use the reduced internal friction angle which produces the same shear resistance of soil under a reduced confining pressure (Ebeling and Morrison, 1992). This approach may work if reasonable engineering judgments are used in determining the level of excess pore water pressure. Many laboratory tests indicate, however, that the pore water pressure generally will not remain constant once the soil begins to deform. Dilatancy of sand at large strain levels complicates the behaviour of saturated sand under cyclic loading. Effects of the initial deviator stress, anisotropic stress conditions, and the rotation of the principal stress axis also complicates the cyclic behaviour of the sand. All of these issues may be best addressed in the context of seismic analysis of the soil-structure interaction phenomena based on cyclic behaviour of sand using effective stress analysis (Iai and Ichii, 1997).

(7) Hydrodynamic pressure

During seismic shaking, the free water in front of the structure exerts a cyclic dynamic loading on the wall; the critical mode occurs during the phase when suction pressure is applied on the wall. The resultant load can be approximated by (Westergaard, 1933):

$$P_{dw} = \frac{7}{12} k_h \gamma_w H_w^2 \qquad (T7.8)$$

where

γ_w: unit weight of seawater

H_w: water depth

The point of application of this force lies $0.4H_w$ above mudline.

The sea water level in front of the wall can be that most conservative in design. The most conservative water level can be a low water level, a high water level, or the level between them, depending on the design conditions of the wall.

(8) Equivalent seismic coefficient

For simplicity, it is often assumed in design practice that $k_v = 0$ (e.g. Ministry of Transport, Japan, 1989). The factor of safety against earthquake for induced deformation is defined by

$$F_s = \frac{k_t}{k_e} \qquad \qquad (T7.9)$$

where k_t: threshold seismic coefficient from pseudo-static analysis

k_e: equivalent seismic coefficient based on design level ground motions

The threshold seismic coefficient k_t is the limit defined for a structure, beyond which a structure-soil system will no longer maintain the pseudo-static equilibrium. For example, the seismic stability of a gravity quay wall is evaluated for sliding, overturning and bearing capacity failure. By increasing the magnitude of the seismic coefficient k_h for use in the pseudo-static approach, the first failure mode to yield a safety factor of unity is identified as the most critical. The seismic coefficient representing this threshold for the most critical failure mode defines the threshold seismic coefficient k_t.

The equivalent seismic coefficient k_e is a pseudo-static acceleration, specified as a fraction of gravity, for use in the pseudo-static analysis to represent the effects of an actual earthquake motion. The equivalent seismic coefficient is not always equal to the design level PGA/g due to the transient nature of the earthquake motions. The ratio of k_e to PGA/g used in practice has been based on empirical data and engineering judgement. For example, the threshold seismic coefficients obtained by back-analyses of a damaged and non-damaged structure at sites of non-liquefiable soils, provide a lower and upper bound estimate for an equivalent seismic coefficient. The threshold seismic coefficient of a damaged structure provides a lower bound estimate because the effective seismic coefficient of the earthquake should have exceeded the threshold seismic coefficient for damage to occur. Conversely, the threshold seismic coefficient of a non-damaged structure provides an upper bound estimate because the effects of the earthquake motion represented by the effective seismic coefficient should have remained below the instability threshold.

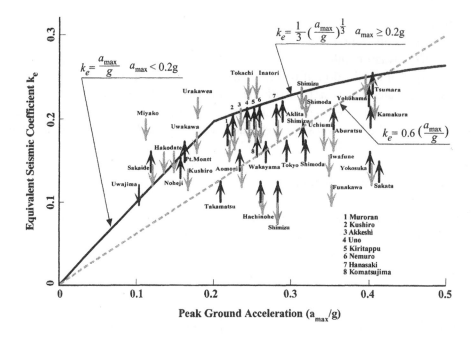

Fig. T7.2. Equivalent seismic coefficient k_e for retaining walls at non-liquefied waterfront sites (after Noda et al., 1975).

Figure T7.2 shows a summary of the lower and upper bound estimates based on the case histories of 129 gravity type quay walls during 12 earthquakes (Noda et al., 1975). The roots of the arrows in the figure, rather than the points, show the exact values. The arrows pointing up indicate lower bound estimates; those pointing down indicate upper bound estimates. The equation for an upper bound envelope was given by Noda et al. (1975) as

$$k_e = \frac{a_{max}}{g} \qquad a_{max} < 0.2g$$

$$k_e = \frac{1}{3}\left(\frac{a_{max}}{g}\right)^{1/3} \qquad a_{max} \geq 0.2g \tag{T7.10}$$

A similar study for anchored steel sheet pile quay walls was performed using data from 110 case histories. It was concluded that the above relation is also applicable for the anchored sheet pile quay walls (Kitajima and Uwabe, 1979).

As shown in Fig. T7.2, the relation in Eqn. (T7.10) is an envelope. An average relationship between the effective seismic coefficient and the peak ground acceleration may be obtained as

$$k_e = 0.6\left(\frac{a_{max}}{g}\right) \tag{T7.11}$$

T7.1.2   Displacement of walls

(1) Non-liquefiable sites

On the basis of earthquake case histories at non-liquefaction sites, a set of empirical equations were derived through regression analysis to define horizontal displacement, settlement, and displacement ratio (i.e. horizontal displacement normalized by wall height). The variable used in the empirical equations is the safety factor F_s defined by Eqns. (T7.9) and (T7.10), which defines the threshold level for the first failure mode among sliding, overturning and bearing capacity failure. The empirical equations are defined as a linear function of $(1/F_s)$ (see Fig. T7.3 and Tables T7.1 and T7.2) (Uwabe, 1983) as

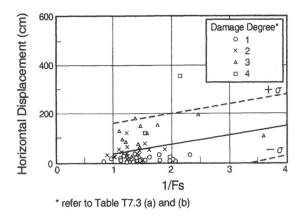

(a)

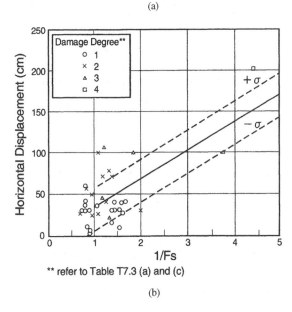

(b)

Fig. T7.3. Statistical summary of case history data for non-liquefaction sites (after Uwabe, 1983).
(a) Gravity quay wall.
(b) Sheetpile quay wall.

Table T7.1. Empirical equations for a gravity quay wall (Uwabe, 1983).

Deformation	Empirical equations**	Correlation coefficient	Standard deviation
Max. horizontal displacement (cm)	$d = -74.2 + 98.2(1/F_s)$	0.34	130
Settlement of gravity wall (cm)	$s = -16.5 + 32.9(1/F_s)$	0.50	30
Normalized displacement* (%)	$r = -7.0 + 10.9(1/F_s)$	0.38	13

* Maximum horizontal displacement normalized with a wall height H.
** F_s: Factor of safety against earthquake = k_t/k_e.

Table T7.2. Empirical equations for a sheet pile quay wall (Uwabe, 1983).

Deformation	Empirical equations**	Correlation coefficient	Standard deviation
Max. horizontal displacement (cm)	$d = -1.6 + 34.9(1/F_s)$	0.68	27
Settlement of sheet pile wall (cm)	$s = -5.3 + 14.7(1/F_s)$	0.40	20
Normalized displacement* (%)	$r = -1.5 + 5.8(1/F_s)$	0.65	5

* Maximum horizontal displacements normalized with a wall height H.
** F_s: Factor of safety against earthquake = k_t/k_e.

$$d = a + b/F_s \tag{T7.12}$$

where

　　　d: horizontal displacement, settlement, or displacement ratio
　　　a, b: empirical constants
　　　F_s: factor of safety during an earthquake

Field performance data from numerous sites in Japan compiled by Kitajima and Uwabe (1979) have been used to derive another empirical method for estimating the approximate degree of damage to anchored sheet pile walls during earthquakes (Gazetas et al., 1990). Two non-dimensional factors, the effective anchor index (*EAI*) and the embedment participation index (*EPI*) have been related to the degree of damage recorded for the sheet pile quay walls. Referring to Fig. T7.4, *EAI* is used to quantify the amount of available anchor capacity with

$$EAI = \frac{D_{anc}}{H} \tag{T7.13}$$

The *EPI* provides the contribution of the wall embedment as

$$EPI = \frac{P_{pe}}{P_{ae}}\left(1 + \frac{D_f}{D_f + H}\right) \tag{T7.14}$$

where D_f is the depth of the effective point of rotation of the sheet pile wall below the mudline. The effective point of rotation is approximated, using the effective seismic coefficient, k_e', and the angle of internal friction of the backfill soil, ϕ, as

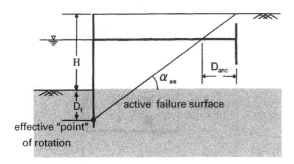

Fig. T7.4. Relevant parameters for defining effective anchor index, *EAI*, and embedment participation index, *EPI* (after Gazetas et al., 1990).

$$D_f = [0.5(1 + k'_e) - 0.02(\phi - 20)]H \qquad (T7.15)$$

The effective point of rotation, D_f, can be conservatively estimated as the embedment length of the sheet pile below the mudline, D_{emb}.

For uniform soils, *EPI* can be approximated as

$$EPI = \frac{K_{pe}}{K_{ae}}\left(\frac{D_f}{D_f + H}\right)^2\left(1 + \frac{D_f}{D_f + H}\right) \qquad (T7.16)$$

The seismic active and passive earth pressure coefficients can be readily estimated from Eqn. (T7.2). The angle, a_{ae}, from a horizontal plane to the active failure plane is approximated as

$$\alpha_{ae} = 45 + \phi/2 - 135(k'_e)^{1.75} \qquad (T7.17)$$

The relationships between the *EAI* and *EPI* factors were plotted for 75 Japanese case histories in Fig. T7.5, with the descriptions of the degrees of damage given in Table T7.3.

(2) Liquefiable sites

When sites contain saturated loose to medium sandy soils, it is necessary to consider the consequences of liquefaction. The simplified geometry of the loose saturated sand relative to the cross section of the wall shown in Figs. T7.6 and T7.7 was used for classifying the case histories for gravity and sheet pile quay walls. Table T7.4 shows the catalogue of the case histories with supplemental information shown in Fig. T7.8 (Iai, 1998a). A summary of the normalized displacements of walls at liquefaction sites are shown in Tables T7.5 and T7.6 for gravity and anchored sheet pile walls. The intensity of the earthquake motions can be categorized in two levels. The first level (E1), is defined as the earthquake motion level being equivalent to the design seismic coefficient. The second level (E2), is defined for those about 1.5 to 2.0 times larger. It should be noted that these earthquake levels are defined for sites of non-liquefiable soils. If a site consists of liquefiable soils, a nearby site of non-liquefiable soil is used for defining the earthquake levels. For evaluating the levels of earthquake motions

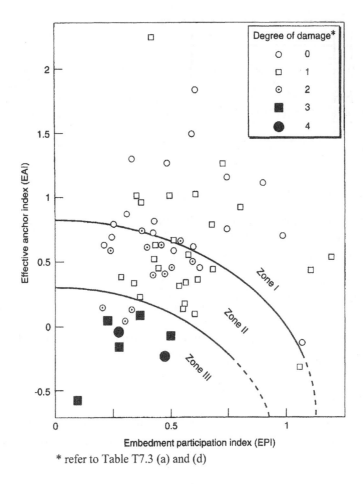

Fig. T7.5. Developed seismic design chart (Gazetas et al., 1990).

relative to the seismic coefficient used for design, the average relationship shown in Eqn. (T7.11) is used. The normalized displacement d/H (%) is defined as the ratio of a horizontal displacement of a quay wall over a wall height from mudline.

From the numbers shown in Tables T7.5 and T7.6, it is possible to obtain a rough estimate of displacements based on the wall height. Case history data of anchored sheet pile walls at liquefied sites are limited, and a statistical summary for E2 motion has not yet been available.

T7.1.3   Pile-supported wharves

In the simplified analysis of pile-deck systems of pile-supported wharves, seismic inertia force is applied to the deck, where the mass of the pile-deck system is

Table T7.3. Qualitative and quantitative description of the reported degrees of damage (after Uwabe, 1983; Gazetas et al., 1990).

(a) Qualitative description of the reported degrees of damage (Uwabe, 1983; Gazetas et al., 1990)

Degree of damage	Description of damage
0	no damage
1	negligible damage to the wall itself; noticeable damage to related structures (i.e. concrete apron)
2	noticeable damage to the wall itself
3	general shape of anchored sheet pile preserved, but significantly damaged
4	complete destruction, no recognizable shape of wall remaining

(b) Quantitative description of the reported degrees of damage for gravity quay wall (after Uwabe, 1983)

Degree of damage	Maximum residual displacement at top of wall (cm)	Average permanent displacement at top of wall (cm)
0	0	0
1	< 25	< 25
2	25 to 70	25 to 40
3	70 to 200	40 to 200
4	> 200	> 200

(c) Quantitative description of the reported degrees of damage for sheet pile quay wall (after Uwabe, 1983)

Degree of damage	Maximum residual displacement at top of sheet pile (cm)	Permanent displacement at top of sheet pile (cm)
0	0	0
1	<30	<10
2	30 to 100	10 to 60
3	100 to 200	60 to 120
4	>200	>120

(d) Quantitative description of the reported degrees of damage for sheet pile quay wall (after Gazetas et al., 1990)

Degree of damage	Permanent displacement at top of sheet pile (cm)
0	<2
1	10
2	30
3	60
4	120

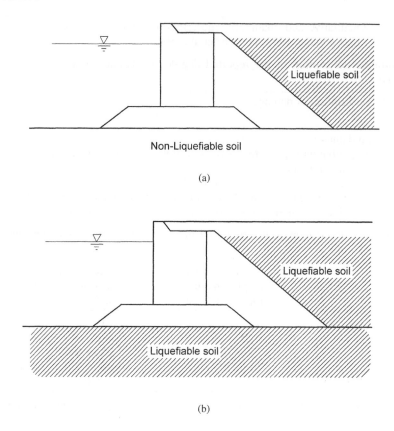

(a)

(b)

Fig. T7.6. Generalized soil conditions for gravity quay wall at liquefiable site.
         (a) Loose sand in backfill only.
         (b) Loose sand in both backfill and foundation.

concentrated. Stresses in the pile-deck system computed in the analysis are compared with the limit stresses at initiation of yield and at plastic hinge formation in
the piles (see Fig. T7.9). The seismic inertia force used in the simplified analysis
is typically a fraction of gravity, often specified in codes and standards.
Comparison of computed stresses in the pile-deck system with limit stresses may
result in an approximate estimate of the threshold level for pile-supported
wharves.

The embedded portion of the piles are typically idealized as a beam on a
Winkler foundation described by

$$EI\frac{d^4\Delta}{d\xi^4} = -P = -pD_p$$
(T7.18)

where
    EI: flexural rigidity (kNm^2)
    ξ: depth from the ground surface (m)

Δ: lateral displacement of a pile at the depth ξ (m)

P: subgrade reaction per unit length of a pile at the depth ξ (kN/m)

p: ($= P/D_p$) subgrade reaction per unit area of a pile at the depth ξ (kN/m^2)

D_p: pile diameter (or equivalent pile width) (m)

In North American practice, the relationships between p and Δ outlined in American Petroleum Institute (API) (1993), called p–y curves, have been widely

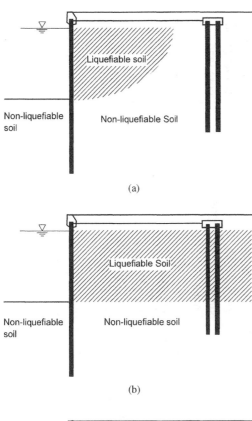

(a)

(b)

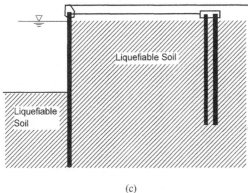

(c)

Fig. T7.7. Generalized soil conditions for sheet pile quay wall at liquefiable site.
(a) Loose sand behind the wall only.
(b) Loose sand backfill.
(c) Loose sand in both backfill and foundation.

Table T7.4. Case histories of seismic performance of retaining walls at liquefied sites.

Structure type/dimensions				Earthquake levels			Soil conditions/displacements		
Type wall	Port/quay	Seismic coefficient k_h	Water depth (m)	Earthquake/magnitude/year	PGA* (Gal)	Acceleration level	Soil conditions	Displacements d** (m)	d/H*** (%)
Gravity	Akita Port Gaiko	0.10	−13.0	Nihonkai-Chubu $M = 7.7$ 1983	205	E1	Loose sand at backfill only	1.50	9
Gravity	Kushiro Port West No. 1	0.20	−9.0	Kushiro-Oki $M = 7.8$ 1993	315	E1	Loose sand at backfill only	0.75	6
Gravity	Kushiro Port East Kita wharf	0.15	−8.0	Kushiro-Oki $M = 7.8$ 1993	315	E2	Loose sand at backfill only	2.00	18
Gravity	Kobe Port Rokko Island RC5	0.15	−14.0	Hyogoken-Nambu (Kobe) $M = 7.2$ 1995	502	E2	Loose sand at both backfill and foundation	5.23 (see Fig. T7.8)	29
Anchored sheet pile	Akita Port Ohama No. 2	0.10	−10.0	Nihonkai-Chubu $M = 7.7$ 1983	205	E1	Loose sand behind the wall only	1.72	14
Anchored sheet pile	Akita Port Ohama No. 3	0.10	−10.0	Nihonkai-Chubu $M = 7.7$ 1983	205	E1	Loose sand behind the wall only	0.82	7
Anchored sheet pile	Ishinomaki Port Shiomi wharf	0.10	−4.5	Miyagiken-Oki $M = 7.4$ 1978	281	E1	Loose sand at backfill	1.16	16
Anchored sheet pile	Hakodate Port Benten wharf	0.15	−8.0	Hokkaido-Nansei-Oki $M = 7.8$ 1993	111	E1	Loose sand at both backfill and foundation	5.21	46

* PGA: Peak ground acceleration converted to SMAC-B2 type equivalent.

** d: Horizontal displacement.

*** d/H: Horizontal displacement normalized with respect to the wall height from mudline.

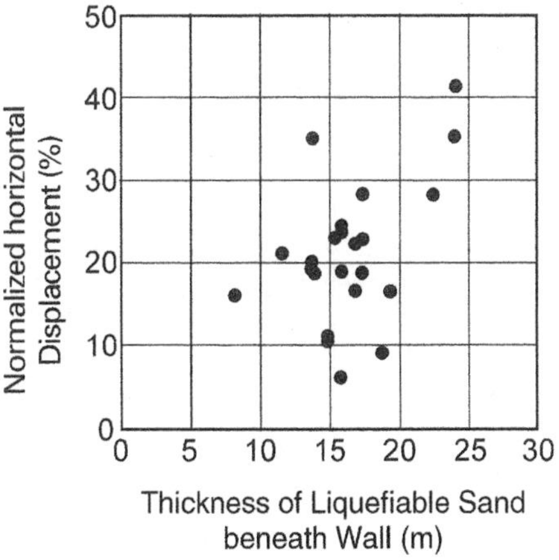

Fig. T7.8. Horizontal displacement normalized with respect to wall height, correlated with thickness of loose soil deposit below the wall (after Iai, 1998a).

Table T7.5. Normalized displacement of gravity walls at liquefied sites.

	Normalized displacement *d/H* (%)			
	0–5	5–10	10–20	20–40
During Design Level Earthquake Motion (E1)				
Non-liquefaction	■			
Loose sand at backfill only		■		
Loose sand at both backfill and foundation			■	
During 1.5 or 2.0 as high as E1 Motion (E2)				
Non-liquefaction		■		
Loose sand at backfill only			■	
Loose sand at both backfill and foundation				■

Table T7.6. Normalized displacement of anchored sheet pile walls at liquefied sites.
During Design Level Earthquake Motion (E1)

	Normalized displacement *d/H* (%)			
	0–5	5–15	15–25	25–50
Non-liquefaction	■			
Loose sand behind the wall only		■		
Loose sand at backfill including anchor			■	
Loose sand at both backfill and foundation				■

adopted for use at ports. The API equations evolved from Reese et al. (1974). The *p–y* curves for soft clay are given as shown in Fig. T7.10(a) and (b) and the relevant parameters, p_u, Δ_c, and ξ_r, are evaluated from standard laboratory tests or empirical correlations with soil parameters obtained from in-situ tests. The required geotechnical information consists of buoyant unit weight, γ_b (for a

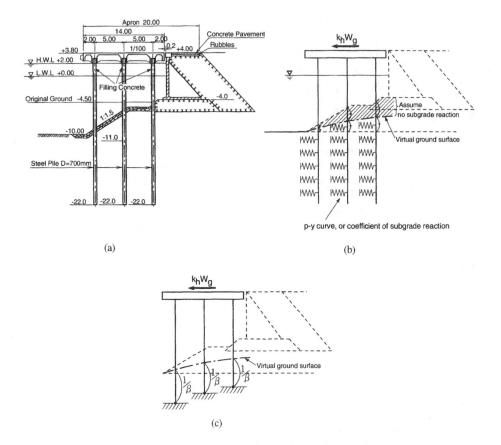

Fig. T7.9. Modelling of pile-deck system of a pile-supported wharf.
(a) Pile-supported wharf.
(b) Modelling by a frame structure supported by Winkler foundation.
(c) Modelling by a frame structure having equivalent lengths.

completely submerged soil deposit), undrained shear strength, c_u, from vane shear or triaxial tests, and the strain at 50% peak shear resistance, ε_c. The ultimate resistance, p_u, increases from $3c_u$ to $9c_u$ as ξ increases from 0 to ξ_r as shown in Fig. T7.10(c). The parameters Δ_c and ξ_r are given by

$$\Delta_c = 2.5\varepsilon_c D_p \tag{T7.19}$$

$$\xi_r = \frac{6D_p}{\dfrac{\gamma_b D_p}{c_u} + J} \tag{T7.20}$$

where J is a dimensionless empirical constant with values ranging from 0.25 to 0.5 determined from field testing.

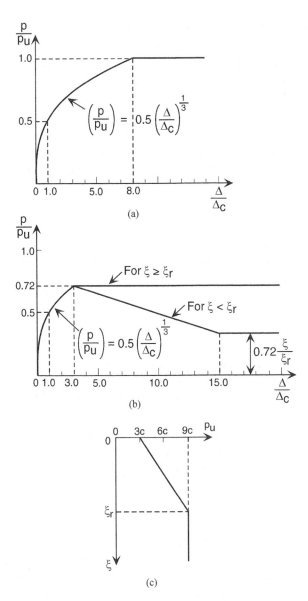

Fig. T7.10. *p–y* curves for soft
clay (API, 1993).
(a) Static loading.
(b) Cyclic loading.
(c) Variation of p_u
with depth.

The API *p–y* curve for sand is defined by

$$p = Ap_u \tanh\left(\frac{k_{sub}\xi}{AP_u}\Delta\right) \qquad\qquad\text{(T7.21)}$$

where

$A = 0.9$ for cyclic loading

$A = \left(3.0 - 0.8\dfrac{\xi}{D_p}\right) \geq 0.9$ for static loading

The required soil parameters include the buoyant unit weight of soil, γ_b, and the angle of internal friction, ϕ. The angle of internal friction is commonly estimated using empirical relationships based on in-situ penetration resistances (SPT or CPT). The relevant parameters, including the ultimate resistance, p_u, initial modulus of subgrade reaction k_{sub}, are then determined from relationships with ϕ (API, 1993).

In Japanese practice, (equivalent) linear or non-linear relationships outlined in Ministry of Transport, Japan (1989) have been widely adopted. The linear relationship, widely known as Chang's method in Japan, is defined by (Chang, 1937)

$$P = E_s \Delta = k_{\text{h-sub}} D_p \Delta \tag{T7.22}$$

where $E_s(= k_{\text{h-sub}} D_p)$: equivalent subgrade elastic modulus (kN/m^2) is assumed constant.

The non-linear relationships evolved from Kubo (1962) are defined as

$$p = k_{\text{s-type}} \xi \Delta^{0.5} \qquad \text{(for S-type subsoil)} \tag{T7.23}$$

$$p = k_{\text{c-type}} \Delta^{0.5} \qquad \text{(for C-type subsoil)} \tag{T7.24}$$

where S-type subsoil refers to that with linearly increasing SPT N-values with depth and C-type subsoil refers to that with constant SPT N-values with depth. The coefficients $k_{\text{h-sub}}$, $k_{\text{s-type}}$ and $k_{\text{c-type}}$ are correlated with SPT N-values (Ministry of Transport, Japan, 1989).

In both the North American and Japanese practices, these pseudo-static methods are applicable for cases involving piles embedded in (a) horizontal soil layers with negligibly small deformation relative to the displacement of the piles due to the inertia load applied at the deck, and (b) soils that do not lose appreciable strength during cyclic loading.

It is quite common to have pile-supported structures that are founded on sloping embankments. In order to take into account the effect of sloping embankments, a virtual (or a hypothetical) embankment surface, which is set below the actual embankment surface, may be assumed for evaluating the lateral resistance of piles embedded in an embankment as shown in Fig. T7.9(b) (Ministry of Transport, Japan, 1989).

The stresses in a pile-deck system can be computed either as a frame structure supported by linear or non-linear Winkler foundations (Fig. T7.9(b)), or a frame structure having 'equivalent fixity' piles (Fig. T7.9(c)). The lengths of 'equivalent fixity' piles from the ground or (virtual) embankment surface is typically given by $1/\beta$ (Ministry of Transport, Japan, 1989), where

$$\beta = \sqrt[4]{\frac{k_{\text{h-sub}} D_p}{4EI}} \tag{T7.25}$$

For a wharf with steel piles, an approximate elastic limit (i.e. threshold level), P_y, at which plastic hinges occur at the pile head in about half of all piles may be obtained based on the equivalent leg lengths of a pile-deck system, $1/\beta$, as (Yokota et al., 1999):

$$P_y = 0.82 P_u \qquad\qquad\qquad (T7.26)$$

$$P_u = \sum \frac{2M_{pi}}{l_i + 1/\beta_i} \qquad\qquad\qquad (T7.27)$$

where

    P_u: the limit lateral load when plastic hinges form at both the top and embedded portion of all piles

    M_p: pile bending moment at plastic hinge state

    l: unsupported length of a pile between the deck and virtual ground surface

    subscript i: i^{th} pile.

T7.1.4   Breakwaters, embankments and slopes

Sloping ground conditions exist throughout ports as natural and engineered embankments (e.g. rubble mounds for breakwaters, dikes for pile-supported wharves, dredged channel slopes, river levees). On-shore and submarine slopes in ports have been often found to be vulnerable to earthquake-induced deformations. Two of the primary causes for this are high water levels and weak foundation soils often encountered in port areas. In addition to waterfront slopes, several recent cases involving failures of steep, natural slopes along marine terraces located in backland areas have resulted in damage to coastal ports. Large-scale deformations of these slopes can impede shipping and damage adjacent foundations and buried structures, thereby limiting port operations following earthquakes.

In most cases involving soils that do not exhibit considerable strength loss due to shaking, common pseudo-static rigid body methods assuming slip surfaces will generally suffice for evaluating the stability of slopes in ports. These methods of evaluation are well established in the technical literature (Kramer, 1996). The seismic stability of slopes is evaluated by modifying static limit equilibrium analyses with the addition of a lateral seismic body force, that is the product of a seismic coefficient k_h similar to k_e previously defined and the mass of the soil bounded by the potential slip surface (see Fig. T7.11). This mass of soil should be modified with an added mass in order to account for the hydrodynamic force due to the presence of sea water. This increase can be derived, as a first approximation, by assuming a Westergaard's pressure distribution along the vertical face. The seismic coefficient k_h is specified as a fraction, roughly ½, of the peak

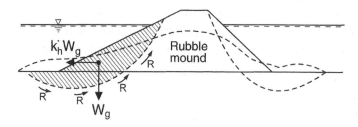

Fig. T7.11. Schematic figure for pseudo-static analysis of an embankment.

horizontal acceleration ratio a_{max}/g at the crest of an earth dam because the lateral inertial force is applied for only a short time interval during transient earthquake loading (Hynes–Griffin and Franklin, 1984; California Division of Mines and Geology, 1997).

The pseudo-static methods outlined above are simple to use, but require a high level of engineering skill and judgement when dealing with the following issues: (a) the influence of excess pore pressure generation on the strength of the soils, including cyclic progressive deformation and strain softening; (b) the range or pattern of slope deformations that may be associated with various factors of safety; and (c) the effect of the elasticity and permeability of the materials used for breakwaters and embankments, that produces hydrodynamic pressures deviating from the classical Westergaard's solution. With the reasonable application of engineering judgement, the pseudo-static methods are routinely used for screening purposes, and can be useful in evaluating the approximate threshold level of slope stability.

T7.2   SIMPLIFIED DYNAMIC ANALYSIS

T7.2.1   Gravity quay walls

(1) Sliding block analysis

Simple, straightforward methods have been developed for evaluating the permanent displacements of gravity walls during earthquakes (e.g., Newmark, 1965; Franklin and Chang, 1977; Richards and Elms, 1979; Whitman and Liao, 1985). These methods are based on the sliding block model, adopting similar assumptions to those made for the pseudo-static analysis outlined in Section T7.1.1.

First, the stability of the wall and the backfill are evaluated using the lateral earth pressure theory based on Mononobe-Okabe's equations. The threshold acceleration is determined by the value resulting in a factor of safety of unity for sliding of the wall-backfill system. For example, the threshold acceleration, a_t, for a vertical retaining wall is given by the expression (Richards and Elms, 1979)

$$a_t = \left(\mu_b - \frac{P_{ae} \cos \delta - \mu_b P_{ae} \sin \delta}{W_g} \right) g \qquad \text{(T7.28)}$$

where μ_b is the coefficient of interface friction between the wall and the foundation rubble or soil, P_{ae} is the active earth thrust computed using the Mononobe-Okabe method, δ is the wall-backfill interface friction angle, W_g is the weight of the wall per unit width, and g is the acceleration of gravity. It should be noted that a_t must be known in order to calculate P_{ae}, therefore an iterative procedure is necessary.

Once the threshold acceleration has been determined, then a set of acceleration time histories are selected for sliding block analysis. The use of more than one acceleration time history is necessary because displacements calculated through sliding block analysis are sensitive to the characteristics of acceleration time history used in the analysis. The acceleration time histories should be representative of the design level ground motions in both duration and frequency content. When the ground motion acceleration exceeds the threshold acceleration, a_t, the wall-backfill system begins to move by translation along the base of the wall and the failure plane through the backfill. By double integrating the area of the acceleration time history that exceeds a_t, and continuing the time integration until the sliding stops, the displacement of the wall relative to the firm base below the failure plane can be determined as shown in Fig. T7.12. This computation can be easily performed using common spreadsheet routines or a simple computer code.

Numerical studies based on the sliding block method of analysis have led to the development of simplified relationships between ground motion intensity and seismically-induced deformations. Based on a large database of nearly 200 earthquake records, conservative envelopes for evaluating sliding displacements were obtained by Franklin and Chang (1977) as shown in Fig. T7.13. These results were obtained for earthquake motions with a peak acceleration and velocity of $0.5g$ and 76 cm/s, respectively.

The sliding block theory suggests that the vertical component of input acceleration time histories should affect the sliding displacement. However, a parameteric study, using an earthquake motion recoded at Kobe Port Island site (PGA$_H$ = 544 cm/s^2, PGA$_V$ = 200 cm/s^2), suggests that the effect of the vertical component is not significant, and it either increases or decreases the sliding displacement less than about 10 percent (Nagao et al., 1995).

A simplified expression to approximate these envelopes was proposed by Richards and Elms (1979) as

$$d = 0.087 \frac{v_{max}^2 a_{max}^3}{a_t^4} \qquad \text{(T7.29)}$$

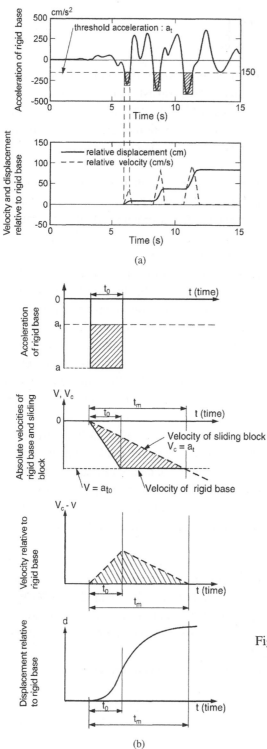

(a)

(b)

Fig. T7.12. Computing displacement in the sliding block analysis.
(a) Example of sliding block analysis.
(b) Acceleration, velocities and displacement from sliding block analysis.

where d is the permanent lateral displacement (cm), v_{max} is the peak ground velocity (cm/s), and a_{max} is the peak ground acceleration (cm/s^2). This expression is considered valid for cases where the ratio a_t/a_{max} is greater than or equal to 0.3.

As mentioned earlier, the computed displacement from sliding block analysis is sensitive to a number of key parameters, including those for evaluating threshold acceleration and the nature of the earthquake motion. Indeed, the scatter of displacements computed by Franklin and Chang (1977) covers a range which spans to about a factor of ten. The sensitivity is evaluated further by Whitman and Liao (1985), taking into account the effects of factors such as the dynamic response of the backfill, kinematics of backfill wedge, tilting of the wall and vertical accelerations on the computed wall deformations. Based on the results of sliding block analyses of 14 ground motions, the following relation was derived to define the mean value:

$$\overline{d} = \left(\frac{37 v_{max}^2}{a_{max}} \right) \exp\left(-\frac{9.4 a_t}{a_{max}} \right) \tag{T7.30}$$

This expression forms the basis for a statistical analysis of the likely sources of error, leading to an alternative recommended design line based on a safety factor of 4, equivalent to a 95% probability of non-exceedance (see Fig. T7.13)

$$\frac{a_t}{a_{max}} = 0.66 - \frac{1}{9.4} \ln \frac{d a_{max}}{v_{max}^2} \tag{T7.31}$$

The design recommendation that has been proposed in the tentative Eurocode (CEN, 1994) provides the horizontal and vertical seismic coefficients through the relations

$$k_h = \frac{a_{design}/g}{r_{EC}}$$
$$k_v = 0.5 k_h \tag{T7.32}$$

where a_{design} is the design acceleration. Values of the reduction factor r_{EC} are related to the allowable displacements given in Table T7.7.

For the designer, these expressions for evaluating the likely sliding displacement of a retaining wall have considerable merit. However, some cautionary notes are necessary. Actual seismic performance of gravity quay walls during earthquakes often does not meet the assumptions inherent in sliding block analysis. Where the movement of the wall is associated with significant deformation in the foundation soils, the displacements computed by the sliding block approach were substantially smaller than the displacements observed in the field (Iai, 1998a; Yang, 1999). Where the foundation is firm but rocking type response of the wall is involved, the design curves discussed earlier are again found to be unconservative (Steedman and Zeng, 1996). For liquefiable backfill, other techniques

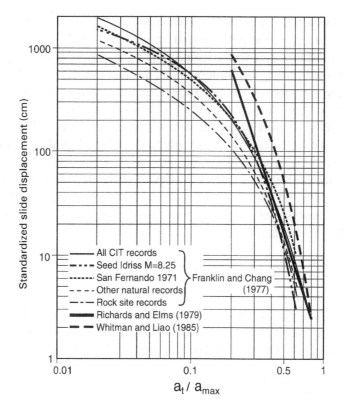

Fig. T7.13. Proposed simplified results for evaluating slide displacement (after Franklin and Chang, 1977; Richards and Elsm, 1979; Whitman and Liao, 1985; as reported by Steedman, 1998).

Table T7.7. Sliding displacement (CEN, 1994).

Type of retaining structure	r_{EC}
Free gravity wall that can accept a displacement up to $30(a_{design}/g)$ (cm)	2
As above with a displacement up to $20(a_{design}/g)$ (cm)	1.5
Other retaining structures without allowing a displacement	1

that enhance limit equilibrium-based methods for predicting seismic displacements are recommended (e.g., Byrne et al., 1994). Consequently, it is important to verify that the design conditions of interest meet the assumptions inherent in sliding block analysis, e.g. rigid firm ground, wall sliding without tilt, and rigid wedge-like backfill movement.

(2) Simplified chart based on parametric study

In order to enhance the applicability of the simplified dynamic analysis for gravity quay walls for general geotechnical conditions, seismic performance of gravity quay walls was studied through effective stress analysis by varying structural and geotechnical parameters of a quay wall under various levels of shaking (Iai et al., 1999). Major parameters studied are width to height ratio of a gravity wall, W/H, thickness of soil deposit below the wall, $D1$, and geotechnical conditions represented by SPT N-values of subsoil below and behind the wall (see Fig. T7.14). The effective stress analysis was based on the multiple shear mechanism using the computer program FLIP (Finite Element Liquefaction Program), which has been demonstrated as applicabile to the seismic analysis of gravity quay walls during the 1995 Hyogoken-Nambu (Kobe) earthquake (Iai, 1998b).

Some of the details assumed for simplicity in the analysis include:

a) The geotechnical conditions of the soil deposits below and behind the wall were assumed to be identical, represented by an equivalent SPT N-value N_{65} (SPT N-value corrected for the effective vertical stress of 65 kPa). The equivalent SPT N-value has been widely used for assessment of liquefaction potential in Japanese port areas (e.g. Port and Harbour Research Institute, 1997). SPT practice in Japan typically corresponds to the energy ratio of 73%, about 20% higher than the practice in the USA, where the SPT N-value is typically normalized to the energy ratio of 60% as in $(N_1)_{60}$. With the additional difference in the effective vertical stresses of 65 kPa and 100 kPa for correction, $(N_1)_{60}$ is about 1.5 times the equivalent SPT N-value.

b) Model parameters for the effective stress analysis were determined from the equivalent SPT N-values based on a simplified procedure (Morita et al., 1997).

c) The peak acceleration of the input seismic excitation was assigned at the base layer as an incident wave (as of $2E$ rather than $E + F$). The time history of the earthquake excitation was that of the incident wave at the Port Island (Kobe) vertical seismic array site at a depth of -79 m during the 1995 Hyogoken-Nambu earthquake.

d) For simplicity, the thickness of backfill, $D2$, was assumed the same as the wall height, H.

The results of the parametric study are summarized in terms of residual horizontal displacement of the wall, as shown in Figs. T7.15 through T7.18. The most sensitive parameter was the SPT N-value of subsoil below and behind the wall. The second was the thickness of the soil deposit below the wall. Although the width to height ratio of a gravity wall is a sensitive parameter for a quay wall with firm competent foundation, the effect of this parameter becomes less obvious when the soil deposit below the wall becomes thick or soft.

Based on the results of the parametric study, a simplified procedure was developed for evaluating the residual horizontal displacement of a gravity quay wall under seismic excitations. In this procedure, the displacement may be evaluated based on the relevant parameters as follows:

1) Normalized residual horizontal displacement, d/H, is evaluated based on equivalent SPT N-value for a prescribed level of shaking at the base by referring to Fig. T7.15 or Fig. T7.16.

2) Effects of thickness of soil deposit below the wall, $D1/H$, is corrected by referring to Fig. T7.17.

3) Effects of width to height ratio, W/H, is corrected by referring to Fig. T7.18.

The applicability of this procedure was confirmed with the case histories from
the 1993 Kushiro-Oki earthquake and the 1995 Hyogoken-Nambu (Kobe) earthquake (Iai et al., 1999). As is the case with other simplified procedures, the engineer should confirm the applicability of these charts based on relevant case
history data and modify the charts accordingly before use in design practice.

(3) Evaluation of liquefaction remediation based on parametric study

The heightened awareness of the potential impact of liquefaction-induced ground
failures on waterfront retaining structures and the corresponding economic impact
of such damage has led to the increased utilization of soil improvement for mitigating liquefaction hazards. All other factors being equal, the effectiveness of the
ground treatment is a function of the level of densification or cementation and the
volume of soil that is treated. Soil improvement techniques have been used at
numerous ports throughout the world (e.g. Iai et al., 1994). Although few case histories exist for the performance of improved soils subjected to design level earthquake motions, past experience at waterfront sites has shown that caissons in
improved soils have performed much more favorably than have adjacent caissons
at unimproved sites which experienced widespread damage (e.g. Iai et al., 1994;
Mitchell et al., 1995).

Design guidelines are available for specifying the extent of soil improvement
adjacent to retaining walls (Port and Harbour Research Institute, 1997). The
method is based on the standard pseudo-static analysis familiar to most engineers
(e.g. Ebeling and Morrison, 1992). In addition to the lateral seismic coefficient, a
dynamic pressure is added to account for earth pressures due to liquefied soil acting at the boundary of the improved soil zone. The improved soil zone can be

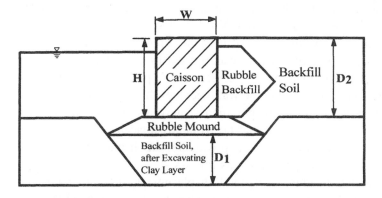

Fig. T7.14. Typical cross section of a gravity quay wall for parameter study.

determined by accounting for the width of the active and passive failure surfaces which pass through the backfill or soils below the dredge line, and adding additional width of improvement as a margin of safety. For structures underlain by liquefiable soils, the improved zone must extend deep enough to preclude failures due to loss of soil bearing capacity or deep seated rotational failure.

In order to develop a simplified technique for estimating seismically-induced deformations of gravity walls (specifically gravity caissons), a parametric study was conducted using a 2D numerical non-linear effective stress model, validated with well-documented case histories (Dickenson and Yang, 1998; Yang, 1999). The results of the study have been synthesized into practice-oriented design charts for estimating the lateral deformations of gravity quay walls. The walls were designed with heights ranging from 12 m to 19 m, and height-to-width (H:W) ratios of 0.7 and 1.2. The H:W ratios correspond to horizontal seismic coefficients of approximately 0.1 to 0.25, the range of interest for most port engineers. The vertical seismic coefficient was not incorporated into the analyses. The configuration of the model caisson and the surrounding soil is shown in Fig. T7.19. The foundation and backfill soils are both cohesionless materials, each layer modelled with uniform density. The range of densities used can be correlated with the normalized SPT N-values, $(N_1)_{60}$, of 25 in the foundation soil, 10 to 20 in unimproved backfill, and 30 in improved backfill.

Five earthquake motions covering the magnitude range of engineering interest (M 6 to 8) were selected for the parametric study. Each record was scaled to three

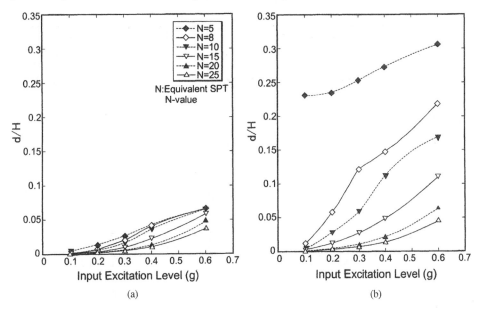

Fig. T7.15. Effects of input excitation level (for $W/H = 0.9$).
      (a) $D1/H = 0.0$.
      (b) $D1/H = 1.0$.

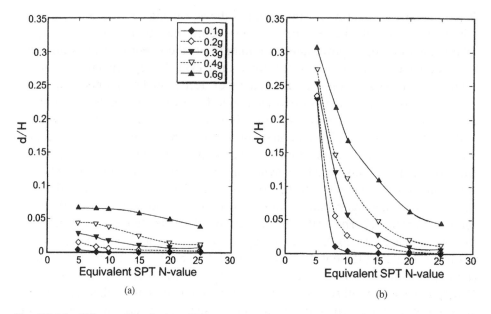

Fig. T7.16. Effects of equivalent SPT *N*-value (for *W/H* = 0.9).
       (a) *D1/H* = 0.0.
       (b) *D1/H* = 1.0.

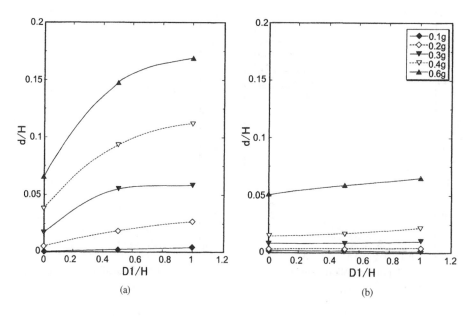

Fig. T7.17. Effects of thickness of soil deposit below the wall (for *W/H* = 0.9).
       (a) Equivalent SPT *N*-value (10).
       (b) Equivalent SPT *N*-value (20).

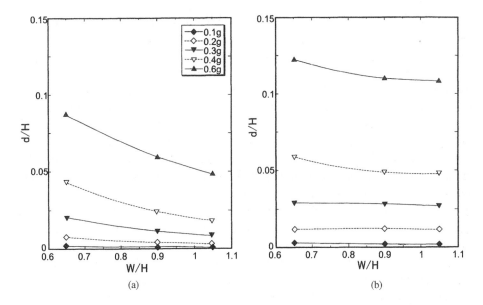

Fig. T7.18. Effects of width to height ratio W/H (for equivalent SPT N-value of 15).
         (a) $D1/H = 0.0$.
         (b) $D1/H = 1.0$.

different maximum acceleration values ranging from $0.1g$ to $0.4g$. The lateral and vertical deformations of the wall and surrounding soil, as well as the excess pore pressures in the backfill were monitored during seismic loading. Notable assumptions in the parametric study include; (a) no pore pressure generation in the foundation or improved backfill soils with $(N_1)_{60} \sim 30$, and (b) the soil improvement extends to the base of the backfill (i.e. the same elevation as the dredge line) depth.

In order to account for the duration of the earthquake motions, a normalized ground motion intensity has been used. This parameter is defined as the maximum horizontal acceleration within the backfill at the elevation of the dredge line $(a_{\max(d)})$ divided by the magnitude scaling factor (MSF) listed in Table T7.8 (Arango, 1996; Youd and Idriss, 1997). It is recommended that, if a site specific seismic study is not performed to determine $a_{\max(d)}$, the peak ground surface acceleration be reduced using the reduction factor, r_d, developed for estimating the variation of cyclic shear stress (or acceleration) with depth (e.g. Makdisi and Seed, 1978). The values of r_d for 15 and 7.5 m walls are approximately 0.78 and 0.95, respectively. It should be noted that the reduction factor was developed using one-dimensional dynamic soil response methods, which will yield approximate acceleration values for the two-dimensional soil-structure interaction applications discussed herein.

The results of the parametric study are shown in Fig. T7.20. The lateral deformations at the top of the wall are plotted versus the width of the improved soil, and as functions of backfill density and the H:W ratios of the caissons. This set of curves can be used for estimating the lateral deformations of gravity caissons in unimproved soils, as well as at improved sites.

The numbered triangles superimposed on the chart correspond to field case histories. Information pertaining to the strength of shaking, the wall geometry and the observed displacements are provided. The relationships provided in Fig. T7.20 demonstrate the benefit of mitigating liquefaction hazards to gravity retaining walls. It is also evident that the incremental benefit of a wider zone of ground treatment begins to decline once the soil improvement extends more than about two to three times the total height of the wall. At this point, the cost of additional soil improvement may outweigh the benefits. It is interesting to note that the soil improvement guidelines prepared by the PHRI (1997) based on shake table tests, correspond to a normalized width of soil improvement of roughly 1.7 to 2.0.

The recommended procedures for utilizing the results of the parametric study to estimate the permanent displacement at the top of caissons include:

1) Design the wall using standard pseudo-static limit equilibrium methods to determine the wall geometry, H:W,

2) Determine $a_{max(d)}$ with a site-specific seismic study or approximate with empirical soil amplification factors to yield the peak ground surface acceleration and the reduction factor, r_d, recommended by Makdisi and Seed (1978),

3) Determine the magnitude scaling factor for the specified earthquake magnitude, and multiply by $a_{max(d)}$ to yield the intensity, and

4) Based on the SPT N-values, $(N_1)_{60}$, of the backfill soils, the width of the ground treatment behind the caisson, and the ground motion intensity factor, enter Fig. T7.20 and obtain the normalized lateral displacement, d/H. From this, the lateral displacement at the top of the wall, d, can be estimated. This lateral deformation includes the contribution of both sliding and tilting on the overall movement at the top of the wall.

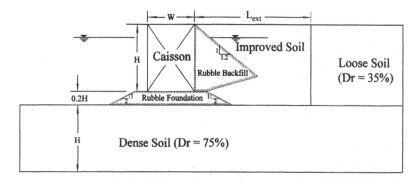

Fig. T7.19. Cross section of model wall.

Table T7.8. Magnitude scaling factors (Arango, 1996).

	Magnitude					
	5.50	6.00	7.00	7.50	8.00	8.25
MSF	3.00	2.00	1.25	1.00	0.75	0.63

T7.2.2 Sheet pile quay walls

(1) Sliding block analysis

In order to evaluate the order-of-magnitude displacement of sheet pile walls based on sliding block analysis, the procedure used for the gravity quay walls discussed in T7.2.1 poses two difficulties. First, the active earth pressure may not be accurate because of the flexible nature of the sheet pile wall. Second, the inertia force on and bottom friction at the wall, i.e. two of the major parameters for gravity quay wall analysis, are not the major parameters in the sheet pile-backfill soil system. These problems are solved by idealizing the movement of the sheet pile-backfill system, assuming united rigid block motion of the wall and backfill (Towhata and Islam, 1987). As shown in Fig. T7.1(b), the driving force is the inertia force acting on the soil wedge and the resistance force includes the passive earth pressures in front of the wall and the shear resistance along the failure surface of the soil wedge.

By expressing the dynamic earth pressure coefficients K_{ae} and K_{pe} and the ultimate anchor resistance T_e as functions of K_a, K_p and T_s before the earthquake (Seed and Whitman, 1970)

$$K_{ae} = K_a + \Delta K_{ae} = K_a + \frac{3}{4} k_h \tag{T7.33}$$

$$K_{pe} = K_p + \Delta K_{pe} = K_p - \frac{17}{8} k_h \tag{T7.34}$$

$$T_e = T_s + \frac{K_{pe} - K_{ae}}{K_p - K_a} \tag{T7.35}$$

the threshold horizontal seismic coefficient is obtained as (Towhata and Islam, 1987)

$$k_t = \frac{a \tan \alpha_{ae} - b + \tan(\phi - \alpha_{ae})(1 + b \tan \alpha_{ae})}{1 + c \tan \alpha_{ae}} \tag{T7.36}$$

where

$$a = \frac{m T_s + P_p + \frac{1}{2} \gamma_w (H_w + D_{emb})^2 + \Delta U_p}{W_m} \tag{T7.37}$$

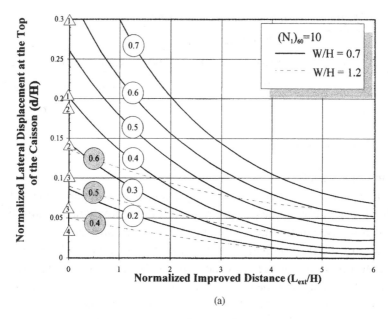

(a)

Case No.	$a_{max(d)}$	$a_{max(d)}/MSF$	d/H	W/H
1. Rokko Island	0.54	0.42	0.21 ~ 0.31	0.7
2. Port Island	0.29	0.22	0.14 ~ 0.19	0.7
3. Akita	0.19	0.19	0.06 ~ 0.10	0.75
4. Kushiro	0.19 (SHAKE)	0.19	0.02 ~ 0.04	1.03

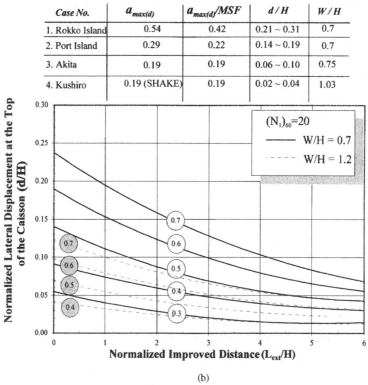

(b)

Fig. T7.20. Normalized lateral displacement at the top of the caisson.
(a) Normalized earthquake intensity ($a_{max(d)}/MSF$) for backfill soil SPT $(N_1)_{60} = 10$.
(b) Normalized earthquake intensity ($a_{max(d)}/MSF$) for backfill soil SPT $(N_1)_{60} = 20$.

$$b = \frac{\frac{1}{2}\gamma_w (H_w + D_{emb})^2 \tan\phi + \Delta U_a \sin\alpha_{ae}}{W_m} \tan\psi \qquad (T7.38)$$

$$c = \frac{1}{W_m} \left[\frac{23mnT_s}{8(K_p - K_a)} + \frac{17P_p\gamma_{sat}}{8K_p\gamma_b} + \frac{7}{12}\gamma_w H_w^2 \right] \qquad (T7.39)$$

$$W_m = \tfrac{1}{2}[\gamma_{sat}(H_w + D_{emb})^2 + \gamma_{wet}(H - H_w)(H + 2D_{emb} + H_w)] \qquad (T7.40)$$

where

m = 0 for no anchor capacity, 1 for full anchor capacity (dependant on the amount of excess pore pressure generation around the anchor)

P_p = static passive earth thrust

ΔU = excess pore water pressure due to cyclic shearing; subscripts a and p denote pressures within the active and passive soil wedges

n = 1 when the anchor block is above the water table, and γ_t/γ_b when the anchor is completely submerged

For definitions of the rest of the parameters, refer to Fig. T7.1(b). Here, it is assumed that the ground water table behind the wall is the same as that of sea water surface (i.e. $H_w = H_{sub}$).

Using the threshold seismic coefficient k_t defined for the sheet pile-backfill system, displacement of the wall is computed based on Newmark's sliding block approach.

(2) Simplified chart based on parametric study

In order to enhance current screening tools for deformation-based design of anchored sheet pile quay walls, a parametric study using a calibrated 2D numerical effective stress model was conducted to examine the performance of sheet pile quay walls (McCullough and Dickenson, 1998; McCullough, 1998). In the parametric study, sheet pile walls were designed with heights ranging from 7.5 to 15 m, and with horizontal seismic coefficients ranging from 0.1 to 0.16. The design procedure compiled by Ebeling and Morrison (1992) was used for sheet pile quay wall design. The coefficients were limited to 0.16, because of the unpracticable anchor requirements required for the 15 m sheet pile walls with coefficients of approximately 0.2 or larger. The vertical seismic coefficient was specified to be zero. The tie-rod length was such that the active failure plane of the wall did not intersect the passive failure plane of the anchor, with the active and passive failure planes originating at the base of the wall and anchor, respectively. The anchor was a continuous sheet pile wall, with the same stiffness as the sheet pile wall. The tie rod spacing was two meters. The distances from the ground surface to the ground water table and the tie rod were 27 percent of the wall height, H.

The foundation and backfill soils were modeled as cohesionless materials, with the engineering properties represented by average SPT N-values. A major aspect of the study was modelling the effectiveness of soil improvement on the computed wall deformation. The normalized SPT N-values $(N_1)_{60}$ for the improved and dense foundation soils were 30 and 25, respectively, while $(N_1)_{60}$ of the backfill was either 10 or 20.

Five earthquake motions were chosen for use in the parametric study in a similar manner as for gravity quay walls, discussed in Section T7.2.1(3). Some assumptions made for the parametric study include (a) no pore pressure generation in the foundation or improved soils, (b) the soil improvement extends to the depth of the backfill layer, and (c) the tie-rod/whale system will not fail. Based on approximately 100 parametric models, sensitivity of five design parameters were evaluated: (a) depth of embedment of the sheet pile wall, D_{emb}, (b) stiffness of the sheet pile wall, EI, (c) length of the tie rod, (d) density of the backfill soil, and (e) extent of soil improvement, L_{ext}. A definition sketch of the modeled geometry is provided in Fig. T7.21.

To increase the applicability of this study, normalized dimensionless factors were developed. A normalized displacement factor shown in Eqn. (T7.41) was developed by normalizing the displacements at the top of the wall, d, by the wall stiffness, EI, total wall height, $H + D_{emb}$, and the buoyant unit weight of the backland soil adjacent to the sheet pile wall.

$$\frac{d \cdot EI}{(H + D_{emb})^5 \cdot \gamma_b} \tag{T7.41}$$

A normalized soil improvement factor, n_{ext}, was also developed by dividing the extent of soil improvement, L_{ext}, by the total wall height, $H + D_{emb}$.

The results of the parametric study are presented in Fig. T7.22. The contour lines indicate various levels of earthquake intensity for backfill soils with $(N_1)_{60}$ of 10 and 20. The normalized earthquake intensity was defined using the same procedure as that for gravity quay walls in Section T7.2.1(3). The effectiveness of soil improvement for minimizing quay wall deformations is clearly demonstrated by the design chart. It is also noted that the incremental benefit of soil improvement beyond n_{ext} values of approximately 2.0 decreases considerably. In comparison, the n_{ext} values as determined from the PHRI (1997) recommendations for the parametric study sheet pile quay walls are approximately 1.9 to 2.5.

There are fourteen case histories plotted on the chart, which are arranged according to the SPT N-values $(N_1)_{60}$ of the backfill soils. Table T7.9 tabulates the case histories. It should be noted that seven case histories are closely predicted by the chart, five case histories are significantly over-predicted and only two of the case histories are significantly under-predicted. These results indicate that the design chart can be conservatively used as a preliminary design chart or screening tool.

The recommended procedures for utilizing the results of the parametric study to estimate the permanent displacement at the top of sheet pile walls include:

1) Design the wall using current pseudo-static methods to determine the wall geometry (H, D_{emb}, EI, and anchor length),
2) Determine $a_{max(d)}$ from a site-specific seismic study or an approximate empirical relationship,
3) Determine the earthquake intensity factor for the specific earthquake by dividing the magnitude scaling factor into $a_{max(d)}$,
4) Based on the density of the backfill and the extent of soil improvement, estimate the permanent lateral displacement at the top of the sheet pile wall, d, using Fig. T7.22 and the normalized displacement equation (Eqn. T7.41).

The results of the parametric study for non-liquefiable sites can be compared to the design chart by Gazetas et al. (1990), previously addressed in Section T7.1.2(1), and shown in Fig. T7.23. The parametric study results for non-liquefiable sites are shown by solid stars in this figure, with numbers beside the solid stars indicating the computed displacements. One group of solid stars (center-right) is for a 7.5 m wall geometry. Two groups (top-right and bottom-right) are from the parametric study varying the length of the tie-rod and anchor, and the last group (left-center) comes from the parametric study varying the depth of embedment. It is evident from Fig. T7.23, that the computed displacements follow a trend that is roughly consistent with the relationship developped by Gazetas et al. (1990); i.e. displacement increases for smaller *EAI* and *EPI*. This indicates that the chart by Gazetas et al. (1990) is adequate as a crude, first-order evaluation of sheet pile walls. However, the parametric study indicates the following limitations in the chart by Gazetas et al. (1990):

1) Although the chart by Gazetas et al. (1990) indirectly includes effects of earthquake motion parameters through *EAI* and *EPI*, the sheet pile wall displacement can be more sensitive to the earthquake motion parameters as indicated by the parametric study.
2) The incremental benefit of sheet pile embedment and tie-rod length beyond the lengths normally specified in seismic design are not as substantial as indicated by the chart by Gazetas et al. (1990).

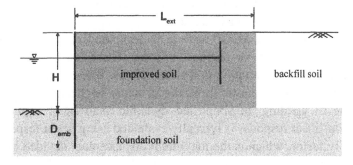

Fig. T7.21. Definition sketch of an anchored sheet pile bulkhead with soil improvement.

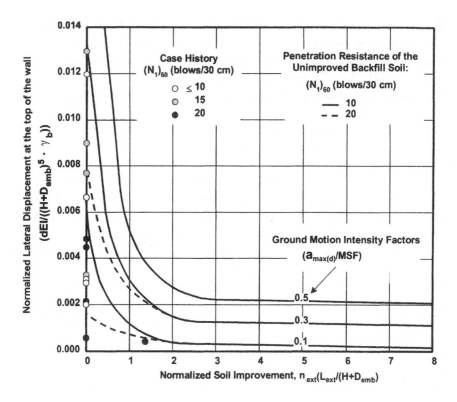

Fig. T7.22. Permanent horizontal displacements at the top of anchored quay walls.

T7.2.3   Pile-supported wharves

(1) Response spectrum analysis

In simplified dynamic analysis, a pile-supported wharf is idealized by a single degree of freedom (SDOF) system. A SDOF system consists of a mass, a spring and a damper as shown in Fig. T7.24. The mass is typically that of the deck and surcharge. The spring is typically computed based on the equivalent pile-deck framework assumed for pseudo-static analysis discussed in Section T7.1.3 or through pushover analysis discussed later.

Dynamic response of a SDOF system is computed using time histories of design earthquake motions or response spectra. These routines are well established in the technical literature (e.g. Kramer, 1996). When the dynamic response of a structure exceeds the elastic limit, the structure begins to exhibit non-linear response due to yielding. In simplified dynamic analysis of a pile-supported wharf, the non-linear response is typically evaluated using linear response spectra and a ductility factor, which is the maximum displacement devided by yield displacement. This approach is widely used for designing buildings and bridges. Some of the fundamental issues relevant to this approach will be reviewed below.

Table T7.9. Case histories of sheet pile quay walls.

Earthquake	$(N_1)_{60}$	$a_{max(d)}/MSF$ (g)	Displacement (cm)	Normalized displacement $d \cdot EI/((H+D_{emb})^5 \cdot \gamma_b)$
1968 Tokachi-Oki	6	0.26	12 to 23	0.0120
1993 Kushiro-Oki	6	0.20	19	0.0020
1964 Niigita	10	0.14	100	0.0031
1983 Nihonkai-Chubu	10	0.10	110 to 160	0.0066
1968 Tokachi-Oki	15	0.23	16 to 57	0.0090
1968 Tokachi-Oki	15	0.23	12 to 19	0.0029
1968 Tokachi-Oki	15	0.12	30	0.0077
1973 Nemuro-Hanto	15	0.16	30	0.0031
1978 Miyagi-Ken-Oki	15	0.10	87 to 116	0.0129
1968 Tokachi-Oki	20	0.26	~60	0.0049
1983 Nihonkai-Chubu	20	0.09	~5	0.0006
1993 Kushiro-Oki	20	0.16	50 to 70	0.0044
1993 Guam	20	0.18	61	0.0019
1993 Kushiro-Oki	30	0.21	no damage (~5)	0.0004

In order to approximate the effect of elasto-plastic response, the following assumption is often made: the energy stored in a SDOF system is the same for linear and elasto-plastic systems. This assumption, called the constant energy assumption, is illustrated in Fig. T7.25. Consider a linear system shaken by an earthquake motion resulting in a response with a maximum displacement d' and a maximum force F'. Consider a response of an elasto-plastic system with elastic limit of F_y undergoing the same earthquake input motion resulting in a response with a maximum displacement d_m. With the constant energy assumption, the maximum displacement d_m of the elasto-plastic system is determined in such a way that the area of the triangle OA'B' is the same as that of OYAB.

Energies stored for linear and elasto-plastic systems are $(1/2)F'd'$ and $(1/2)F_y d_y(2\mu-1)$, where $\mu = d_m/d_y$ denotes displacement ductility. By introducing a spring coefficient k for the linear system, it follows that $F' = kd'$ and $F_y = kd_y$, so that the constant energy assumption demands

$$\frac{1}{2k}(F')^2 = \frac{1}{2k}(F_y)^2(2\mu-1) \qquad (T7.42)$$

This equation reduces to the following relationship.

$$F_y = \sqrt{\frac{1}{2\mu-1}}F' \qquad (T7.43)$$

Since inertia force is proportional to acceleration, Eqn. (T7.43) is converted to a relation between peak response accelerations of linear and elasto-plastic systems a' and a_y as follows.

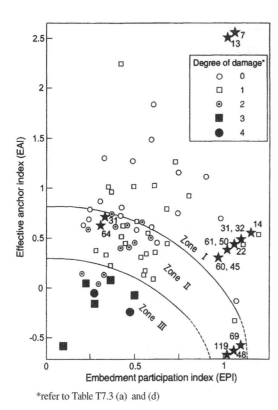

Fig. T7.23. Case history data of sheet pile quay walls, including data from the parametric study for models improved sites (solid stars with computed lateral deformations (cm)).

$$a_y = \sqrt{\frac{1}{2\mu - 1}} a' \qquad (T7.44)$$

A parallel relationship for displacements of yield limit and maximum response are given by the definition of ductility factor as follows.

$$d_m = \mu d_y \qquad (T7.45)$$

Since residual displacement d_r of the elasto-plastic system is given by $d_r = d_m - d_y$, Eqn. (T7.45) demands $d_r = (\mu - 1)d_y$. By introducing a modification factor c to correct for the difference between dynamic response and the constant energy assumption, the residual displacement d_r is given by

$$d_r = c(\mu - 1)d_y \qquad (T7.46)$$

For example, the correction factor for a bridge supported by concrete piers is typically $c = 0.6$ (Japan Road Association, 1996).

The allowable (displacement) ductility factor μ for use in design is generally determined from ultimate displacement d_u with allowing a safety margin; typically $\mu = 2$ to 4 for bending and $\mu = 1$ for shear. For further discussions on ductility criteria, see Section 4.3 in the Main Text.

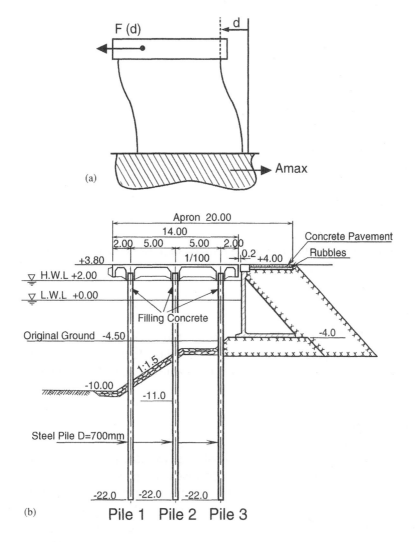

Fig. T7.24. Modelling of a pile-supported wharf for simplified dynamic analysis.
  (a) SDOF system.
  (b) Pile supported wharf.

The ultimate displacement d_u is governed by ultimate curvature ϕ_u for piles having a plastic hinge length L_p with a free pile length L as follows (Priestly et al., 1996):

$$d_u = d_y + 3(\phi_u - \phi_y)L_p(L - 0.5L_p) \tag{T7.47}$$

Plastic hinge length L_p is proportional to a diameter of pile D_p or given as $L_p = 0.1$ to $0.5D_p$ for a concrete pile (Japan Road Association, 1996).

Equation (T7.47) forms the basis to relate the ductility limit specified in curvature with displacement ductility limit μ. The curvature ductility limit is further

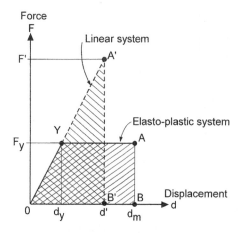

Fig. T7.25. Load-displacement response of elasto-plastic system evaluated with constant energy assumption.

related to the strain limit through the relationship between the extreme fibre strain and the curvature as

$$\varepsilon = \frac{D_p}{2} \phi \qquad (T7.48)$$

Equations (T7.47) and (T7.48) were used as the basis to develop the damage criteria for piles discussed in Section 4.3 in the Main Text. For further details, refer to Priestly et al. (1996), Ferritto (1997a), Ferritto et al. (1999), and Yokota et al. (1999).

(2) Pushover analysis

In simplified dynamic analysis, the load-displacement relationship for the pile-deck system shown in Fig. T7.25, is evaluated by pushover analysis. Pushover analysis is accomplished by performing a multi-stage pseudo-static analyses with an increasing level of external load. In this analysis, the structure is generally idealized by a beam-deck framework with or without the subgrade reaction springs shown in Fig. T7.9 to take into account the reaction on the embedded portion of the structure. With an increasing level of external load, the sequence of yield in the structures and a transition from elastic response to the ultimate state of failure will be identified. The yield generally begins from the pile heads most landward to those seaward and then moves down to the embedded portion of the piles. Figure T7.26 shows an example of the results of a pushover analysis.

Simplified dynamic analysis of pile-supported deck may be accomplished through the following steps:
1) evaluate the elasto-plastic parameters d_y and F_y defining the load-displacement relationship shown in Fig. T7.25 based on the results of pushover analysis or other simplified procedures;
2) evaluate the seismic response of SDOF based on non-linear response analysis using a time integration technique and obtain the ductility factor μ through

Eqn. (T7.45). Alternatively, use the linear response spectra to obtain the peak response acceleration a' and obtain the ductility factor through Eqns. (T7.43) or (T7.44);

3) compare the response ductility with the relevant ductility criteria/strain limit.

In simplified dynamic analysis based on response spectra/pushover analysis, the movement/deformation of the dike and the ground or the soil-structure interaction is not fully taken into account. The effects of these issues should be considered in the final evaluation of the results of the simplified dynamic analysis.

(3) Ductility criteria

Extensive studies have been performed to establish ductility criteria for pile-supported wharves (Ferritto, 1997a; Ferritto et al., 1999, Yokota et al., 1999). The damage criteria defined in terms of strain in piles may be given as shown in Table T7.10. In this table, the design ultimate compression strain of confined concrete may be taken as

$$\varepsilon_{cu} = 0.004 + (1.4\rho_s f_{yh}\varepsilon_{sm}) f'_{cc} \geq 0.005 \tag{T7.49}$$

where

ρ_s: effective volume ratio of confining steel

f_{yh}: yield stress of confining steel

ε_{sm}: strain at peak stress of confining reinforcement, 0.15 for grade 40 and 0.12 for grade 60

f'_{cc}: confined strength of concrete approximated by $1.5f'_c$, where f'_c is 28 days unconfined strength of concrete.

The serviceability limit shown in Table T7.10 corresponds to an essentially elastic state, but it does not imply a rigid requirement that the strains in all parts remain below yield limit. Concrete structures may be considered serviceable, without any significant decrement to structural integrity, provided concrete strains reached during maximum seismic response do not cause incipient spalling of concrete cover, and if residual crack widths after the earthquake are sufficiently small to negate the need for special corrosion protection measures such as crack grouting. This latter requirement implies that significantly larger crack widths might momentarily exist during seismic response, but these have no effect on corrosion potential. Thus the serviceability limits shown in Table T7.10 are based on strain limits sufficiently low so that spalling of cover concrete will not occur, and any residual cracks will be fine enough that remedial grouting will not be needed. The specified incremental strain for prestressing strand is less than for reinforcing steel to ensure that no significant loss of effective prestress force occurs at the serviceability limit state. For either hollow or concrete filled steel shell piles, it is recommended that the maximum tension strain in the shell be also limited to 0.01 to ensure that residual displacements are negligible. For the steel shell pile connected to the deck by a dowel reinforcing bar detail, the limit strains for reinforced concrete apply.

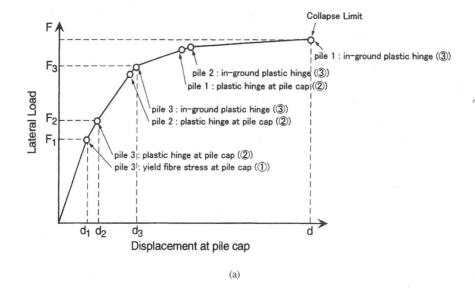

(a)

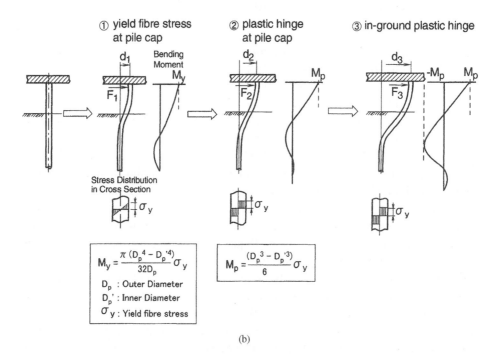

(b)

Fig. T7.26. Pushover analysis of a pile-supported wharf.
(a) Load-displacement relationship.
(b) Sequential change in bending moment in piles.

Table T7.10. Strain ductility criteria for pile-supported wharves (Ferritto et al., 1999; with modification).

Extent of damage	Parameters		Upper limit
Degree I (Serviceable)	Concrete extreme fibre compression strain		0.004
	Reinforcing steel tension strain		0.010
	Prestressing strand incremental strain		0.005
	Structural steel fibre compression strain (pile and concrete filled pipe)		0.008
	Hollow steel pipe pile fibre compression strain		0.008
Degree II (Repairable)	Pile-deck hinge	Concrete extreme fibre compression strain	Eqn. (T7.49), but <0.025
		Reinforcing steel tension strain	0.05
		Prestressing strand	0.04
		Structural steel fibre compression strain (pile and concrete filled pipe)	0.035
		Hollow steel pipe pile fibre compression strain	0.025
	Embedded portion (in-ground hinge)	Concrete extreme fibre compression strain	Eqn. (T7.49), but <0.008
		Reinforcing steel tension strain	0.010
		Prestressing strand	0.015
		Structural steel fibre compression strain (pile and concrete filled pipe)	0.035
		Hollow steel pipe pile fibre compression strain	0.025

For repairable damage state, limited permanent deformation are acceptable. Distinction needs to be made between the pile/deck hinge and the in-ground hinge locations, since access to the latter will frequently be impractical after an earthquake. The criteria shown in Table T7.10 reflect these considerations. For further details on strain ductility criteria, refer to Technical Commentary 5.

Displacement ductility is related to curvature/strain through Eqns. (T7.47) and (T7.48) and is inversely proportional to pile length. Based on the common dimensions of pile-supported wharves, the strain ductility criteria shown in Table T7.10 may be converted to displacement ductility criteria.

In Japanese practice, it is common to use steel piles for pile-supported wharves. Displacement ductility limits shown in Table T7.11 have been adopted for use with the simplified analysis discussed in Section T7.1.3 for typical pile-supported wharves with vertical piles (Ministry of Transport, Japan, 1999). In this table, the displacement ductility criterion for Level 2 earthquake for use with the simplified analysis is given by

$$\mu = 1.25 + 62.5\frac{t_p}{D_p}, \text{ but } \mu \leq 2.5 \tag{T7.50}$$

where

t_p: thickness of steel pipe pile
D_p: pile diameter

For use with the simplified dynamic analysis, including pushover analysis, the strain ductility limit is given by

$$\varepsilon_{max} = 0.44\frac{t_p}{D_p} \tag{T7.51}$$

The upper limit for Degree II damage is defined when the embedded portion of the first few piles reach at the strain limit specified by Eqn. (T7.51).

T7.2.4   Breakwaters, embankments and slopes

In most applications involving waterfront slopes and embankments, it is necessary to estimate the permanent deformations that may occur in response to cyclic loading. Allowable deformation limits for these slopes will often reflect the sensitivity of adjacent structures, foundations and other facilities to these soil movements. Enhancements to traditional pseudo-static limit equilibrium methods of embankment analysis have been developed to estimate embankment deformations for soils which do not lose appreciable strength during earthquake shaking (Ambraseys and Menu, 1988; Makdisi and Seed, 1978; Jibson, 1993).

As dicussed earlier for gravity and sheet pile walls, rigid body, sliding block analyses, which assume that the soil behaves as a rigid, perfectly plastic material, can be used to estimate limited earthquake-induced deformations. Numerical studies based on the sliding block method of analysis have led to the development of useful relationships between ground motion intensity and seismically-induced deformations (e.g. Ambraseys and Menu, 1988; Makdisi and Seed, 1978; Jibson, 1993). The relationship proposed by Makdisi and Seed is shown in Fig. T7.27.

Table T7.11. Displacement ductility factor (Ministry of Transport, Japan, 1999; with modification).

Reference earthquake level	Importance of structure (see Table 3.3)	Limit displacement ductility factor
Level 1	Special class	1.0
	Class A	1.3
	Class B	1.6
	Class C	2.3
Level 2	Special class	Eqn. (T7.50)

Given that the sliding block analyses are based on limit equilibrium techniques, they suffer from many of the same limitations previously noted for pseudo-static analyses. One of the primary limitations with respect to their application for submarine slopes in weak soils is that strain softening behaviour is not directly incorporated in the analysis. The sliding block methods have, however, been applied for liquefiable soils by using the post-liquefaction undrained strengths for sandy soils. A recent method proposed by Byrne et al. (1994) has been developed for contractive, collapsible soils that are prone to liquefaction.

Examples of displacement criteria for use with the sliding block analysis of dikes supporting pile-supported wharves used for Navy Facilities and Oil Terminals are 7 to 15 cm for the serviceablility limit and 15 to 30 cm for repairable state limit (Ferritto, 1997a, 1997b; Ferritto et al., 1999).

T7.3   DYNAMIC ANALYSIS

T7.3.1   Analysis procedures

Dynamic analysis, generally using finite element or finite difference techniques, involves coupled soil-structure interaction, wherein, the response of the foundation and backfill soils is incorporated in the computation of the structural response. A structure is idealized as either a linear or non-linear model, depending on the level of earthquake motion relative to the elastic limit of the structure. The stress-strain behaviour of the soil is commonly idealized with either equivalent linear or effective stress constitutive models, depending on the anticipated strain level within the soil deposit. Fairly comprehensive results can be obtained from soil-structure interaction analysis, possibly including failure modes of the soil-structure systems, extent of displacement, and stress/strain states in soil and structural components.

Issues regarding structural modelling of port structures, either through linear or non-linear models, are similar to those for other structures, including buildings, bridges and offshore structures. For example, modelling of a caisson for gravity quay walls is generally accomplished using a linear model. Modelling of sheet pile walls, cellular walls and pile-supported wharves often requires an elasto-plastic type non-linear model to allow ductile response of the structure beyond the elastic response limit. Special attention is needed for pile-deck connection details for designing pile-supported wharves. For further details, refer to Technical Commentary 5.

Geotechnical modelling of foundation and backfill soil, either through linear (equivalent linear total stress), total stress non-linear or effective stress models, is a primary consideration when evaluating the seismic performance of port structures. Total stress analysis procedures, either equivalent linear or non-linear, do

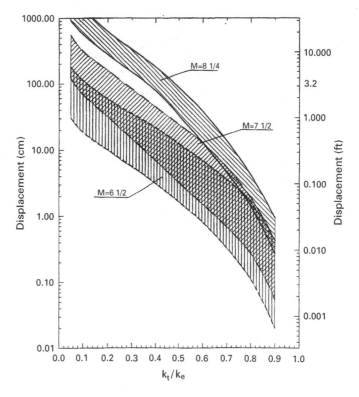

Fig. T7.27. Empirical relationships between permanent displacement of sliding block and
ratios of accelerations (after Makdisi and Seed, 1978).

not include the effects of change in excess pore water pressure or effective stress
during shaking, and therefore changes in the soil stiffness and strength are not
accounted for. Among the total stress analysis procedures, the equivalent linear
procedure has been the most widely used technique in practice for computing the
dynamic response of soil deposits, earth embankments, and soil-structure interaction. This is based on a linear analysis, using strain level dependent shear moduli
and damping. The computer code, FLUSH (Lysmer et al., 1975), is one of the
most widely used. Limitations in the equivalent linear analysis include:
– residual displacements of soil-structure systems are not computed; and
– applicability beyond the shear strain level of about one percent is questionable.
The shear strain level of one percent corresponds, for example, to a state where
level ground with a 10 m deposit deforms 0.1 m. This deformation is considerably
smaller than those discussed for performance criteria for port structures in
Chapter 4 of the Main Text. Those who use the equivalent linear analysis should
be aware of these limitations.

Modifications to a total stress analysis have been applied to overcome
the limitations in the equivalent linear analysis. These modifications relate

to: (1) incorporating the effect of excess pore water pressure increase in terms of reduced shear moduli in the equivalent linear analysis (e.g. FLUSH-L (Ozutsumi et al., 2000)), and (2) using reduced shear strength in the non-linear analysis such as using easily available computer codes (e.g. FLAC (ITASCA, 1995)). These total stress computer codes have an improved range of applicability over equivalent linear analysis. Certain limitations still remain because the effects of a progressive increase in excess pore water pressures are not included.

Non-linear, effective stress analysis methods are the most ideal for analyzing residual displacements and/or evaluating performance of structures beyond the strain level of one percent in soil. Many computer codes have been developed and utilized in practice. However, users should be aware of the fact that most of these computer codes are still under development and useful only as a research tool. It is best to consult with specialists in geotechnical earthquake engineering and effective stress analysis to discuss the stage of development of the computer code and the effective stress model and then decide whether the computer program is appropriate for use in practice. Proceedings of the relevant conferences, including Japanese Geotechnical Society (1989, 1991) and National Science Foundation, USA (Arulanandan and Scott, 1993), might offer relevant information. The final decision on what computer code should be used for the analysis will depend on the following issues:

– availability of solid theoretical background outlined in technical literature;
– availability of solid procedures for determining analysis parameters from commonly used geotechnical investigations;
– applicability of the procedure for well documented case histories of seismic performance of port structures;
– availability of computer code (preferably widely used and tested by non-specialists/consulting engineers other than the special group of researchers who originally developed it).

For more details, refer to Technical Commentary 4.

T7.3.2 Examples

The following are examples of dynamic analyses for typical port structures.

(1) Gravity quay wall

During the 1995 Hyogoken Nambu earthquake in Kobe, Japan, many of the caisson walls suffered damage as shown in Fig. T7.28. These caisson walls were constructed on a loose saturated backfill foundation of decomposed granite, which was used for replacing the soft clayey deposit in Kobe Port to attain the required foundation bearing capacity. Subjected to a strong earthquake motion having peak accelerations of $0.54g$ and $0.45g$ in the horizontal and vertical directions, these caisson walls were displaced an average of 3 m (maximum displacement – 5 m) toward the sea, settled 1 to 2 m and tilted about 4° toward the sea. Although a sliding mechanism could explain the large horizontal displacement of the caisson walls, this mechanism did not explain the large settlement and tilt of the caissons. Reduction in the stiffness of foundation soils due to development of excess pore water pressure was speculated as a main cause of the observed caisson damage at Kobe Port.

This speculation was confirmed by a series of effective stress analyses using the computer code FLIP. The model parameters were evaluated based on in-situ velocity logging, the blow counts of Standard Penetration Tests (SPT N-values) and the results of cyclic triaxial tests. The specimens used for cyclic triaxial tests were undisturbed samples obtained by an in-situ freezing technique. Input earthquake motions were those recorded at the Port Island site about 2 km from the quay wall. The spatial domain used for the finite element analysis covered a cross-sectional area of about 220 m by 40 m in the horizontal and vertical directions.

The effective stress analysis resulted in the residual deformation shown in Fig. T7.29. As shown in this figure, the mode of deformation of the caisson wall was to tilt into and push out the foundation soil beneath the caisson. This was consistent with the observed deformation mode of the rubble foundation shown in Fig. T7.30, which was investigated by divers. The order of wall displacements was also comparable to that observed and shown in Fig. T7.28.

In order to evaluate the overall effect of geotechnical conditions on the displacements of a gravity wall, the following three analyses were performed as a parameter study. The parameter study included a virtual soil model, to be called non-liquefiable soil, which has the same properties as those used in the aforementioned analysis but without the effect of excess pore water

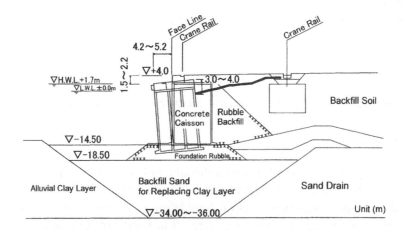

Fig. T7.28. Cross section of gravity quay wall at Kobe Port and deformation/failure during 1995 Great Hanshin earthquake, Japan.

pressures. To distinguish the cases in the parameter study, the case which dealt with the actual quay wall during the earthquake described earlier is designated as Case 1. Cases 2 through 4 are defined depending on the extent of the non-liquefiable soil relative to the caisson wall, as shown in Fig. T7.31.

Major results of the analysis are summarized in Table T7.12, including those of Case 1. These results indicate that the deformation of the gravity wall may be reduced up to about one half of that actually observed, if the excess pore water pressure increase was prevented in the subsoil as in Case 2. In particular, the effect of the pore water pressure increase in the foundation soil beneath the caisson wall is about twice as much as that of the backfill as understood from the comparison of Cases 3 and 4. These results were partly confirmed by the seismic performance of the quay walls at Port Island (Phase II), where a caisson wall had been placed on a foundation improved by the sand compaction pile technique before the earthquake (Iai et al., 1996).

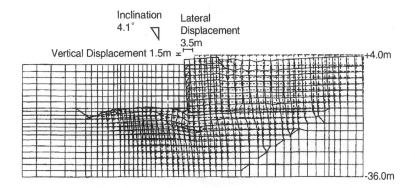

Fig. T7.29. Computed deformation of a gravity quay wall.

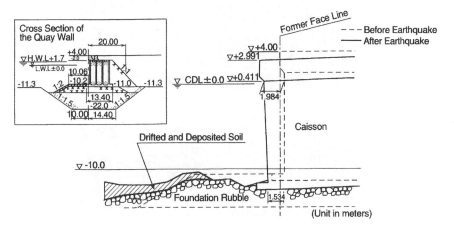

Fig. T7.30. Deformation of rubble foundation of a quay wall investigated by divers.

For variations in the peak acceleration of the earthquake motion input at the base, the horizontal residual displacement at the top of the caisson wall was computed as shown in Fig. T7.32. These response curves constitute the basis for performance based design. For example, let us suppose a caisson quay wall is constructed at a site where seismic hazard analysis resulted in 0.2g and 0.3g for Levels 1 and 2 earthquakes, respectively. For a Case 1 quay wall, horizontal displacements at the top of the caisson will be 1.3 and 2.1 m for Levels 1 and 2 earthquakes, which do not satisfy the performance criteria discussed in Chapter 4 in the Main Text. For a Case 2 quay wall with full liquefaction remediation measures, displacements will be reduced to 0.3 and 0.9 m for Levels 1 and 2 earthquakes, which satisfy the performance criteria.

The effect of the vertical component of input acceleration time histories was studied by varying the peak acceleration of the vertical component of acceleration ranging from zero to 0.40g (Ichii et al., 1997). The peak acceleration of the horizontal component was unchanged from the original value of 0.54g. The phase difference between the vertical and the horizontal components were also varied. This parametric study suggested that the effect of the vertical component is to increase the residual displacement less than about 10%.

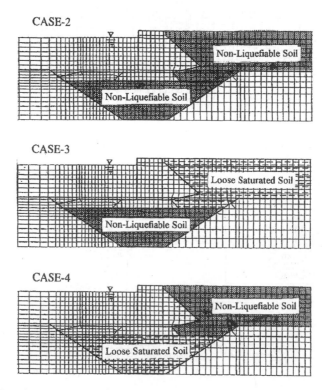

Fig. T7.31. Conditions assumed for parametric study, Case 2 through 4.

Table T7.12. Major results of parametric study for gravity quay wall.

| Case | Residual displacements of caisson | | |
	Horizontal (m)	Vertical (m)	Tilt (°)
Case 1	3.5	1.5	4.1
Case 2	1.6	0.6	2.4
Case 3	2.1	0.7	3.1
Case 4	2.5	1.1	2.2

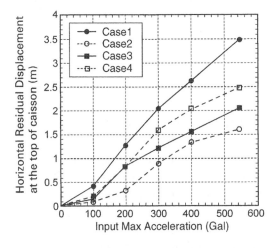

Fig. T7.32. Effects of input acceleration levels on horizontal residual displacement.

(2) Sheet pile quay wall

During the 1983 Nihonkai-Chubu earthquake in Japan, many anchored steel sheet pile walls suffered damage at Akita Port. Most of the walls were constructed by backfilling clean sand dredged from the nearby seabed with placed SPT N-values ranging from 20 to 50. During the earthquake, these walls were shaken by an earthquake motion with peak accelerations of $0.24g$ and $0.05g$ in the horizontal and vertical directions. Figure T7.33 shows a typical cross-section and deformation of an anchored sheet pile wall. The backfill liquefied and the sheet pile wall moved 1.1 to 1.8 m towards to the sea as shown by the broken lines. The sheet pile wall yielded and a crack opened about 6 m below sea level.

An effective stress analysis using FLIP resulted in the deformation shown in Fig. T7.34 (Iai and Kameoka, 1993). Increases in earth pressures and bending moments were also computed as shown in Fig. T7.35. In this analysis, the model parameters were evaluated based on the SPT N-values and cyclic triaxial test results. The wall and anchors were modeled using linear beam elements and, thus, there was a limitation in simulating the seismic response of the wall after yield of the wall. It is to be noted, however, that the displacement computed at the top of the wall was 1.3 m, consistent with that observed. The maximum bending moment computed in the sheet pile wall exceeded the yield level, also consistent with that observed.

In order to discuss the overall effect of the geotechnical conditions in the backfill, a parameter study was performed by varying the SPT N-values of the backfill. The maximum bending moment in the sheet pile wall was reduced for higher SPT N-values as shown in Fig. T7.36. In this figure, N_{65} designates SPT N-values corrected to the effective overburden pressure of 65 kPa, which is widely used in the design practice in Japanese ports (Port and Harbour Research Institute, 1997). The evidence to support this result was available from the case history of an undamaged quay wall at Akita Port (Iai and Kameoka, 1993).

(3) Pile-supported wharf

During the 1995 Hyogoken Nambu earthquake of 1995 in Japan, a pile-supported wharf suffered damage at Takahama Wharf in Kobe Port. The horizontal residual displacement of the wharf ranged from 1.3 to 1.7 m, with a typical example of the cross-section and deformation of the pile-supported wharf shown in Fig. T7.37. As shown in this figure, the wharf was constructed on a firm foundation deposit consisting of alternating layers of Pleistocene clay and sandy gravel. The SPT N-values ranged from 10 to 25 for the clay and 30 to 50 or higher for the sandy gravel. The firm deposit was overlain by an alluvial sand layer with SPT N-values of about 15, the thickness of which was variable, of about two meters on average. Behind the retaining wall made of concrete cellular blocks was a hydraulic

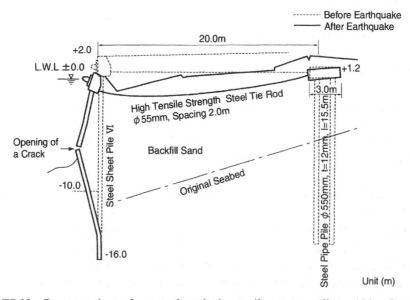

Fig. T7.33. Cross section of an anchored sheet pile quay wall at Akita Port and damage/deformation during 1983 Nihonkai-Chubu earthquake, Japan.

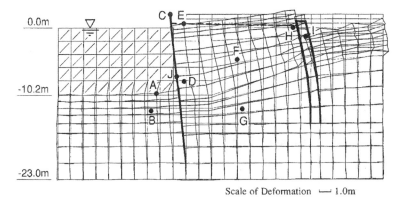

Fig. T7.34. Computed deformation of a sheet pile quay wall.

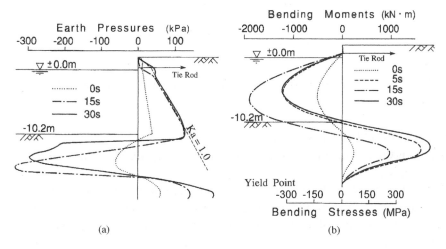

Fig. T7.35. Computed earth pressures and bending moments in a sheet pile quay wall.
            (a) Earth pressures.
            (b) Bending moment.

backfill of decomposed granite with SPT N-values of about 10. The deck of the quay was made of reinforced concrete slabs and beams supported by steel pipe piles having a diameter of 700 mm.

The steel piles buckled at the pile heads, except for the piles located most landward. A crack was found at the pile – concrete beam connection located most landward. The piles, pulled out after the earthquake for investigation, shown in Fig. T7.37(b), also showed buckling below the mudline at the depths shown in Fig. T7.37(a). As shown in this figure, some of the buckling was located close to the boundary between the layers of alluvial sand and Pleistocene gravel. Displacements of the rubble dike, measured by divers at five locations 5 m apart, were about the same as those of the deck as shown in Fig. T7.37(c).

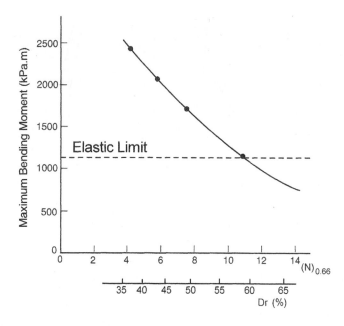

Fig. T7.36. Reduction of maximum bending moment with increasing SPT N-values/density.

The backfill behind the retaining structure settled about 1 m. These measurements indicate a somewhat uniform movement of the dike and the retaining wall toward the sea.

Effective stress analysis using FLIP resulted in the residual deformation shown in Fig. T7.38 (Iai, 1998b). In this analysis, the model parameters were evaluated based on the SPT N-values. The piles were idealized using elasto-plastic (bi-linear) beam elements. Linear springs were included to account for the three-dimensional interaction between the piles and the rubble dike. The stiffness of the springs used for this particular analysis corresponds to the coefficient of horizontal subgrade reaction $k_{h\text{-sub}} = 147$ kN/m^3. The computed deformation of the dike and the retaining wall, shown in Fig. T7.38, is associated with the significant deformation in the alluvial sand layer, which is consistent with the observed deformation of the pile-supported dike shown in Fig. T7.37. The bending moment and curvature of the piles accumulated gradually, reaching the residual values shown in Fig. T7.39. These residual values were close to the peak values during shaking. The computed locations at which the piles yielded (shown in Fig. T7.39) are also consistent with those observed, shown earlier in Fig. T7.37. In particular, the bending moment at the pile heads located most landward was not large, contrary to the results obtained by conventional design practice taking into account only the effect of the inertia force on the deck. The displacement of the dike obviously redistributed the bending moments among the three rows of piles.

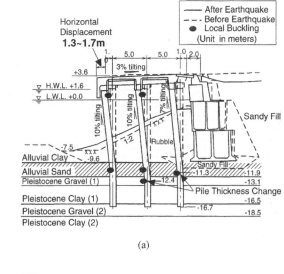

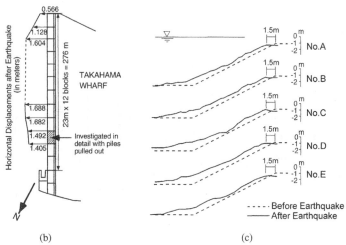

Fig. T7.37. Cross section of a pile-supported wharf at Kobe Port and deformation/failure
during 1995 Great Hanshin earthquake, Japan.
(a) Cross section and deformation/damage.
(b) Plan of pile-supported wharf.
(c) Displacements in rubble dike.

(4) Breakwater

Similar to the gravity walls mentioned earlier, the composite breakwaters in
Kobe Port were constructed on a loose decomposed granite, which was backfilled into the area excavated from the original alluvial clay layer. The thickness
of the clay layer reaches a maximum of 25 m in Kobe Port as shown in
Fig. T7.40. During the 1995 earthquake, the breakwater settled about 1.4 to 2.6 m.

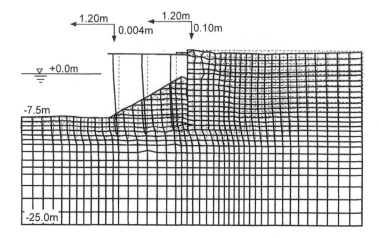

Fig. T7.38. Computed deformation of a pile-supported wharf.

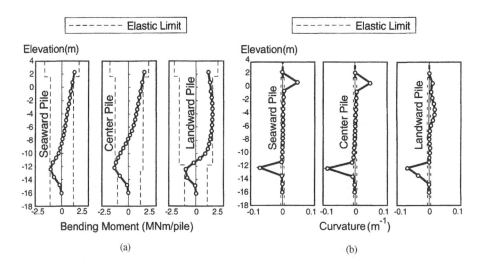

Fig. T7.39. Computed residual bending moments and curvatures in piles.
          (a) Bending moments.
          (b) Curvatures.

The horizontal displacements of the breakwater, however, were less than tens of centimeters. The mode of deformation suggested that the caisson was pushed into the rubble foundation and the rubble was also dragged down and pushed into the loose deposit below it. Although the caissons used for the quay walls and the breakwaters were similar to each other, the modes of deformation were different. The effect of backfill or lack of it was presumed to be the cause for this fundamental difference.

The effective stress analysis using the computer code FLIP resulted in the deformation shown in Fig. T7.41 (Iai et al., 1998). The deformation mode is consistent with that observed and shown in Fig. T7.40. The settlement at the top of the breakwater accumulated gradually and reached 1.3 m while the computed horizontal displacement at the top of the breakwater remains very small. The results are basically consistent with the observed performance.

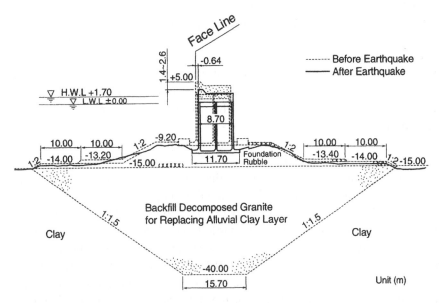

Fig. T7.40. Cross section of a breakwater at Kobe Port and deformation during 1995 Great Hanshin earthquake, Japan.

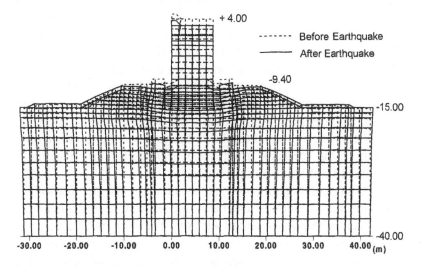

Fig. T7.41. Computed deformation of a breakwater.

T7.3.3 Remarks on dynamic analysis of slopes and walls

In situations where the movement of a slope has an impact on adjacent structures, such as pile-supported structures embedded in dikes, buried lifelines and other soil-structure interaction problems, it is becoming more common to rely on numerical modelling methods to estimate the range of slope deformations which may be induced by design level ground motions (Finn, 1990). The numerical models used for soil-structure interaction problems can be broadly classified based on the techniques that are used to account for the deformations of both the soil and the affected structural elements. In many cases, the movement of the soil is first computed, then the response of the structure to these deformations is determined. This type of analysis is termed uncoupled, in that the computed soil deformations are not affected by the existence of embedded structural components.

A common enhancement to this type of uncoupled analysis includes the introduction of an iterative solution scheme which modifies the soil deformations based on the response of the structure so that compatible strains are computed. An example of this type of analysis is drag loading on piles due to lateral spreading. In an uncoupled analysis, the ground deformations would be estimated using either an empirical relationship (e.g. Bartlett and Youd, 1995) or a sliding block type evaluation (e.g. Byrne et al., 1994). Once the pattern of ground deformations has been established, a model such as LPILE (Reese and Wang, 1994) can be used to determine the loads in the deformed piles. In addition, modifications can be made to the p–y curves to account for the reduced stiffness of the liquefied soil (O'Rourke and Meyerson, 1994). The lateral spread displacement is forced onto the p–y spring (i.e., drag loading).

In coupled numerical analysis, the deformations of the soil and structural elements are solved concurrently. Two-dimensional numerical models such as FLUSH (Lysmer et al., 1975), FLAC (Itasca, 1995), DYSAC2 (Muraleetharan et al., 1988), LINOS (Bardet, 1990), and FLIP (Iai et al., 1992) have been used to model the seismic performance of waterfront components at ports (e.g., Finn, 1990; Roth et al., 1992; Werner, 1986; Wittkop, 1994, Iai, 1998b). The primary differences in these numerical analyses include (a) the numerical formulation employed (e.g. finite element method, finite difference method), (b) the constitutive model used for the soils, and (c) the ability to model large, permanent deformations. Each of the methods listed, have been useful in evaluating various aspects of dynamic soil-structure interaction.

Advanced numerical modelling techniques are recommended for soil-structure interaction applications, such as estimating permanent displacements of slopes and embankments with pile-supported wharves. The primary advantages of these models are as follows: (a) complex embankment geometries can be evaluated, (b) sensitivity studies can be done to determine the influence of various parameters on the seismic stability of the structure, (c) dynamic soil behaviour is much more realistically reproduced, (d) coupled analyses which allow for such factors as excess pore pressure generation in contractive soils during ground shaking and the associated reduction of soil stiffness and strength can be used, (e) soil-structure interaction and permanent deformations can be evaluated.

These advantages may warrant the use of the dynamic analysis for designing high performance grade structures despite the following disadvantages: (a) the engineering time required to construct the numerical model can be extensive, (b) numerous soil parameters are often required, thereby increasing laboratory testing costs (the number of soil properties required is a function of the constitutive soil model employed in the model), (c) very few of the available models have been validated with well documented case studies of the seismic performance of actual retaining structures, therefore the level of uncertainty in the analysis is often unknown. As is the case with other analytical procedures, it may be advisable for the user to confirm the applicability of the analytical procedure based on the relevant well documented case histories before using them for design purposes.

T7.4   INPUT PARAMETERS FOR ANALYSIS

This section further elaborates the input parameters for analysis discussed in Section 5.4 in the Main Text. As mentioned in the Main Text, the reliability of the results depends not only on the type of analysis, but also on the reliability of input parameters. The input parameters elaborated in Tables T7.13 through T7.15 in this section, may be useful in identifying possible sources of uncertainties in input parameters, and setting up an engineering plan to improve the reliability.

Table T7.13. Input parameters for simplified analysis.

Port structures/conditions		Design parameters	Input parameters
Gravity quay wall	Earthquake	k_e: equivalent seismic coefficient	a_{max}: Regional PGA at bedrock Site Category (site amplification factor)
	Structural	k_t: threshold seismic coefficient (Geometrical extent of liquefiable soils relative to the position and dimensions of a wall for a liquefiable site)	W, H & water level: Cross-sectional dimensions of a gravity wall
	Geotechnical		c, ϕ: cohesion and internal friction angle of soils μ_b, δ: friction angles at the bottom and back face of the wall ground water level (SPT/CPT for a liquefiable site)
Sheet pile quay wall	Earthquake	k_e: equivalent seismic coefficient	a_{max}: Regional PGA at bedrock Site Category (site amplification factor)
	Structural	k_t: threshold seismic coefficient (Geometrical extent of liquefiable soils relative to the position and dimensions of a wall for a liquefiable site)	Sheet pile wall: Dimensions (H, D_{emb} & water level) & material parameters (e.g. E, I, A, yield stress) Tie-rod: Length L and material parameters (e.g. E, A, yield stress) Anchor: Type (vertical or batter pile or deadman), Dimensions and material parameters (e.g. E, I, yield stress)
	Geotechnical		c, ϕ: cohesion and internal friction angle of soils δ: friction angle at sheet pile/soil interface ground water level (SPT/CPT for a liquefiable site)
Pile-supported wharf	Earthquake	Pile-deck structure: design response spectra Dike/retaining wall: k_e: equivalent seismic coefficient	a_{max}: Regional PGA at bedrock Site Category (site dependent response spectra)
	Structural	Pile-deck structure: resistance at elastic limit $1/\beta$: depth, from mudline, at the bottom end of an equivalent unsupported pile $$\beta = \sqrt[4]{\frac{k_{h\text{-sub}} D_p}{4EI}}$$	Pile-deck structure: Piles: Dimensions (L, D_p & water level) & material parameters (e.g. E, I, A, D_p, yield/ultimate stress/strain) Deck: Dimensions & material parameters Dike/retaining wall: relevant structural parameters for gravity/ sheet pile walls

	Geotechnical	Dike/retaining wall: k_t: threshold seismic coefficient (Geometrical extent of liquefiable soils relative to the position and dimensions of a wall for a liquefiable site)	Pile-deck structure: $k_{h\text{-sub}}$: coefficient of lateral subgrade reaction (or p–y curve) Dike/retaining wall: relevant geotechnical parameters for dike, gravity/sheet pile walls
Cellular quay wall	Earthquake	k_e: equivalent seismic coefficient	a_{max}: Regional PGA at bedrock Site Category (site amplification factor)
	Structural	for overall stability: k_t: threshold seismic coefficient (Geometrical extent of liquefiable soils relative to the position and dimensions of a wall for a liquefiable site) for structural performance: elastic limit limits	Cell: Dimensions (H, W, D_{emb} & water level) & material parameters (e.g. E, I, A, yield stress)
	Geotechnical		c, ϕ cohesion and internal friction angle of soils δ: friction angle at the cell/soil interface ground water level (SPT/CPT for a liquefiable site)
Crane	Earthquake	design response spectra	a_{max}: Regional PGA at bedrock Site Category (site dependent response spectra)
	Structural	resistance at elastic limit	Dimensions and material parameters (e.g. E, I, A, yield stress)
	Geotechnical	N/A	N/A
Break-water	Earthquake	k_e: equivalent seismic coefficient	a_{max}: Regional PGA at bedrock Site Category (site amplification factor)
	Structural	k_t: threshold seismic coefficient (Geometrical extent of liquefiable soils relative to the position and dimensions of a breakwater for a liquefiable site)	W, H & water level: Cross-sectional dimensions of breakwater
	Geotechnical		c, ϕ: cohesion and internal friction angle of rubble mound and foundation soils μ_b: friction angles at the bottom of a caisson for a composite type break water (SPT/CPT for a liquefiable site)

Table T7.14. Input parameters for simplified dynamic analysis.

		Design parameters	Input parameters
Gravity quay wall	Earthquake	empirical equations: a_{max}: peak acceleration v_{max}: peak velocity time history analysis: time histories of earthquake motions	bedrock earthquake motions
	Structural		W, H & water level: Cross-sectional dimensions of a gravity wall
	Geotechnical	sliding block analysis: a_t: threshold acceleration simplified chart: W, H & water level: Cross-sectional dimensions of a gravity wall SPT N-values (extent of soil improvement)	G–γ & D–γ curves (for site response analysis) c, ϕ: cohesion and internal friction angle of soils μ_b, δ: friction angles at the bottom and back face of the wall ground water level undrained cyclic properties and/or SPT/CPT data
Sheet pile quay wall	Earthquake	empirical equations: a_{max}: peak acceleration v_{max}: peak velocity time history analysis: time histories of earthquake motions	bedrock earthquake motions
	Structural	sliding block analysis: a_t: threshold acceleration simplified chart: W, H & water level: Cross-sectional dimensions of a gravity wall SPT N-values (extent of soil improvement)	Sheet pile wall: Dimensions (H, D_{emb} & water level) & material parameters (e.g. E, I, A, yield/ultimate stress/strain limits, ductility factor) Tie-rod: Length L and material parameters (e.g. E, A, yield/ultimate stress/strain limits) Anchor: Type (vertical or batter pile or deadman), Dimensions and material parameters (e.g. E, I, yield/ultimate stress/strain limits)
	Geotechnical		G–γ & D–γ curves (for site response analysis) c, ϕ: cohesion and internal friction angle of soils δ: friction angle at wall/soil interface ground water level undrained cyclic properties and/or SPT/CPT data

Pile-supported wharf	Earthquake	Pile-deck structure: standard procedure: design response spectra site specific study: time histories of earthquake motions Dike/retaining wall: empirical equations: a_{max}: peak acceleration v_{max}: peak velocity time history analysis: time histories of earthquake motions	bedrock earthquake motions
	Structural	Pile-deck structure: resistance at elastic limit/ultimate states ductility factor/strain limit $1/\beta$: depth, from mudline, at the bottom end of an equivalent unsupported pile $$\left(\beta = \sqrt[4]{\frac{k_{h\text{-}sub} D_p}{4EI}}\right)$$	Pile-deck structure: Piles: Dimensions (L, D_{emb} & water level) & material parameters (e.g. E, I, A, D_p, yield/ultimate bending moment, shear fracture criteria, ductility factor, M–ϕ curve, strain limit) Deck: Dimensions & material parameters (e.g. E, dimensions, shear fracture criteria, strain limit, pile-deck connection) Dike/retaining wall: relevant structural parameters for gravity/sheet pile walls
	Geotechnical	Dike/retaining wall: a_t: threshold acceleration	G–γ & D–γ curves (for site response analysis) undrained cyclic properties and/or SPT /CPT data Pile-deck structure: $k_{h\text{-}sub}$: coefficient of lateral subgrade reaction (or p–y curve) Dike/retaining wall: relevant geotechnical parameters for dike, gravity/sheet pile walls
Cellular quay wall	Earthquake	empirical equations: a_{max}: peak acceleration v_{max}: peak velocity time history analysis: time histories of earthquake motions	bedrock earthquake motions

Table T7.14. Continued.

		Design parameters	Input parameters
	Structural	for overall stability: a_t: threshold acceleration for structural performance: yield/ultimate stress/strain limits	Cell: Dimensions (H, W & water level) & material parameters (e.g. E, I, A, yield/ultimate stress/strain limits)
	Geotechnical		$G-\gamma$ & $D-\gamma$ curves (for site response analysis) c, ϕ: cohesion and internal friction angle of soils δ: friction angles at cell/soil interface ground water level undrained cyclic properties and/or SPT /CPT data
Crane	Earthquake	standard procedure: design response spectra site specific study: time histories of earthquake motions	bedrock earthquake motions
	Structural	resistance at elastic limit/ultimate strain limit ductility factor	Dimensions and material parameters (e.g. E, I, A, B, yield and ultimate bending moment, shear fracture criteria, ductility factor, $M-\phi$ curve, strain limit)
	Geotechnical	N/A	N/A
Break-water	Earthquake	empirical equations: a_{max}: peak acceleration v_{max}: peak velocity time history analysis: time histories of earthquake motions	bedrock earthquake motions
	Structural	a_t: threshold acceleration	W, H & water level: Cross-sectionals dimensions of breakwater
	Geotechnical		$G-\gamma$ & $D-\gamma$ curves (for site response analysis) c, ϕ: cohesion and internal friction angle of rubble mound and foundation soils μ_b: friction angles at the bottom of a caisson for a composite type break water undrained cyclic properties and/or SPT /CPT data

Table T7.15. Input parameters for dynamic analysis.

		Design parameters	Input parameters
Gravity quay wall	Earthquake	time histories of earthquake motions at the bottom boundary of analysis domain	If the bottom boundary of analysis domain differs from the bedrock; for moderate shaking: G–γ & D–γ curves bedrock earthquake motions for strong shaking: the same as geotechnical parameters for non-linear analysis (below)
	Structural	W, H & water level: Cross-sectional dimensions of a gravity wall	
	Geotechnical	equivalent linear analysis: G–γ & D–γ curves non-linear analysis: parameters for elastic properties parameters for plastic properties: for shear and dilatancy	G, K: shear and bulk modulus c, ϕ: cohesion and internal friction angle of soils μ_b, δ: friction angles at the bottom and back face of the wall ground water level undrained cyclic properties and/or SPT/CPT data
Sheet pile quay wall	Earthquake	time histories of earthquake motions at the bottom boundary of analysis domain	If the bottom boundary of analysis domain differs from the bedrock; for moderate shaking: G–γ & D–γ curves bedrock earthquake motions for strong shaking: the same as geotechnical parameters for non-linear analysis (below)
	Structural	linear analysis: Sheet pile wall: Dimensions (H, D_{emb} & water level) & material parameter (EA, yield stress) Tie-rod: Length L and material parameter (EA, yield stress) Anchor: Type (vertical or batter pile or deadman), non-linear analysis: in addition to those for linear analysis, M–ϕ curve, yield/ultimate stress/strain limits	Dimensions (H, D_{emb} & water level) & material parameters (E, I, A, yield stress) Dimensions and material parameters (E, I, yield stress)

Table T7.15.   Continued.

		Design parameters	Input parameters
	Geotechnical	equivalent linear analysis: $G-\gamma$ & $D-\gamma$ curves	
		non-linear analysis: parameters for elastic properties parameters for plastic properties: for shear and dilatancy	G, K: shear and bulk modulus c, ϕ: cohesion and internal friction angle of soils δ: friction angle at sheet pile/soil interface ground water level undrained cyclic properties and/or SPT/CPT data
Pile-supported wharf	Earthquake	time histories of earthquake motions at the bottom boundary of analysis domain	If the bottom boundary of analysis domain differs from the bedrock; for moderate shaking: $G-\gamma$ & $D-\gamma$ curves bedrock earthquake motions for strong shaking: the same as geotechnical parameters for non-linear analysis (below)
	Structural	linear analysis: pile-deck structure: Piles: Dimensions (L, D_{emb} & water level) & linear material parameters Deck: Dimensions & material parameters dike/retaining wall: relevant structural parameters for gravity/sheet pile walls non-linear analysis: in addition to those for linear analysis, $M-\phi$ curve, shear fracture criteria, yield/ultimate stress/strain limits	linear material parameters (E, I, A, D_p, yield stress)
	Geotechnical	equivalent linear analysis: $G-\gamma$ & $D-\gamma$ curves	
		non-linear analysis: parameters for elastic properties parameters for plastic properties: for shear and dilatancy	G, K: shear and bulk modulus c, ϕ: cohesion and internal friction angle of soils μ_b, δ: friction angles at retaining wall ground water level undrained cyclic properties and/or SPT/CPT data

Cellular quay wall	Earthquake	time histories of earthquake motions at the bottom boundary of analysis domain	If the bottom boundary of analysis domain differs from the bedrock; for moderate shaking: G–γ & D–γ curves bedrock earthquake motions for strong shaking: the same as geotechnical parameters for non-linear analysis (below)
	Structural	linear analysis: Cell: Dimensions (H, W & water level) & material parameters (e.g. E, I, A, yield stress) non-linear analysis: in addition to those for linear analysis, M–ϕ curve, yield/ultimate stress/strain limits	
	Geotechnical	equivalent linear analysis: G–γ & D–γ curves	G, K: shear and bulk modulus c, ϕ: cohesion and internal friction angle of soils δ: friction angle at cell/soil interface ground water level undrained cyclic properties and/or SPT/CPT data
		non-linear analysis: parameters for elastic properties parameters for plastic properties: for shear and dilatancy	
Crane	Earthquake	time histories of earthquake motions at the bottom boundary of analysis domain	If the bottom boundary of analysis domain differs from the bedrock; for moderate shaking: G–γ & D–γ curves bedrock earthquake motions for strong shaking: the same as geotechnical parameters for non-linear analysis (below)
	Structural	linear analysis: Dimensions and material parameters for crane (and quay wall, if crane-quay wall interaction is analyzed) Non-linear analysis: in addition to those for linear analysis, M–ϕ curve, yield/ultimate stress/strain limits, shear fracture criteria	
	Geotechnical	The same as those for quay wall analysis if crane-quay wall interaction is analyzed	

Table T7.15. Continued.

		Design parameters	Input parameters

Break-water	Earthquake	time histories of earthquake motions at the bottom boundary of analysis domain	If the bottom boundary of analysis domain differs from the bedrock;
			for moderate shaking:
			  G–γ & D–γ curves
			  bedrock earthquake motions
			for strong shaking:
			  the same as geotechnical parameters for non-linear analysis (below)

	Structural	W, H & water level: Cross-sectional dimensions of breakwater	
	Geotechnical	equivalent linear analysis: G–γ & D–γ curves	G, K: shear and bulk modulus of soils/rubbles c, ϕ: cohesion and internal friction angle of soils/rubbles μ_b: friction angles at the bottom of a caisson for a composite type break water undrained cyclic properties and/or SPT/CPT data

		non-linear analysis:
		  parameters for elastic properties
		  parameters for plastic properties: for shear and dilatancy	

TC8: Examples of Seismic Performance Evaluation

T8.1   Gravity Quay Wall (Grade A)
    T8.1.1   Performance requirements and design conditions
    T8.1.2   Simplified analysis
    T8.1.3   Simplified dynamic analysis
    T8.1.4   Simplified dynamic analysis for evaluation of liquefaction remediation
    T8.1.5   Dynamic analysis
T8.2   Sheet Pile Quay Wall (Grade B)
    T8.2.1   Performance requirements and design conditions
    T8.2.2   Simplified analysis (North American method)
    T8.2.3   Simplified analysis (Japanese method)
    T8.2.4   Simplified dynamic analysis (deformations)
T8.3   Sheet Pile Quay Wall (Grade S)
    T8.3.1   Performance requirements and design conditions
    T8.3.2   Dynamic analysis
T8.4   Pile-Supported Wharf (Grade B)
    T8.4.1   Performance requirements and design conditions
    T8.4.2   Simplified analysis
    T8.4.3   Simplified dynamic analysis
T8.5   Pile-Supported Wharf (Grade S)
    T8.5.1   Performance requirements and wharf conditions
    T8.5.2   Dynamic analysis

TECHNICAL COMMENTARY 8

Examples of Seismic Performance Evaluation

This technical commentary presents examples of seismic performance evaluations for five port structures and illustrates the basic procedures employed. These examples are based on field case studies, modified slightly to fit, where necessary, to the seismic guidelines addressed in the Main Text. Thus, the examples shown are hypothetical. These design examples are intended to illustrate the application of the various analysis procedures outlined in this report.

T8.1 GRAVITY QUAY WALL (GRADE A)

T8.1.1 Performance requirements and design conditions

A gravity quay wall with a water depth of 14.5 m was proposed for construction. The performance grade of this quay wall was determined to be Grade A over the life span of 50 years (see Section 3.1 in the Main Text). Based on seismic hazard analysis for the quay wall site, the PGA of L1 and L2 earthquake motions with a probability of exceedance of 50% and 10% were $0.15g$ and $0.3g$ at the bedrock, respectively. One of the simplified dynamic analysis methods requires input data on the earthquake moment magnitudes for the L1 and L2 earthquakes. In this example, these magnitudes were 6.0 and 7.0, respectively. Damage criteria for this quay wall were determined as shown in Table T8.1. These criteria were established based on the criteria discussed in Section 4.1 in the Main Text, with modification for tilting and differential settlement due to the specific performance requirements of the quay wall.

The proposed cross-section of the gravity quay wall is shown in Fig. T8.1. It was proposed to remove the clay layer below the caisson wall and replace it with sand backfill, in order to attain sufficient bearing capacity. The four possible geotechnical cross sections with respect to liquefaction remediation of backfill sand were considered for seismic performance evaluation as shown in Fig. T8.2.

Table T8.1. Damage criteria for a gravity quay wall.*

Extent of damage		Degree I	Degree II	Degree III	Degree IV
Gravity wall	Normalized residual horizontal displacement (d/H)**	Less than 1.5%	1.5~5%	5~10%	Larger than 10%
	Residual tilting towards the sea	Less than 2°	2~5°	5~8°	Larger than 8°
Apron	Differential settlement between apron and non-apron areas	Less than 0.3 m	N/A	N/A	N/A

* Modified from the general criteria shown in Table 4.1 in the Main Text, due to the specific requirements of the quay wall.
** d: Residual horizontal displacement; H: height of gravity wall.

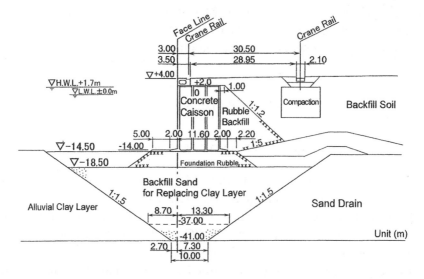

Fig. T8.1. Cross section of a gravity quay wall.

Since the required performance was Grade A, simplified and simplified dynamic analyses were performed in the preliminary stage, and dynamic analysis was performed in the final stage (see Section 5.1 in the Main Text). The simplified analysis was based on pseudo-static analysis and the simplified dynamic analyses were based on parametric studies (see Technical Commentary 7). The dynamic analysis utilized the effective stress based finite element method program FLIP (see Section 5.3 in the Main Text and Technical Commentary 7). Input data necessary for the analysis were acquired from geotechnical investigations performed at a nearby site, including SPT, PS-logging, and laboratory cyclic tests of in-situ frozen samples (see Section 5.4 in the Main Text and Technical Commentary 4). SPT N-values and the location of in-situ freezing sampling are shown in Fig. T8.3.

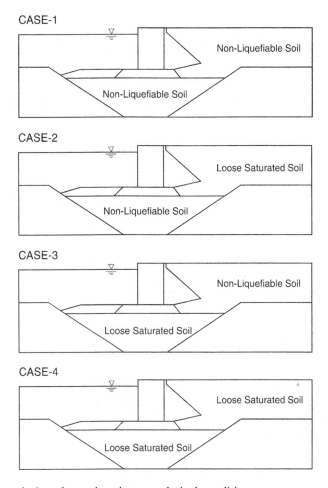

Fig. T8.2. Four design alternatives in geotechnical conditions.

The structural parameters of the caisson and geotechnical parameters for backfill soils and rubble are shown in Tables T8.2 and T8.3, respectively

For the seismic performance evaluation, the limiting curves were defined as shown in Fig. T8.4 for a wall height of 18.5 m based on the damage criteria shown in Table T8.1.

T8.1.2   Simplified analysis

Pseudo-static analysis of the gravity quay wall was performed for L1 earthquake motion. Using the L1 acceleration of $0.15g$ at bedrock, site response was performed, and a ground surface acceleration of $0.25g$ was obtained. A seismic coefficient of 0.15 used in the pseudo-static analysis was determined using Eq. (T7.11) as follows:

$$k_h = 0.6 \times 0.25 = 0.15$$

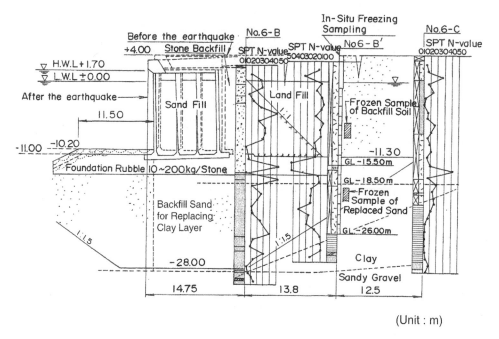

(Unit : m)

Fig. T8.3. SPT *N*-values and location of in-situ freezing sampling.

Cross-sections of the structure with all relevant and simplified parameters required for the analysis are shown in Figs. T8.5 and T8.6.

(1) Active earth pressures and thrust

The active angle of failure from the horizontal direction for a level surface condition is computed as

$$\alpha_{ae} = \phi + \arctan\left[\frac{-\tan\phi + \sqrt{\tan\phi(\tan\phi + \cot\phi)(1 + \tan\delta\cot\phi)}}{1 + \tan\delta(\tan\phi + \cot\phi)}\right]$$

$$= 40° + \arctan\left[\frac{-\tan 40° + \sqrt{\tan 40°(\tan 40° + \cot 40°)(1 + \tan 15°\cot 40°)}}{1 + \tan 15°(\tan 40° + \cot 40°)}\right]$$

$$= 63°$$

Thus, the active soil wedge is defined by the angle of failure measured from the vertical direction as

$$90° - 63° = 27°$$

The numbers with the square marks shown in Fig. T8.6 within the soil wedge are areas in m² of various portions used in the analysis. Using these areas, a representative unit weight of the material above the water table, including the

Table T8.2. Parameters for concrete caisson and joint elements.

Parameters	density (t/m³)	Young's modulus (kPa)	Shear modulus (kPa)	Poisson's ratio
Caisson	2.10	3.0×10^7	1.3×10^7	0.20

Joint elements: Friction angle $\delta = 31°$ (bottom of caisson); 15° (behind caisson)

Table T8.3. Parameters for soils and rubble.

Parameters	Density (t/m³)	Initial shear modulus (kPa)	Poisson's ratio	Internal friction angle (°)	Phase transformation angle (°)
Foundation soil*	1.8	$5660 \times (\sigma'_{mo})^{0.5}$	0.3	37	28
Backfill soil*	1.8	$10000 \times (\sigma'_{mo})^{0.5}$	0.3	36	28
Clay	1.7	$6270 \times (\sigma'_{mo})^{0.5}$	0.3	30	–
Foundation rubble and rubble backfill	2.0	$18200 \times (\sigma'_{mo})^{0.5}$	0.3	40	–

* Refer to Iai et al. (1998b) on the parameters for dilatancy.

capping material and backfill soil, is (1 kN = 0.1 tf)

$$(20 \times 0.19 + 18 \times 28.92)/29.1 = 18 \text{ kN/m}^3$$

Then, Eqn. (T7.6) gives for the equivalent unit weight of the backfill

$$\gamma_e = 18\left[1 - \left(\frac{15.1}{18.5}\right)^2\right] + 10 \times \left(\frac{15.1}{18.5}\right)^2 = 12.7 \text{ kN/m}^3$$

A representative internal friction angle for the backfill can be determined by direct averaging based on the corresponding cross-sectional areas (see Fig. T8.6)

$$\phi = (0.19 \times 40 + 28.92 \times 36 + 14.4 \times 36 + 43.69 \times 40)/87.2$$

$$= (7.6 + 1041.12 + 518.4 + 1747.6)/87.2 = 38°$$

Assuming that half the operational surcharge is present during the earthquake, the modified seismic coefficient is derived by Eqn. (T7.7)

$$k'_h = k_h \cdot \frac{10 \times 18.5 + \frac{1}{2} \times 18 \times 3.40^2 + 18 \times 3.40 \times 15.10 + \frac{1}{2} \times 20 \times 15.1^2}{10 \times 18.5 + \frac{1}{2} \times 18 \times 3.40^2 + 18 \times 3.40 \times 15.10 + \frac{1}{2} \times 10 \times 15.1^2}$$

$$= 0.15 \times \frac{185 + 104 + 924.1 + 2280.1}{185 + 104 + 924.1 + 1140.1} = 0.15 \times 1.48 = 0.22$$

The alternative procedure using Eqn. (T7.7′) will be shown in the example of Section T8.2.3. Use of Eqn. (T7.1) now gives

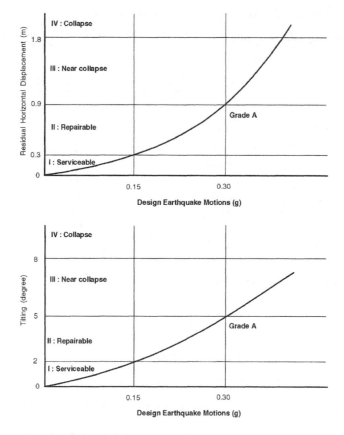

Fig. T8.4. Limiting curve for defining wall displacement.

$$\psi = \arctan(k'_h) = 12.4°$$

and the dynamic active earth pressure coefficient of the Mononobe-Okabe method is found by Eqn. (T7.2)

$$K_{ae} = \frac{\cos^2(38° - 12.4°)}{\cos 12.4° \cos(15° + 12.4°)\left[1 + \sqrt{\dfrac{\sin(38° + 15°)\sin(38° - 12.4°)}{\cos(15° + 12.4°)}}\right]^2} = 0.356$$

The total earth thrust is now

$$P_{ae} = 0.356 \times \frac{1}{2} \times \left(\gamma_{eq} + \frac{\frac{1}{2} q_{sur}}{H}\right) \times 18.50^2 = 0.356 \times \frac{1}{2}\left(12.7 + \frac{10}{18.50}\right) \times 18.50^2$$

$$= 806.6 \text{ kN/m}$$

This gives a horizontal force

$$P_{ae}\cos\delta = 806.6 \times \cos 15° = 779.1 \text{ kN/m}$$

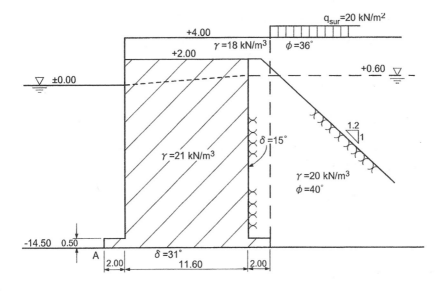

Fig. T8.5. Simplified cross-section for pseudo-static analysis.

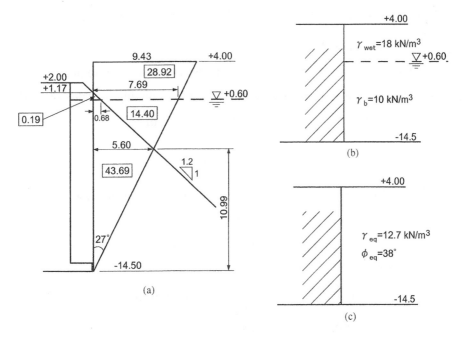

Fig. T8.6. Diagram for evaluating average values.
      (a) Soil wedge geometry.
      (b) Original parameters.
      (c) Equivalent parameters.

at about $0.45 \times 18.50 = 8.33$ m above the -14.50 m elevation, and a vertical contribution

$$P_{ae}\sin\delta = 806.6 \times \sin 15° = 208.8 \text{ kN/m}$$

at the interface between structure and soil.

(2) Hydrodynamic force

The hydrodynamic force in front of the wall gives a horizontal force according to Westergaard's expression

$$P_{dw} = \frac{7}{12} \times 0.15 \times 10.2 \times 14.5^2 = 187.6 \text{ kN/m}$$

at $0.4 \times 14.5 = 5.8$ m above the -14.50 m elevation.

(3) Inertia and other driving forces

The wall inertia forces are estimated as follows:

    cap $11.60 \times 2 \times 18 \times 0.15 = 62.6$ kN/m at 17.50 m above bed (-14.5 m)

    caisson $11.60 \times 16.00 \times 21 \times 0.15 = 584.6$ kN/m at 8.50 m above bed

    footing $15.60 \times 0.5 \times 21 \times 0.15 = 24.6$ kN/m at 0.25 m above bed

The backfill inertia force

    $2.0 \times 18.0 \times 18 \times 0.15 = 97.2$ kN/m at 9.50 m above -14.50 m

The static bollard pull can be taken into account by reducing its value by 50%

    $50\% \times 20 = 10$ kN/m at 19.0 m above -14.50 m

(4) Vertical forces

The resisting forces and their points of application are determined as follows:

    cap $11.60 \times 2.0 \times 18 = 417.6$ kN/m

             at 7.8 m from A (see Fig. T8.5)

    caisson $11.60 \times 1.70 \times 21 = 414.1$ kN/m

$$\text{at } \frac{11.6}{3} \times \frac{2 \times 1.40 + 2}{1.40 + 2.0} + 2 = 7.46 \text{ m from A}$$

            $11.60 \times 14.30 \times 11 = 1824.7$ kN/m

             at 7.8 m from A

    footing $15.60 \times 0.50 \times 11 = 85.8$ kN/m

             at 7.8 m from A

Using the backfill stone area above the water table $= \frac{1}{2} \times 0.83(1+2) + 2 \times 0.57$
$= 2.39 \text{ m}^2$, the backfill equivalent unit weight above water table is determined
as

$\{20 \times 2.39 + 18 \times (2 \times 3.40 - 2.39)\} / 6.80 = 18.7 \text{ kN/m}^3$

$2.0 \times 3.40 \times 18.7 = 127.2 \text{ kN/m}$ at 14.60 m from A

$2.0 \times 13.40 \times 10 = 268 \text{ kN/m}$ at 14.60 m from A

The hydrostatic pressure due to the variation of water table gives the following
forces:

$\frac{1}{2} \times 0.6^2 \times 10.2 = 1.8 \text{ kN/m}$ at $\frac{1}{3} \times 0.60 + 14.50 = 14.70$ m above -14.50 m

$14.50 \times 0.60 \times 10.2 = 88.7 \text{ kN/m}$ at 7.25 m above -14.50 m

Uplift $\frac{1}{2} \times 0.60 \times 10.2 \times 15.60 = 47.7 \text{ kN/m}$ at $\frac{2}{3} \times 15.60 = 10.40$ m from A

It is noted that the hydrostatic pressure distribution depends on the permeability
of the foundation and backfill rubble. If the latter has a much lower permeability
than the former, then no uplift force is generated and the pressure diagram at the
back face dies off linearly to the foundation level AB.

(5) Force and moment balance

The various forces mentioned previously act on the structure as shown in Fig. T8.7.
Now, the stabilizing moments with respect to point A are calculated as follows:

stabilizing	395.2	×	14.60	=	5769.9 kNm/m
	208.8	×	13.60	=	2839.7
	414.1	×	7.46	=	3089.2
	2,328.1	×	7.80	=	18,159.2
	3,346.2				29,858.0 kNm/m
	− 47.7				
	3,298.5 kN/m				

The overturning moments with respect to the same point are

	10.0	×	19.00	=	190.0
	62.6	×	17.50	=	1,095.5
	584.6	×	8.50	=	4,969.1
	24.6	×	0.25	=	6.2
	187.6	×	5.80	=	1,088.1
(negative)	47.7	×	10.40	=	496.1
	779.1	×	8.33	=	6,489.9
	1.8	×	14.70	=	26.5
	88.7	×	7.25	=	643.1
	97.2	×	9.50	=	923.4
	1,836.2 kN/m				15,927.9 kNm/m

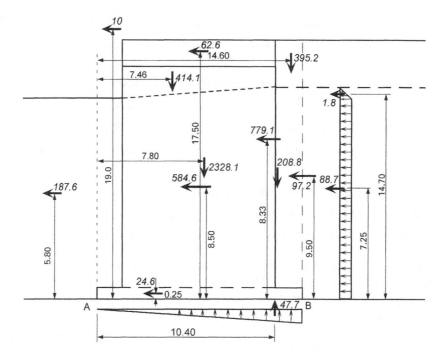

Fig. T8.7. Loads acting on structure during earthquake.

The factor of safety against overturning is

$$F_{so} = 29{,}858.0/15{,}927.9 = 1.87$$

Depending on the definition of this factor of safety and when a non-zero angle of friction between the backfill and the wall back face is assumed, one can take the overturning moment of the earth thrust before analysing it into vertical and horizontal components.

The factor of safety against sliding is

$$F_{ss} = 3{,}298.5 \times \mu_b / 1{,}836.2$$
$$\text{where} \quad \mu_b \approx \tan 31° = 0.6$$
$$\text{Thus,} \quad F_{ss} = 1.08$$

Thus, the proposed cross section is likely to satisfy the seismic performance requirements at L1 earthquake motion.

The pressures at foundation are derived as follows:
Eccentricity

$$e_{ex} = (15{,}927.9 + 414.1 \times 0.34 - 208.8 \times 5.80 - 395.2 \times 6.80)/3{,}298.5$$
$$= 3.69 \text{ m} > 15.60/6 = 2.60 \text{ m}$$

Therefore, the vertical stress applied at the both ends of footing A (sea side) and B (land side) are given as

$\sigma_B = 0$

$\sigma_A = 2 \times (\text{total vertical load})/(3 \times (\text{width}))$

$\qquad = 2 \times 3{,}298.5/(3 \times (7.80 - 3.69)) = 535 \text{ kN/m}^2$

add surcharge at half its value $535 + \frac{1}{2} \times 20 = 545 \text{ kN/m}^2$

T8.1.3 Simplified dynamic analysis

The preliminary analysis includes the use of a simplified method to predict the approximate range of deformations expected. This method utilizes non-dimensional parameters with respect to caisson geometry, thickness of soil deposit below the caisson, and geotechnical conditions represented by SPT N-values of the subsoil below and behind the wall (see Section T7.2.1(2) in Technical Commentary 7). The displacement at the top of the caisson under the prescribed earthquake motion was estimated.

Among the four alternatives shown in Fig. T8.2, Cases 1 and 4 were chosen as examples. Case 1 represents a quay wall with non-liquefiable soil deposits and, in this example, equivalent SPT N-value, N_{65}, was assumed to be 14 and 20. The N_{65} is an SPT N-value measured with a Japanese SPT energy efficiency of 72% and normalized to an overburden pressure of 65 kPa (T4.4.2 in Technical Commentary 4). Case 4 represents a quay wall with liquefiable soil deposits and in this example, equivalent SPT N-value, N_{65}, was assumed to be 7.

For a wall height, H, of 18.5 m, and a width, W, of 15.6 m, the normalized caisson geometry is

$$\frac{W}{H} = \frac{15.6\,\text{m}}{18.5\,\text{m}} = 0.84$$

The normalized thickness of soil deposit below the caisson is (see Fig. T7.14 for definitions)

$$\frac{D1}{H} = \frac{18.5\,\text{m}}{18.5\,\text{m}} = 1.0$$

For the L1 earthquake motion ($a_{\max} = 0.15g$ at the bedrock), the residual lateral displacement was estimated through Fig. T7.16(b), referring to $W/H = 0.9$. This approximation ($W/H = 0.84 \approx 0.9$) was quite reasonable as can be inferred by the graph of Fig. T7.18(b).

Thus, we have

Case 1 (Non-liquefiable):

$\qquad$ For $N_{65} = 14$: $\dfrac{d}{H} \cong 0.005$, hence, $d \approx 0.005 \times 18.5 = 0.09$ m

$\qquad$ For $N_{65} = 20$: $\dfrac{d}{H} \cong 0$, hence, $d \approx 0$ m

Case 4 (Liquefiable):

$$\text{For } N_{65} = 7: \frac{d}{H} \cong 0.04, \text{ hence, } d \approx 0.04 \times 18.50 = 0.74 \text{ m}$$

For the L2 earthquake motion ($a_{max} = 0.30g$ at the bedrock), the displacements become Case 1 (Non-liquefiable):

$$\text{For } N_{65} = 14: \frac{d}{H} \cong 0.03, \text{ hence, } d \approx 0.03 \times 18.5 = 0.56 \text{ m}$$

$$\text{For } N_{65} = 20: \frac{d}{H} \cong 0.0075, \text{ hence, } d \approx 0.0075 \times 18.5 = 0.14 \text{ m}$$

Case 4 (Liquefiable):

$$\text{For } N_{65} = 7: \frac{d}{H} \cong 0.18, \text{ hence, } d \approx 0.18 \times 18.50 = 3.33 \text{ m}$$

The requirements for Grade A performance were that the residual horizontal displacements at the top of the wall are less than 0.3 m and 0.9 m, for the L1 and L2 earthquake motions, respectively (Fig. T8.4). Table T8.4 compares the estimated displacements with the allowable displacements. Based on the simplified dynamic analysis, it was seen that if the soil deposit was non-liquefiable with N_{65} larger than about 10 (i.e. Case 1), this quay wall satisfied the required performance criteria.

T8.1.4   Simplified dynamic analysis for evaluation of liquefaction remediation

The preliminary analysis includes the use of a simplified method to predict the approximate range of deformations expected with respect to the area of soil improvement as remediation of liquefiable soils. This method utilizes non-dimensional parameters with respect to caisson geometry, and area of soil improvement to estimate displacement at the top of the caisson under the prescribed earthquake motion (see Section T7.2.1(3) in Technical Commentary 7).

Among the four alternatives shown in Fig. T8.2, Case 2 was chosen as the example. Using the above parameters, it is possible to enter the design chart and obtain estimates for the permanent lateral deformation at the top of the caisson.

Table T8.4. Comparison of estimated and allowable displacement for a caisson wall.

Earthquake levels	Estimated displacement (m)			Allowable displacement (m)
	Case 4	Case 1		
	$N_{65} = 7$	$N_{65} = 14$	$N_{65} = 20$	
Level 1	0.74	0.09	0	0.3
Level 2	3.33	0.56	0.14	0.9

Similarly to Section T8.1.3, the normalized caisson geometry is

$$\frac{W}{H} = \frac{15.6\,\text{m}}{18.5\,\text{m}} = 0.84$$

Assuming that the rubble fill is non-liquefiable, it is treated as a zone of soil improvement, with an equivalent width (L) equal to approximately half of the rubble width (9.9 m). Therefore, the normalized soil improvement is

$$\frac{L}{H} = \frac{9.9\,\text{m}}{18.5\,\text{m}} = 0.54$$

For the L1 earthquake motion (M_w 6.0 and $a_{\text{max(rock)}}$ of 0.15g), the site response study indicated that the maximum acceleration at the elevation of the dredge line within the backfill ($a_{\text{max(d)}}$) is equal to 0.25g. The magnitude scaling factor (*MSF*) was determined as 2.0 from Table T7.8 in Technical Commentary 7. The normalized earthquake motion is

$$\frac{a_{\text{max(d)}}}{MSF} = \frac{0.25g}{2.0} = 0.13g$$

The loose material has SPT $(N_1)_{60}$-values approximately equal to 10. Therefore, Fig. T7.20a in Technical Commentary 7 was used to determine the normalized and dimensional lateral displacements at the top of the caisson as

$$\frac{d}{H} \cong 0.03$$

$$d = (0.03)(18.5\ \text{m}) = 0.56\ \text{m}$$

If the soil in the back land was improved, so that the SPT $(N_1)_{60}$-values were approximately 20, Fig. T7.20b in Technical Commentary 7 would be used and the displacement would be estimated as follows:

$$\frac{d}{H} \cong 0.01$$

$$d = (0.01)(18.5\ \text{m}) = 0.19\ \text{m}$$

For the L2 earthquake motion (M_w 7.0 and $a_{\text{max(rock)}}$ of 0.30g), the site response study indicated that the maximum acceleration at the elevation of the dredge line within the backfill ($a_{\text{max(d)}}$) is equal to 0.35g. The magnitude scaling factor (*MSF*) was determined as 1.25 from Table T7.8 in Technical Commentary 7. The normalized earthquake motion is

$$\frac{a_{\text{max(d)}}}{MSF} = \frac{0.35g}{1.25} = 0.28g$$

The loose material has SPT $(N_1)_{60}$-values that are approximately equal to 10, and therefore, Fig. T7.20a in Technical Commentary 7 was used to determine that the lateral displacement at the top of the caisson was

$$\frac{d}{H} \cong 0.08 \text{ to } 0.09$$

$$d = (0.08 \text{ to } 0.09)(18.5 \text{ m}) = 1.5 \text{ to } 1.7 \text{ m}$$

If the soil in the back land was improved, so that the SPT $(N_1)_{60}$-values were approximately 20, Fig. T7.20b in Technical Commentary 7 would be used and the displacement would be estimated as follows:

$$\frac{d}{H} \cong 0.04$$

$$d = (0.04)(18.5 \text{ m}) = 0.74 \text{ m}$$

As mentioned earlier, the requirements for Grade A performance are that the residual horizontal displacements at the top of the wall are less than 0.3 m and 0.9 m, for the L1 and L2 earthquake motions, respectively (Fig. T8.4). Based on the simplified analysis, it is necessary to improve the backfill soil sufficiently to preclude liquefaction from occurring. It should be noted that this simplified analysis method is not able to estimate the tilting of the caisson.

T8.1.5   Dynamic analysis

As mentioned earlier, the limiting curves for seismic performance evaluation were defined, as shown in Fig. T8.4, for a wall height of 18.5 m based the damage criteria shown in Table T8.1.

The cross section of the gravity quay wall was idealized into plane strain finite elements shown in Fig. T8.8. Element size was defined small enough to allow seismic wave propagation of up to 2 Hz throughout the analysis, including the liquefaction state. In order to simulate the incoming and outgoing waves through the boundaries of the analysis domain, equivalent viscous dampers were used at the boundaries. The effect of the free field motions was also taken into account by performing one dimensional response analysis

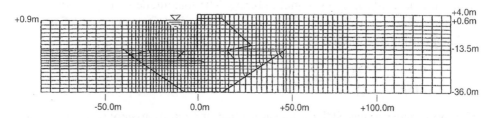

Fig. T8.8. Finite element mesh for analysis.

at the outside fields, and assigning the free field motion through the viscous dampers. Before the earthquake response analysis, a static analysis was performed with gravity under drained conditions to simulate the stress conditions before the earthquake. The results of the static analyses were used for the initial conditions for the earthquake response analyses. The input earthquake motions were those recorded at the Port Island site during the Great Hanshin earthquake.

Concrete caisson was modelled by linear elements, with soil-structure interfaces modelled by joint elements, as shown in Fig. T8.9. The parameters are shown in Table T8.2. The seawater in front of the caisson was modelled as an incompressible fluid.

Parameters for soil elements were calibrated based on the results of the geotechnical investigation as shown in Table T8.3. In particular, undrained cyclic properties of sand were calibrated as shown in Fig. T8.10. It should be noted that appropriate characterization of pore water pressure build-up and the increase in shear strain amplitude are important for parameter calibration, since these undrained cyclic properties significantly affect the degree of deformation of the soil-structure system.

Seismic analysis was performed for the four alternatives shown in Fig. T8.2 over L1 and L2 earthquake motions. Computed residual deformations of the quay wall for Cases 1 and 4 for L2 earthquake motion are shown in Fig. T8.11. The results of the analysis are summarized in Fig. T8.12. Although all the alternatives satisfy the criteria for tilting as shown in Fig. T8.12(b), only Case 1 satisfied the horizontal displacement criteria as shown in Fig. T8.12(a). The remaining criteria with for differential settlement between apron and non-apron areas for L1, as shown in Table T8.1, are also satisfied by Case 1. Consequently, Case 1 was proposed as the recommended design of a Grade A gravity quay wall.

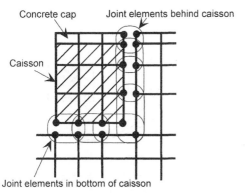

Fig. T8.9. Modelling of caisson structure.

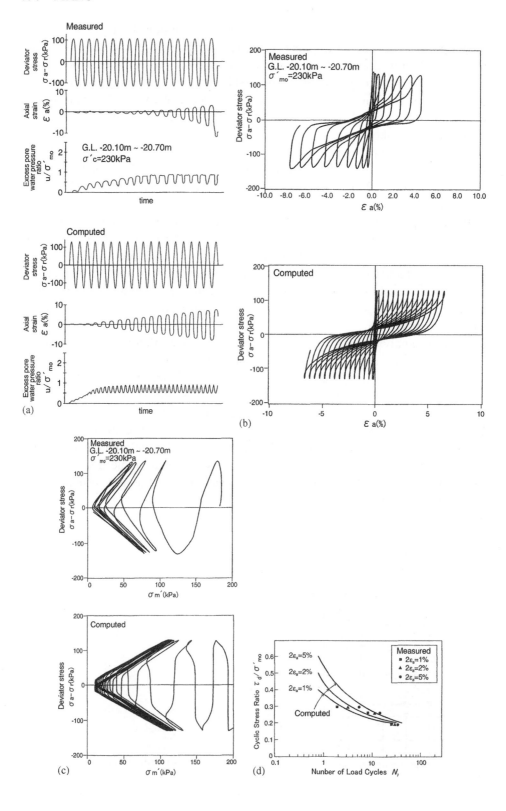

(a)

(b)

(c)

(d)

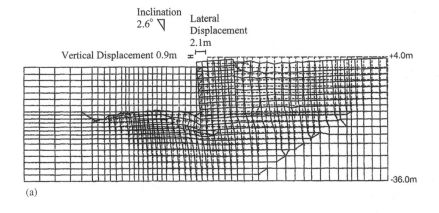

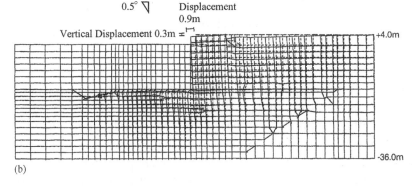

Fig. T8.11. Computed residual deformation.
          (a) Alternative case 4.
          (b) Alternative case 1.

T8.2   SHEET PILE QUAY WALL (GRADE B)

T8.2.1   Performance requirements and design conditions

A new sheet pile quay wall with a 6 m water depth was being constructed for a small harbour in a seismically active region. The seismic performance grade of this relatively small quay wall was designated as Grade B. Simplified analysis was performed as a preliminary stage to determine the depth of sheet pile embedment and the structural parameters of the sheet pile wall and tie rod. Based on Grade B requirements, a simplified dynamic analysis was also performed to estimate the

Fig. T8.10. Undrained cyclic properties: laboratory test and computed results.
          (a) Time histories.
          (b) Stress-strain relationships.
          (c) Stress paths.
          (d) Cyclic resistance.

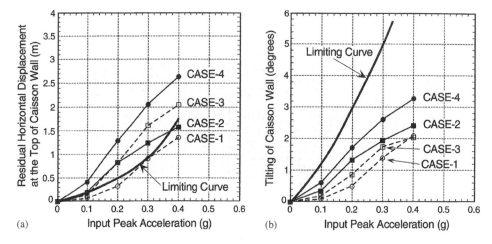

Fig. T8.12. Seismic response and limiting curve for acceptable damage.
(a) Residual horizontal displacement.
(b) Tilting.

deformations of the sheet pile quay walls and the required extent of soil improvement to limit the permanent deformations.

The simplified analysis was based on the pseudo-static analysis using Mononobe-Okabe's method (see Section T7.1.1 in Technical Commentary 7). When applied to sheet pile wall analysis, the pseudo-static analysis differs depending on the regional design practice. The North American practice (Ebeling and Morrison, 1992) and the Japanese practice (Ministry of Transport, Japan, 1999) were used to illustrate the differences. The simplified dynamic analysis to estimate the wall deformations was based on the method discussed in Section T7.2.2 (McCullough, 1998; McCullough and Dickenson, 1998).

The proposed geometry of the sheet pile quay wall is shown in Fig. T8.13. The wall was backfilled with hydraulically placed granular materials that were improved adjacent to the wall due to soil liquefaction concerns, which have been noted to cause serious damage to sheet pile quay walls (Kitajima and Uwabe, 1979; Gazetas, et al., 1990). The foundation soils were fairly dense granular materials, with relatively high SPT $(N_1)_{60}$-values. The anchor system for the wall consisted of a continuous sheet pile wall. It was assumed during this analysis, that no excess pore pressures were generated adjacent to the wall, a reasonable assumption since the SPT $(N_1)_{60}$-values in the foundation and improved backfill are approximately 20. The Mononobe-Okabe method requires an initial estimate of the depth of embedment (D_{emb}). From several iterations, this was chosen as 3.75 meters. For simplicity, the water table behind the wall was assumed to be the same as that of the sea water level. Surcharge on the quay wall was not considered in this example problem.

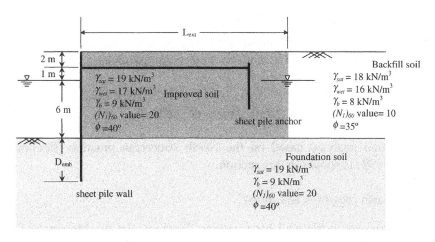

Fig. T8.13. A proposed sheet pile quay wall.

Table T8.5. Qualitative and quantitative description of the reported degrees of damage (after Kitajima and Uwabe, 1979).

Degree of damage (description of damage)	Maximum permanent displacement at top of sheet pile (cm)	Average permanent displacement at top of sheet pile (cm)
No damage	0	0
Negligible damage to the wall itself; noticeable damage to related structures (i.e. concrete apron)	Less than 30	Less than 10
Noticeable damage to the wall itself	30 to 100	10 to 60
General shape of anchored sheet pile preserved, but significantly damaged	100 to 400	60 to 120
Complete destruction, no recognizable shape of wall remaining	Larger than 400	Larger than 120

Grade B performance requires that, for an L1 earthquake motion, the structure will be serviceable, and for an L2 earthquake motion, the structure will not collapse (Fig. 3.2 in the Main Text). For a sheet pile quay wall, these damage criteria can be approximately quantified by the relationships proposed by Kitajima and Uwabe (1979), and presented in Table T8.5. These relationships indicate that for the wall to be serviceable, the average deformations should be approximately 10 cm or less and that to prevent collapse of the wall, the average deformations should be approximately 120 cm or less. The specific Grade B requirement is that the deformation factor (d/H) be less than 1.5% (Table 4.2 in the Main Text), which for this 9 m high wall, is 13.5 cm.

A site response analysis was performed at the quay wall site for the L1 and L2 motions of $M_w = 6.5$ and 7.5 earthquakes. The peak ground accelerations at the ground surface for L1 and L2 motions were obtained as $a_{max(surface)} = 0.45g$ and $0.60g$, respectively.

T8.2.2   Simplified analysis (North American method)

The simplified analysis based on the North American practice (Ebeling and Morrison, 1992) is shown in this section.

(1) Active earth pressures and thrust

The soils adjacent to the wall have varying densities, yet the Mononobe-Okabe analysis method requires that the densities be uniform. It was therefore necessary to estimate equivalent densities assuming the wedge failure plane (Fig T8.14), where γ_{wet} is the unit weight of the improved backfill above the water table, γ_{sat} is the unit weight of the improved backfill beneath the water level and the unit weight of the foundation soil (since they are the same). The equivalent buoyant unit weight of the backfill soil is equal to

$$\gamma_e = \frac{\gamma_{wet}((H + D_{emb})^2 - (H_{sub} + D_{emb})^2) + \gamma_b (H_{sub} + D_{emb})^2}{(H + D_{emb})^2}$$

$$= \frac{(17 \text{ kN/m}^3)((12.75 \text{ m})^2 - (9.75 \text{ m})^2) + (9 \text{ kN/m}^3)(9.75 \text{ m})^2}{(12.75 \text{ m})^2}$$

$$= 12.3 \text{ kN/m}^3$$

The equivalent saturated unit weight is similarly obtained as

$$\gamma_{e\text{-sat}} = \frac{(17 \text{ kN/m}^3)((12.75 \text{ m})^2 - (9.75 \text{ m})^2) + (19 \text{ kN/m}^3)(9.75 \text{ m})^2}{(12.75 \text{ m}^2)}$$

$$= 18.2 \text{ kN/m}^3$$

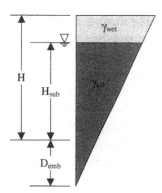

Fig. T8.14. Diagram used in estimating an equivalent soil unit weight.

The vertical seismic coefficient (k_v) was assumed to be zero and the horizontal seismic coefficient (k_h) was calculated as (see Section T7.1.1 in Technical Commentary 7)

$$k_h = \frac{1}{3}\left(\frac{a_{max}}{g}\right)^{1/3} = \frac{1}{3}\left(\frac{0.45g}{g}\right)^{1/3} = 0.26$$

The Mononobe-Okabe equations were originally derived for dry backfill. With the inclusion of water in this example, the horizontal seismic coefficient is modified using Eqn. (T7.5) as

$$k'_{h1} = k_h \cdot \frac{\gamma_{e\text{-sat}}}{\gamma_e} = 0.26 \cdot \frac{18.2 \text{ kN/m}^3}{12.3 \text{ kN/m}^3} = 0.38$$

The above equation is essentially the same as Eqn. (T7.7). The seismic inertia angle (Eqn. (T7.1) in Technical Commentary 7) is then

$$\psi_1 = \tan^{-1}(k'_{h1}) = \tan^{-1}(0.38) = 20.6°$$

The friction angle between the wall and soil (δ) was assumed to be 20°, half of the internal friction angle of the soil (ϕ).

For level backfill, as shown in the definition sketch, the coefficient of active earth pressure is calculated using Eqn. (T7.2) in Technical Commentary 7

$$K_{ae} = \frac{\cos^2(\phi - \psi_1)}{\cos\psi_1 \cos(\psi_1 + \delta)\left[1 + \sqrt{\frac{\sin(\phi + \delta)\ \sin(\phi - \psi_1)}{\cos(\delta + \psi_1)}}\right]^2}$$

$$= \frac{\cos^2(40° - 20.6°)}{\cos 20.6° \cos(20.6° + 20°)\left[1 + \sqrt{\frac{\sin(40° + 20°)\ \sin(40° - 20.6°)}{\cos(20° + 20.6°)}}\right]^2} = 0.48$$

The horizontal dynamic active thrust acting on the wall is calculated using Eqn. (T7.3) in Technical Commentary 7

$$
\begin{aligned}
(P_{ae})_x &= K_{ae} \cdot \tfrac{1}{2} \cdot \gamma_e \cdot (H + D_{emb})^2 \cdot \cos\delta \\
&= 0.48 \cdot \tfrac{1}{2} \cdot 12.3 \text{ kN/m}^3 \cdot (12.75 \text{ m})^2 \cdot \cos 20° \\
&= 452 \text{ kN per meter of wall}
\end{aligned}
$$

This force includes both the static, $(P_a)_x$, and incremental dynamic, $(\Delta P_{ae})_x$, forces

$$(P_{ae})_x = (P_a)_x + (\Delta P_{ae})_x$$

It is now necessary to determine the point of action of the dynamic force. It is known that the incremental dynamic portion acts at an elevation ($\xi_{\Delta P_{ae}}$) above the pile toe that is 60% of ($H + D_{emb}$). Using this information, and the calculation of the static component force $(P_a)_x$ and its location (ξ_{P_a}), it is possible to estimate the location of the dynamic earth pressure force ($\xi_{P_{ae}}$) as

$$\xi_{P_{ae}} = \frac{(P_a)_x \cdot \xi_{P_a} + (\Delta P_{ae})_x \cdot \xi_{\Delta P_{ae}}}{(P_{ae})_x}$$

In order to find the static earth pressure force, Coulomb's earth pressure theory is used, with the coefficient of active earth pressure (K_a) being

$$K_a = \frac{\cos^2 \phi}{\cos \delta \left[1 + \sqrt{\dfrac{\sin(\phi + \delta)\sin \phi}{\cos(\delta)}} \right]^2}$$

$$= \frac{\cos^2 40°}{\cos 20° \left[1 + \sqrt{\dfrac{\sin(40° + 20°)\sin 40°}{\cos(20°)}} \right]^2} = 0.20$$

Using the active earth pressure coefficient and the areas represented by 1, 2 and 3 in Fig. T8.15, the horizontal components of the static active earth pressure forces and the location of these forces (above the pile toe) is given by

$$(P_{a1})_x = K_a \cdot [\tfrac{1}{2} \cdot 17 \text{ kN/m}^3 \cdot (3 \text{ m})^2] \cdot \cos \delta$$
$$= 0.20 \cdot \tfrac{1}{2} \cdot 17 \text{ kN/m}^3 \cdot (3 \text{ m})^2 \cdot \cos 20°$$
$$= 14.3 \text{ kN} \text{ per meter of wall}$$

$$\xi_{pa1} = 3.75 \text{m} + 6 \text{m} + \tfrac{1}{3} \cdot 3 \text{m} = 10.75 \text{m}$$

$$(P_{a2})_x = K_a \cdot \left[17 \text{ kN/m}^3 \cdot 3 \text{ m} \cdot 9.75 \text{ m} \right] \cdot \cos \delta$$
$$= 0.20 \cdot 17 \text{ kN/m}^3 \cdot 3 \text{ m} \cdot 9.75 \text{ m} \cdot \cos 20°$$
$$= 93.2 \text{ kN} \text{ per meter of wall}$$

$$\xi_{pa2} = \tfrac{1}{2} \cdot 9.75 \text{ m} = 4.88 \text{ m}$$

$$(P_{a3})_x = K_a \cdot \left[\tfrac{1}{2} \cdot 9 \text{ kN/m}^3 \cdot 9.75 \text{ m} \cdot 9.75 \text{ m} \right] \cdot \cos \delta$$
$$= 0.20 \cdot \tfrac{1}{2} \cdot 9 \text{ kN/m}^3 \cdot 9.75 \text{ m} \cdot 9.75 \text{ m} \cdot \cos 20°$$
$$= 80.2 \text{ kN} \text{ per meter of wall}$$

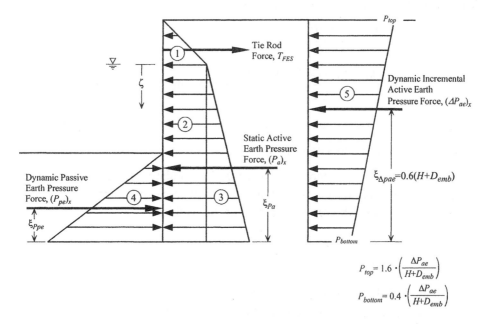

Fig. T8.15. Resultant earth pressure forces.

$$\xi_{pa3} = \tfrac{1}{3} \cdot 9.75\,\text{m} = 3.25\,\text{m}$$

Therefore, the total static active earth pressure is

$$(P_a)_x = (P_{a1})_x + (P_{a2})_x + (P_{a3})_x$$

$$= 14.3\ \text{kN} + 93.2\ \text{kN} + 80.2\ \text{kN} = 188\ \text{kN per meter of wall}$$

The total static active earth pressure force acts at the following distance above the pile toe:

$$\xi_{pa} = \frac{(P_{a1})_x \cdot \xi_{pa1} + (P_{a2})_x \cdot \xi_{pa2} + (P_{a3})_x \cdot \xi_{pa3}}{(P_a)_x}$$

$$= \frac{14.3\ \text{kN} \cdot 10.75\ \text{m} + 93.2\ \text{kN} \cdot 4.88\ \text{m} + 80.2\ \text{kN} \cdot 3.25\ \text{m}}{188\ \text{kN}} = 4.6\ \text{m}$$

The incremental dynamic active earth pressure force is then equal to

$$(\Delta P_{ae})_x = (P_{ae})_x - (P_a)_x = 452\ \text{kN} - 188\ \text{kN} = 264\ \text{kN per meter of wall}$$

The location of the incremental active earth pressure force is

$$\xi_{\Delta P_{ae}} = 0.6 \cdot (3.75\ \text{m} + 6\ \text{m} + 3\ \text{m}) = 7.65\ \text{m}$$

Therefore, the dynamic active earth pressure acts at

$$\xi_{P_{ae}} = \frac{(P_a)_x \cdot \xi_{Pa} + (\Delta P_{ae})_x \cdot \xi_{\Delta P_{ae}}}{(P_{ae})_x}$$

$$= \frac{188 \text{ kN} \cdot 4.6 \text{ m} + 265 \text{ kN} \cdot 7.7 \text{ m}}{452 \text{ kN}} = 6.4 \text{ m above the pile toe}$$

i.e. at about 50% of the total wall height.

(2) Passive earth pressures and thrust

Following the same procedures as those used in calculating the active earth pressures behind the wall, the equivalent horizontal seismic coefficient for use in estimating the passive earth pressure in front of the wall is

$$k'_{h2} = k_h \cdot \frac{\gamma_{sat}}{\gamma_b} = 0.26 \cdot \frac{19 \text{ kN/m}^3}{9 \text{ kN/m}^3} = 0.55$$

The seismic inertia angle (Eqn. T7.1) is then

$$\psi_2 = \tan^{-1}(k'_{h2}) = \tan^{-1}(0.55) = 28.8°$$

A factor of safety of 1.2 is typically applied to the passive soil strengths to take into account the partial mobilization of passive resistance at the design condition (Ebeling and Morrison, 1992). Factoring the angle of internal friction for the soil strength

$$\phi^* = \arctan\left(\frac{\tan(\phi)}{1.2}\right) = \arctan\left(\frac{\tan(40°)}{1.2}\right) = 35.0°$$

The factored angle of friction between the wall and soil is

$$\delta^* = \arctan\left(\frac{\tan(\delta)}{1.2}\right) = \arctan\left(\frac{\tan(20°)}{1.2}\right) = 16.9°$$

The passive earth pressure coefficient is calculated as

$$K_{pe} = \frac{\cos^2(\phi^* - \psi_2)}{\cos\psi_2 \cos(\psi_2 + \delta^*)\left[1 - \sqrt{\frac{\sin(\phi^* + \delta^*)\sin(\phi^* - \psi_2)}{\cos(\delta^* + \psi_2)}}\right]^2}$$

$$= \frac{\cos^2(35.0° - 28.8°)}{\cos 28.8° \cos(28.8° + 16.9°)\left[1 - \sqrt{\frac{\sin(35.0° + 16.9°)\sin(35.0° - 28.8°)}{\cos(16.9° + 28.8°)}}\right]^2} = 3.9$$

The horizontal passive earth pressure force is then given by

$$(P_{pe})_x = K_{pe} \cdot \tfrac{1}{2}\gamma_b \cdot (D_{emb})^2 \cdot \cos\delta'$$
$$= 3.9 \cdot \tfrac{1}{2} \cdot 9 \text{ kN/m}^3 (3.75 \text{ m})^2 \cdot \cos 16.9° = 234 \text{ kN per meter of wall}$$

The horizontal passive earth pressure force acts at approximately one-third the depth of embedment from the pile toe (Ebeling and Morrison, 1992)

$$\xi_{Ppe} \cong \tfrac{1}{3} \cdot 3.75 \text{m} = 1.25 \text{m above the pile toe}$$

(3) Hydrodynamic force

Since the fluid elevation is equal on both sides of the wall, the net hydrostatic fluid pressure is zero, but there is a hydrodynamic fluid force (P_{wd}), that is estimated using the seismic coefficient, and the depth of water in front of the wall (H_w), using Eqn. (T7.8)

$$P_{wd} = \tfrac{7}{12} \cdot k_h \cdot \gamma_w \cdot (H_w)^2 = \tfrac{7}{12} \cdot 0.26 \cdot 9.81 \text{ kN/m}^3 \cdot (6 \text{ m})^2$$
$$= 52.6 \text{ kN per meter of wall}$$

The hydrodynamic force acts at

$$\xi_{P_{wd}} = 0.4 \cdot H_w + D_{emb} = 0.4 \cdot 6 \text{ m} + 3.75 \text{ m} = 6.15 \text{ m above the pile toe}$$

(4) Moment balance about the tie-rod

The depth of embedment is calculated by summing the moments about the elevation of the tie-rod/anchor, with the distance from the pile toe to the tie rod being 10.75 m.

$$\sum M_{\text{tie-rod}} = (P_{pe})_x \cdot (10.75 \text{ m} - \xi_{P_{pe}}) - (P_{ae})_x \cdot (10.75 \text{ m} - \xi_{P_{ae}}) - P_{wd} \cdot (10.75 \text{ m} - \xi_{P_{wd}})$$
$$= (234 \text{ kN}) \cdot (10.75 \text{ m} - 1.25 \text{ m}) - (452 \text{ kN}) \cdot$$
$$(10.75 \text{ m} - 6.40 \text{ m}) - (52.6 \text{ kN}) \cdot (10.75 \text{ m} - 6.15 \text{ m})$$
$$= 16.2 \text{ kN} \cdot \text{m out of balance}$$

Using an embedment depth of 3.75 meters, results in a relatively small conservative moment imbalance, and therefore, will be used as the seismic design depth of embedment. It should also be noted that the depth of embedment may need to be increased depending on environmental and operational conditions (such as scour, propeller wash, etc.).

(5) Tie-rod force (T_{FES})

The tie-rod force (T_{FES}) is calculated by summing the forces in the horizontal direction

$$T_{FES} = (P_{ae})_x + P_{wd} - (P_{pe})_x$$
$$= 452 \text{ kN} + 52.6 \text{ kN} - 234 \text{ kN} = 270 \text{ kN per meter of wall}$$

(6) Maximum Bending Moment (M_{FES} and M_{design})

The maximum bending moment occurs at the same elevation as zero shear within the sheet pile wall. The zero shear is calculated at elevation ζ beneath the fluid level (Fig. 8.15). Since the dynamic active earth pressure distribution is unknown, it is separated into two parts, the static active earth pressure and the incremental dynamic active earth pressure. Above the dredge level, the contribution due to the static active earth pressure is (refer to Fig. T8.15)

$$(P_a)_x = [[\tfrac{1}{2} \cdot 17 \text{ kN/m}^3 \cdot (3 \text{ m})^2] + [17 \text{ kN/m}^3 \cdot 3 \text{ m} \cdot \zeta] + [\tfrac{1}{2} \cdot 9 \text{ kN/m}^3 \cdot \zeta^2]]$$
$$\cdot K_a \cdot \cos \delta$$
$$= [76.5 \text{ kN/m} + 51 \text{ kN/m}^2 \cdot \zeta + 4.5 \text{ kN/m}^3 \cdot \zeta^2] \cdot 0.20 \cdot \cos 20$$
$$= 14.4 \text{ kN} + 9.6 \text{ kN/m} \cdot \zeta + 0.84 \text{ kN/m}^2 \cdot \zeta^2 \text{ per meter of wall}$$

The incremental dynamic active earth pressures is shown as area 5 in Fig. T8.15 (Ebeling and Morrison, 1992).

$$p_{top} = 1.6 \cdot \left(\frac{(\Delta P_{ae})_x}{H + D_{emb}} \right) = 1.6 \cdot \left(\frac{265 \text{ kN}}{9 \text{ m} + 3.75 \text{ m}} \right)$$
$$= 33.2 \text{ kN/m per meter of wall}$$

$$p_{bottom} = 0.4 \cdot \left(\frac{(\Delta P_{ae})_x}{H + D_{emb}} \right) = 0.4 \cdot \left(\frac{265 \text{ kN}}{9 \text{ m} + 3.75 \text{ m}} \right)$$
$$= 8.3 \text{ kN/m per meter of wall}$$

$$(\Delta P_{ae})_x = \tfrac{1}{2} \cdot (3 \text{ m} + \zeta) \left[2 p_{top} - \left[\frac{p_{top} - p_{bottom}}{H + D_{emb}} \right] (3 \text{ m} + \zeta) \right]$$

$$\frac{p_{top} - p_{bottom}}{H + D_{emb}} = 1.95 \text{ kN/m}^2 \text{ per meter of wall}$$

$$(\Delta P_{ae})_x = \left(-\frac{1.95 \text{ kN/m}^2}{2} \right) \cdot \zeta^2 + (-3 \text{ m} \cdot 1.95 \text{ kN/m}^2 + 33.2 \text{ kN/m}) \cdot \zeta$$
$$+ \left(3 \text{ m} \cdot 33.2 \text{ kN/m} - 9 \text{ m}^2 \cdot \frac{1.95 \text{ kN/m}^2}{2} \right)$$

$$= -0.98 \text{ kN/m}^2 \cdot \zeta^2 + 27.3 \text{ kN/m} \cdot \zeta + 90.8 \text{ kN per meter of wall}$$

The distribution of the hydrodynamic fluid pressure is given by

$$P_{wd} = \tfrac{7}{12} \cdot k_h \cdot \gamma_w \cdot \zeta^2 = \tfrac{7}{12} \cdot 0.26 \cdot 9.81 \text{ kN/m}^3 \cdot \zeta^2$$
$$= 1.46 \text{ kN/m}^2 \cdot \zeta^2 \text{ per meter of wall}$$

At the point of zero shear, the summation of the forces in the horizontal direction is equal to zero

$$(P_a)_x + (\Delta P_{ae})_x + P_{wd} - T_{FES} = 0$$

Substituting in the values and solving for ζ

$$(0.84 - 0.98 + 1.46) \text{ kN/m}^2 \cdot \zeta^2 + (9.6 + 27.3) \text{ kN/m} \cdot \zeta$$
$$+ (14.4 + 90.8) \text{ kN} - 270 \text{ kN} = 0$$
$$\zeta = 3.93 \text{ m beneath the fluid level}$$

It is now necessary to sum the moments an elevation of ζ. This will provide the maximum moment in the sheet pile wall. Table T8.6 provides the data for summing the moments about the tie rod. From the table, the maximum moment in the sheet pile wall (M_{FES}) is approximately 450 kN per meter of wall, which occurs at approximately 3.93 m beneath the water table.

The maximum moment is reduced using Rowe's (1952) moment reduction procedure (Fig. T8.16). This procedure utilizes a flexibility number (ρ), and estimates the reduction in moment for various sheet pile sections.

$$\rho = \frac{(H + D_{emb})^4}{E \cdot I} \tag{T8.1}$$

Table T8.6. Forces and lever arms for calculating the maximum moment.

Horizontal force designation	Horizontal force (kN per meter of wall)	Lever arm about tie rod (m)	Moment about tie rod (kN·m per meter of wall)
$(P_{a1})_x$	14.4	0	0
$(P_{a2})_x$	$9.6 \cdot \zeta = 37.6$	$\zeta/2 + 1 = 3.0$	113
$(P_{a3})_x$	$0.84 \cdot \zeta^2 = 13.02$	$2/3 \cdot \zeta + 1 = 3.62$	47.1
$(\Delta P_{ae})_x$	$-0.98 \cdot \zeta^2 + 27.3 \cdot \zeta + 90.8 = 183$	1.169*	214
P_{wd}	$1.49 \cdot \zeta^2 = 22.6$	$\zeta + 1 - 0.4 \cdot \zeta = 3.36$	75.8
		$\Sigma (M_{FES}) =$	450

*Using the stress distribution from Fig. T8.15.

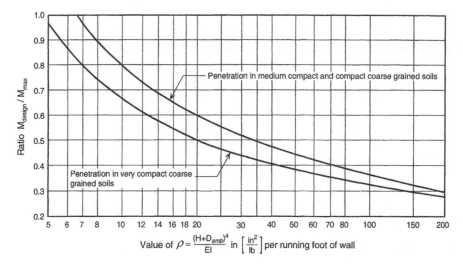

Fig. T8.16. Moment reduction curves (after Rowe, 1952; as reported in U.S. Navy, 1982).

where

E: Young's modulus (200 GPa)

I: moment of inertia (m^4)

The flexibility number is then calculated by (and has been converted to English units for the use of Rowe's chart)

$$\rho = \frac{(6\text{ m} + 1\text{ m} + 2\text{ m} + 3.75\text{ m})^4}{200 \times 10^9 \text{ Pa} \cdot I} = \frac{26427 \text{ m}^4}{200 \times 10^9 \text{ Pa} \cdot I} = \frac{2189 \text{ in}^4}{\text{psi} \cdot I}$$

The allowable stress (σ_{allow}) of the sheet pile is commonly taken as approximately 87% of the yield stress (σ_y) for seismic design, where the yield stress for steel is taken as 250 MPa.

$$\sigma_{allow} = 0.87 \cdot \sigma_y = 0.87 \cdot 250 \times 10^6 \text{ Pa} = 218 \times 10^6 \text{ Pa}$$

The allowable moment (M_{allow}) can be calculated as

$$M_{allow} = Z_e \cdot \sigma_{allow} \text{ per meter of wall}$$

where Z_e is the elastic section modulus of the sheet pile wall (m^3)

The sheet pile properties, and the design and allowable moments are shown in Table T8.7 for several standard Bethlehem Steel sheet pile sections. From this table, it can be seen that section PZ22 is the sheet pile wall section best suited for this specific design scenario. The design bending moment is

$$M_{design} = 203 \text{ kNm} \text{ per meter of wall}$$

Table T8.7. Calculations of the design and allowable sheet pile wall moments.

Sheet pile (Bethlehem Steel)	I (in^4 per ft of wall)	Z_e (cm^3 per m of wall)	$\rho*$ (1/psi per ft of wall)	M_{design}/M_{FES} (Fig. T8.16)	M_{design} (kNm per m)	M_{allow} (kNm per m)
PZ22	84.4	973	25.93	0.45	203	212
PZ27	184.2	1624	11.88	0.61	275	354
PZ35	361.2	2607	6.06	0.86	387	568
PZ40	490.8	3263	4.46	1.0	450	711

* Rowe's flexibility number.

(7) Design of the tie rods (T_{design})

The seismic design of the tie rods normally utilize a factor of safety of 1.3, therefore the design tie-rod force is

$$T_{design} = 1.3 \cdot T_{FES} = 1.3 \cdot 270 \text{ kN/m} = 351 \text{ kN per meter of wall}$$

Assuming a 2 meter spacing of the tie-rods, each tie-rod has a design load of

$$T_{2m\text{-design}} = 2 \text{ m} \cdot 351 \text{ kN/m} = 702 \text{ kN}$$

Assuming a standard allowable stress in the tie-rods of 60% of the yield stress, the minimum diameter of the tie-rod is calculated as

$$\text{Diameter} = \sqrt{4 \cdot \frac{\left(\dfrac{T_{2\,m\text{-design}}}{0.6 \cdot \sigma_y} \right)}{\pi}} = \sqrt{4 \cdot \frac{\left(\dfrac{702 \text{ kN}}{0.6 \cdot 250 \times 10^6 \text{ Pa}} \right)}{3.14159}} = 7.7 \text{ cm}$$

(8) Design of the anchor

The anchor design follows the same procedures as outlined for the design of the sheet pile wall, with the seismic active pressures behind the anchor and T_{FES} being balanced by the seismic passive resistance in front of the anchor. An anchor design has not been performed for this example, and the reader is referred to Ebeling and Morrison (1992) for a complete design example.

(9) Location of the anchor

It is commonly recommended (Ebeling and Morrison, 1992) that the location of the anchor be such that the active failure plane behind the sheet pile wall does not intersect the passive failure plane in front of the anchor, as shown in Fig. T8.17. The seismic active (α_{ae}) and passive (α_{pe}) angles of failure for a level surface

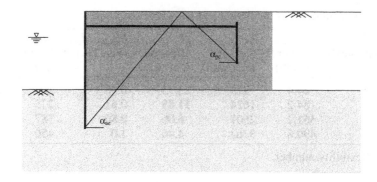

Fig. T8.17. Recommended anchor location (North American Practice).

condition are calculated using

$$\alpha_{ae} = \phi - \psi + \arctan\left[\frac{-\tan(\phi - \psi) + c_1}{1 + [\tan(\delta + \psi)(\tan(\phi - \psi) + \cot(\phi - \psi))]}\right] \qquad \text{(T8.2a)}$$

$$\alpha_{pe} = -\phi + \psi + \arctan\left[\frac{\tan(\phi - \psi) + c_2}{1 + [\tan(\delta + \psi)(\tan(\phi - \psi) + \cot(\phi - \psi))]}\right] \qquad \text{(T8.2b)}$$

where

$$c_1 = \sqrt{\tan(\phi - \psi)[\tan(\phi - \psi) + \cot(\phi - \psi)][1 + \tan(\delta + \psi)\cot(\phi - \psi)]} \qquad \text{(T8.2c)}$$

$$c_2 = \sqrt{\tan(\phi - \psi)[\tan(\phi - \psi) + \cot(\phi - \psi)][1 + \tan(\delta + \psi)\cot(\phi - \psi)]} \qquad \text{(T8.2d)}$$

T8.2.3   Simplified analysis (Japanese method)

The method adopted in Japanese design practice is based on the analysis of a beam embedded in Winkler foundation (Takahashi et al., 1993; Ministry of Transport, Japan, 1999). This method is essentially the same as Rowe's approach to sheet pile wall (1952, 1955). In the following example, the first trial for embedment depth was 4.0 m.

(1) Active earth pressures and thrust

The case of partially submerged soil is treated by idealizing the effect of overlaying soil layer as a surcharge to the soil layer of interest: Eqn. (T7.7′) is used instead of Eqn. (T7.7). For simplicity, design surcharge q_{sur} will be omitted in this example (i.e. $q_{sur} = 0$). Active earth pressures at designated elevations are

Table T8.8. Seismic active earth pressures at designated elevations.

Elevation (m)	Thickness h_i (m)	γ_i (kN/m³)	ϕ (°)	δ (°)	$\Sigma\gamma_i h_i$ (kN/m²)	k'_h	$K_{ae}\cos\delta$	p_{ae} (kN/m²)
+ 3.0	3.0	17	40	15	0	0.26	0.35	0
− 0.0					51			18
0.0	6.0	9	40	15	51	0.41	0.49	25
− 6.0					105			51
− 6.0	4.0	9	40	15	105	0.46	0.55	58
− 10.0					141			78

computed as shown in Table T8.8 using the equation

$$p_{ae} = K_{ae}\cos\delta(\Sigma\gamma_i h_i + q_{sur}) \tag{T8.3}$$

where γ_i is $(\gamma_{wet})_i$ and $(\gamma_b)_i$ above and below the water table, respectively.

The active earth thrust is (see Table T8.9)

$$P_{ae} = \Sigma\tfrac{1}{2}\{(p_{ae})_{top} + (p_{ae})_{bottom}\}_i\, h_i = 527 \text{ kN per meter of wall}$$

(2) Passive earth pressures and thrust

Passive earth pressures and thrust are calculated through a procedure similar to that shown in Section T.8.2.2. Using typical wall friction angle in Japanese design practice (i.e. $\delta = 15°$), the passive earth pressure coefficient is calculated as follows:

$$k'_{h2} = k_h \frac{\gamma_{sat}}{\gamma_b} = 0.26 \cdot \frac{19}{9} = 0.55$$

$$K_{pe} = \frac{\cos^2(\phi - \psi_2)}{\cos\psi_2 \cos(\psi_2 + \delta)\left[1 - \sqrt{\dfrac{\sin(\phi + \delta)\,\sin(\phi - \psi_2)}{\cos(\delta + \psi_2)}}\right]^2}$$

$$= \frac{\cos^2(40° - 28.8°)}{\cos 28.8° \cos(28.8° + 15.0°)\left[1 - \sqrt{\dfrac{\sin(40° + 15.0°)\,\sin(40° - 28.8°)}{\cos(15.0° + 28.8°)}}\right]^2}$$

$$= 5.43$$

The horizontal passive earth pressure at the wall toe is given by

$$p_{pe} = K_{pe} \cos\delta \cdot \gamma_i D_{emb} = 5.43 \times \cos 15.0° \times 9.0 \times 4.0 = 189 \text{ kN/m}^2$$

The horizontal passive earth thrust is then given by

$$P_{pe} = \tfrac{1}{2} 189 \times 4.0 = 378 \text{ kN per meter of wall}$$

acting at

$$z_{P_{pe}} \cong 7.0 + \tfrac{2}{3} \cdot 4.0 = 9.7 \text{ m below the tie-rod level}$$

(3) Hydrodynamic forces

The hydrodynamic forces are the same as shown in Section T8.2.2 and given as

$$P_{wd} = \tfrac{7}{12} \cdot k_h \cdot \gamma_w \cdot (H_w)^2 = \tfrac{7}{12} \cdot 0.26 \cdot 9.81 \text{ kN/m}^3 \cdot (6 \text{ m})^2$$
$$= 52.6 \text{ kN per meter of wall}$$

The hydrodynamic force acts at

$$z_{P_{wd}} = 7.0 - 0.4 \cdot H_w = 7.0 - 0.4 \cdot 6 \text{ m} = 4.6 \text{ m below the tie-rod}$$

(4) Moment balance about the tie-rod

The moment balance about the tie-rod is

$$\sum M_{\text{tie-rod}} = P_{pe} \cdot z_{P_{pe}} - \sum \tfrac{1}{2} \{(p_{pe})_{\text{top}}(z)_{\text{top}} + (p_{pe})_{\text{bottom}}(z)_{\text{bottom}}\}_i h_i - P_{wd} \cdot z_{P_{wd}}$$

$$= (378 \text{ kN}) \cdot (9.7 \text{ m}) - (\text{see Table T8.9} = 3466 \text{ kN} \cdot \text{m}) - (52.6 \text{ kN}) \cdot (4.6 \text{ m})$$

$$= 41 \text{ kN} \cdot \text{m out of balance}$$

The depth of embedment of 4 meters results in a small conservative moment imbalance. In the Japanese design practice, a safety factor of 1.2 is multiplied for driving moment. If this practice is followed, the embedment depth of 4 meters is unconservative.

(5) Simple beam analysis (M_{simple} and T_{simple})

Bending moments in the sheet pile and forces in the tie-rods are initially evaluated using the simplifying assumption that the contra flexure point is at the mudline level (Tschebotarioff, 1949). Therefore, the sheet pile wall is assumed to be a simple beam spanning between the tie-rod and the mudline levels for the purpose of calculating both bending moments and the tie-rod force. These results are later corrected for sheet pile flexure and foundation rigidity based on a beam analysis in Winkler foundation. This correction is essentially the same as the use of Rowe's reduction factor (Rowe, 1952, 1955).

For the simple beam supported at the tie-rod and the mudline levels, active earth thrust above the mudline and driving moment about the tie-rod is computed as (see Table T8.9)

$$(P_{ae})_{\text{mudline}} = \sum_{\text{above mud line}} \tfrac{1}{2}\{(p_{pe})_{\text{top}} + (p_{pe})_{\text{bottom}}\}_i h_i$$

$$= 255 \text{ kN per meter of wall}$$

$$M_{\text{mudline}} = \sum_{\text{above mud line}} \frac{1}{2}\{(p_{pe})_{\text{top}}(z)_{\text{top}} + (p_{pe})_{\text{bottom}}(z)_{\text{bottom}}\}_i h_i + P_{\text{wd}} \cdot z_{Pwd}$$

$$= 990 + 242 = 1232 \text{ kN} \cdot \text{m}$$

The equivalent reaction force exerted by the foundation (i.e. mudline support) is

$$P_{\text{foundation}} = M_{\text{mudline}} / H_{\text{simple}} = 1232/7.0 = 176 \text{ kN per meter of wall}$$

where H_{simple}: a span of the simple beam (m) (i.e. sheet pile wall span between tie-rod and mudline levels)

The tie-rod force T_{simple} is obtained by summing the forces in the horizontal direction

$$T_{\text{simple}} = (P_{ae})_{\text{mudline}} + P_{\text{wd}} - P_{\text{foundation}}$$

$$= 255 \text{ kN} + 53 \text{ kN} - 176 \text{ kN} = 132 \text{ kN per meter of wall}$$

The maximum moment occurs at the same elevation as zero shear within the sheet pile. The zero shear is calculated as elevation z beneath the tie-rod level. Using the tie-rod force T_{simple}, earth thrusts in Table T8.9, and hydrodynamic pressures, zero shear is found to occur between the elevation 0.0 and -6.0 m. At the zero shear location z:

$$(P_{ae})_{\text{above } y} + (P_{\text{wd}})_{\text{above } z} - T_{\text{simple}} = 0$$

Table T8.9. Seismic active earth pressures at designated elevations along full height of wall.

Elevation (m)	Thickness h_i (m)	$(p_{ae})_{\text{top}}$ $(p_{ae})_{\text{bottom}}$ (kN/m²)	$(1/2)(p_{ae})_{\text{top}} h_i$ $(1/2)(p_{ae})_{\text{bottom}} h_i$ (kN/m²)	Distance below tie-rod elevation z_i (m)
+3.0	3.0	0	0	$1.0 - 2/3 \times 3.0 = -1.0$
0.0		18	27	$1.0 - 1/3 \times 3.0 = 0.0$
0.0	6.0	25	75	$1.0 + 1/3 \times 6.0 = 3.0$
-6.0		51	153	$1.0 + 2/3 \times 6.0 = 5.0$
-6.0	4.0	58	116	$7.0 + 1/3 \times 4.0 = 8.3$
-10.0		78	156	$7.0 + 2/3 \times 4.0 = 9.7$

$$\sum_{\text{above } z} \frac{1}{2}\{(p_{pe})_{\text{top}} + (p_{pe})_{\text{bottom}}\}_i h_i + \frac{7}{12} \cdot k_h \cdot \gamma_w \cdot H_w^{0.5}(z-1)^{1.5} - T_{\text{simple}}$$

$$= 27 + \frac{1}{2}\left[25 + \left\{25 + (51-25)\frac{z-1}{6}\right\}\right](z-1) + 3.7(z-1)^{1.5} - 132$$

$$= 2.17(z-1)^2 + 3.7(z-1)^{1.5} + 25(z-1) - 105$$

$$= 0$$

$z = 3.9$ m beneath the tie-rod level

It is now necessary to sum the moments about the tie rod, at an elevation of z. This will provide the maximum moment in the sheet pile wall. Table T8.10 provides the data for summing the moments about z. By multiplying the data shown on the second and fourth columns in this table, and summing these moments, the maximum moment in the sheet pile wall is

$M_{\text{simple}} = 269$ kNm per meter of wall,

which occurs at 3.9 m beneath the tie-rod level.

(6) Corrections by Winkler beam analysis (M_{design} and T_{design})

The results of the simple beam analysis are corrected based on Winkler beam analysis (Takahashi et al., 1993; Ministry of Transport, Japan, 1999). This procedure is analogous to the use of Rowe's reduction factor and utilizes a non-dimensionalized Rowe's flexibility number, called similarity number, defined by

$$\omega = \rho k_{\text{h-sub}} = \frac{H_{\text{simple}}^4 k_{\text{h-sub}}}{EI} \tag{T8.4}$$

Table T8.10. Seismic active earth pressures at designated elevations.

Elevation (m)	Thickness h_i (m)	$(1/2)(p_{ae})_{\text{top}} h_i$ $(1/2)(p_{ae})_{\text{bottom}} h_i$ (kN/m^2)	Distance below the tie-rod level z_i (m)	Distance above the elevation at z=3.9 m
+ 3.0	3.0	0	−1.0	4.9
0.0		27	0.0	3.9
0.0	2.9	36	$1.0 + 1/3 \times 2.9 = 2.0$	1.9
− 2.9		54	$1.0 + 2/3 \times 2.9 = 2.9$	1.0
+ 1.0	Tie rod	$T_{\text{simple}} = -133$ kN	0.0	3.9
	Fluid Pressure	$P_{\text{wd}} = 18$ kN	$1.0 + 2.9 \times 0.6 = 2.7$	1.2

where

ρ: Rowe's flexibility number (see Eqn. (T8.1))

$k_{h\text{-sub}}$:   coefficient of subgrade reaction (MPa/m)

H_{simple}: a span of the simple beam (m) (i.e. sheet pile wall span between tie-rod and mudline levels)

E: Young's modulus (MPa)

I: moment of inertia (m^4/m)

The Winkler beam analysis results in the following equations:

$$\frac{M_{design}}{M_{simple}} = 4.5647\omega^{-0.2} + 0.1329 \tag{T8.5a}$$

$$\frac{T_{design}}{T_{simple}} = 2.3174\omega^{-0.2} + 0.5514 \tag{T8.5b}$$

$$\frac{D_{emb}}{H_{simple}} = 5.0916\omega^{-0.2} - 0.2591 \tag{T8.5c}$$

The coefficient of subgrade reaction is determined based on Fig. T8.18. For SPT N-value of 20, $k_{h\text{-sub}} \approx 30$ Mpa/m. The similarity number is then calculated by

$$\omega = \frac{7^4 \times 30}{I \cdot 2 \times 10^5} = \frac{0.36}{I}$$

The Japanese steels of SY295 and SY390 have $\sigma_{allow} = 180 \times 10^3$ kPa and 235×10^3 kPa, respectively. The allowable bending moment can be calculated as

$$M_{allow} = Z_e \cdot \sigma_{allow} \text{ per meter of wall}$$

where Z_e is the section modulus of the sheet pile wall.

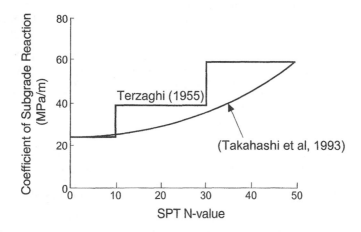

Fig. T8.18. Coefficient of subgrade reaction.

The relevant parameters and design bending moments computed are shown in Table T8.11 for typical sheet pile sections. From this table, it can be seen that SP-III with SY390 steel or SP-IV with SY295 steel is the sheet pile wall section that is suited for this specific design scenario. Design bending moments are

$$M_{design} = 300 \text{ or } 339 \text{ kNm per meter of wall}$$

The design tie-rod force is also computed using similarity number, as shown in Table T8.12. The design tie-rod force is

$$T_{design} = 140 \text{ or } 149 \text{ kN per meter of wall}$$

(7) Embedment depth and anchor location

Design embedment depth should be deep enough to satisfy the fixed earth support condition. Using Eqn. (T8.5c) for sheet pile SP-III with SY295 steel, the design embedment depth is determined as

$$D_{emb} = 7.0 \times 0.833 = 5.8 \text{ m}$$

The recommended location of anchor is determined as shown in Fig. T8.19. The active failure plane behind the sheet pile wall is assumed to originate from the mudline level because the design/analysis is performed for fixed earth support condition.

Table T8.11. Sheet piles and design bending moment.

Sheet pile type (Japanese)	I (m^4/m)	Z_e (m^3/m)	Similarity number ω	M_{design}/M_{simple} Eqn. (T8.5a)	M_{design} kNm/m	M_{allow} kNm/m	
						SY295	SY390
SP-II	0.87×10^{-4}	0.87×10^{-3}	4.1×10^3	1.00	269	157	204
SP-III	1.64×10^{-4}	1.31×10^{-3}	2.2×10^3	1.11	300	235	308
SP-IV	3.19×10^{-4}	2.06×10^{-3}	1.1×10^3	1.26	339	371	484

Table T8.12. Design tie-rod forces.

Sheet pile (Japanese)	I (m^4/m)	Z_e (m^3/m)	Similarity number ω	T_{design}/T_{simple} Eqn. (T8.5b)	T_{design} kN/m
SP-II	0.87×10^{-4}	0.87×10^{-3}	4.1×10^3	0.99	132
SP-III	1.64×10^{-4}	1.31×10^{-3}	2.2×10^3	1.05	140
SP-IV	3.19×10^{-4}	2.06×10^{-3}	1.1×10^3	1.12	149

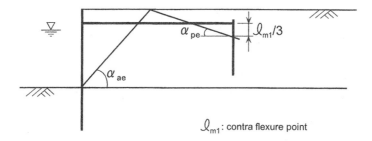

Fig. T8.19. Location of anchor (Japanese practice).

(8) Comparison of North American and Japanese methods

Major design parameters obtained through the simplified analyses in Sections T8.2.2 and T8.2.3 are shown in Table T8.13. These results indicate the following:

a) Sheet pile bending moment are computed in both design practices based on Winkler foundation analysis using Rowe's reduction (correction) factor but differences are noted because of the different earth pressure computations.

b) Design tie-rod forces significantly differ from each other because of the different conditions adopted for design (i.e. free earth support condition for North American practice and Winkler foundation (Rowe's) condition for Japanese practice).

c) Embedment depth and location of the anchor is also different because of the different design conditions mentioned above.

   The reader should be aware of these differences in the pseudo-static analysis of sheet pile walls.

T8.2.4   Simplified dynamic analysis (deformations)

The sheet pile wall determined in Section T8.2.2 was used for the simplified dynamic analysis. The simplified dynamic analysis was performed to determine the extent of soil improvement necessary to meet Grade B performance requirements stated in Section T8.2.1.

   This example utilizes the empirically developed design chart proposed by McCullough (1998) and McCullough and Dickenson (1998) (see Section T7.2 in Technical Commentary 7). The design chart shows non-dimensional parameters with respect to a wall displacement and a soil improvement area for a quay wall under prescribed earthquake motions. The backfill conditions are specified by the SPT N-values equal to 10 or 20.

Table T8.13. Comparison of simplified analysis results.

Analysis method	M_{design} (kNm/m)	T_{design} (kN/m)	D_{emb} (m)
North American practice	203	351	3.75
Japanese practice	300 to 339	140 to 149	5.8

By referring to the sheet pile wall section shown in Fig. T8.13, the sheet pile wall parameters for a PZ27 sheet pile section are

$$E = 2 \times 10^8 \text{ kPa}$$

$$I = 2.51 \times 10^{-4} \text{ m}^4/\text{m}$$

The wall height (H) is 9 m, and the wall depth (D_{emb}) is 3.75 m. The soil buoyant unit weight (γ_b) directly adjacent to the wall is 9 kPa.

Substituting these values into the normalized lateral displacement factor yields

$$\frac{d \cdot EI}{(H + D_{emb})^5 \cdot \gamma_b} = \frac{d \cdot (2 \times 10^8 \text{ kPa}) \cdot (2.51 \times 10^{-4} \text{ m}^4/\text{m})}{(9 \text{ m} + 3.75 \text{ m})^5 \cdot (9 \text{ kPa})} = \frac{d}{60.4}$$

where d is the displacement at the top of the wall.

The ground motion intensity factor is determined for the earthquake motions using Table T7.8 in Technical Commentary 7 to determine the magnitude scaling factor (MSF) and Fig. T8.20 to determine the depth reduction factor (r_d), using the following equation:

$$\frac{a_{max(d)}}{MSF} = \frac{a_{max} \cdot r_d}{MSF}$$

The magnitude scaling factor values for both earthquake motions (L1 and L2) are tabulated in Table T8.14.

The normalized lateral displacement factor is then determined as shown in Table T8.15 as a function of soil improvement (n_{ext}) using Fig. T7.22 in Technical Commentary 7 for uniform normalized blowcounts of 20.

By plotting the values tabulated above, it is possible to develop a figure that relates the extent of soil improvement to lateral displacements (Fig. T8.21). It is interesting to note that for this example, it is practically impossible to limit the deformations to 10.3 cm for the L1 motion and 13.9 cm for the L2 motion, utilizing densification of the sandy backfill. Other forms of mitigation (e.g. soil treatment by cementation/grouting or strengthening, and structural mitigation measures) may be necessary to limit earthquake-induced deformations to smaller values. It can also be noted that soil improvement beyond approximately 30 meters has little effect on the earthquake-induced lateral displacements.

Based on Grade B requirements, and the estimated displacements highlighted above, the length of soil improvement needs to be approximately 30 meters to meet

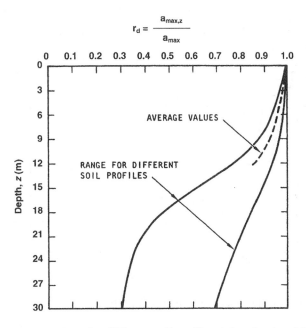

Fig. T8.20. Range of r_d values for different soil profiles (after Seed and Idriss, 1982).

Table T8.14. Magnitude scaling factors.

	$a_{max(surface)}$	r_d (at 9 m)	$a_{max(d)}$	MSF	$a_{max(d)}$/MSF
L1	0.45	0.92	0.41	1.63	0.25
L2	0.60	0.92	0.55	1.00	0.55

Table T8.15. Lateral displacements.

	Normalized soil improvement, n_{ext}	Length of soil improvement, L_{ext} (m)	Normalized lateral displacement, d/H	Lateral displacement, d (m)
	0	0	0.011	0.66
	1	15	0.0035	0.21
L1	2	30	0.0021	0.13
	3	45	0.0019	0.12
	8	120	0.0017	0.10
	0	0	Exceeds limits of design chart	
	0.5	7.5	0.014	0.85
	1	15	0.006	0.36
L2	1.5	22.5	0.0035	0.21
	2	30	0.003	0.18
	3	45	0.0025	0.15
	8	120	0.0023	0.14

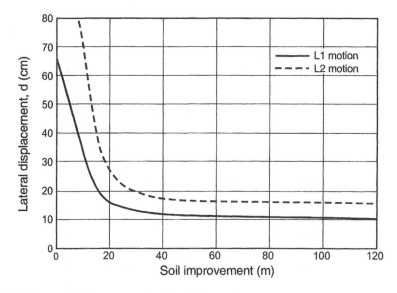

Fig. T8.21. Lateral displacement for the design problem.

Grade B requirements of 1.5% deformation (d/H) during the L1 event (deformation of 13.5 cm). Note that these same criteria cannot be met for the L2 event.

T8.3   SHEET PILE QUAY WALL (GRADE S)

T8.3.1   Performance requirements and design conditions

An existing sheet pile quay wall with a water depth of -11.5 m was proposed for upgrading into a quay wall with a water depth of -16.6 m. The performance grade of this new quay wall was designated as Grade S (see Section 3.1 in the Main Text). Based on seismic hazard analysis, the earthquake motions from the active seismic fault located nearby was dominant and determined as L2 motion with an estimated earthquake magnitude of $M_{JMA} = 7.5$. PGA at the bedrock level was evaluated as $0.53g$. Based on the needs of the owners/users of the quay wall, extremely rigorous damage criteria were determined as shown in Table T8.16. In particular, 'elastic' in this table was defined as the state that the extreme fibre stress is less than the allowable stress.

Proposed cross section of the sheet pile quay wall is shown in Fig. T8.22. Two different types of anchorage, i.e. battered and vertical pile types, were installed with two sets of tie rods connected separately. Profiles of natural ground soil layers are summarized in Fig. T8.23. The bottom boundary for the analysis was chosen at -46 m where the bedrock layer was encountered.

Table T8.16. Damage criteria for a sheet pile quay wall.

Extent of damage		Degree I
Residual displacements	Normalized residual horizontal displacement (*d/H*)*	Less than 1.5%
Peak response stresses	Sheet pile wall	Elastic**
	Tie rod	Elastic**
	Anchor	Elastic**

d: Residual horizontal displacement; *H*: height of sheet pile wall from mudline.
**Less than allowable stress.

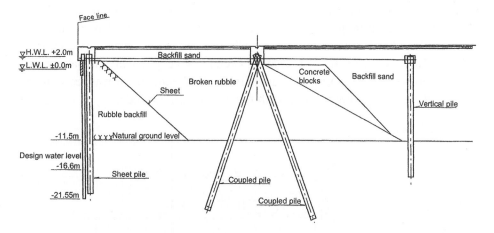

Fig. T8.22. An example of sheet pile type quay wall.

D.L-11.5m

-19.5m	Soil γ_b=10 kN/m³ ϕ =35°
-28.5m	Clay γ_b=7.4 kN/m³ c=515kPa
-33.0m	Clay γ_b=5.1 kN/m³ c=390kPa
-46~-52m	Clay γ_b=6.5 kN/m³ c=475kPa

Fig. T8.23. Natural ground soil layers.

T8.3.2   Dynamic analysis

Since the required performance was Grade S, dynamic analysis was performed for seismic performance evaluation for L2 earthquake motion (see Section 5.1 in the Main Text). The method used and the evaluation of input parameters were similar to those for the gravity quay wall analysis shown in Section T8.1.5.

Cross section of the sheet pile quay wall was idealized into a set of finite element mesh shown in Figs. T8.24 and T8.25. Parameters for structural materials are summarized in Table T8.17.

Computed deformation and computed distribution of excess pore water pressure ratios are shown in Figs. T8.26 and T8.27. The computed displacement at the top of the sheet pile was 13 cm, and was confirmed to satisfy the displacement criterion of less than 30 cm (equal to 1.5% of the wall height of 20.4 m). Computed bending moments of the sheet pile wall and vertical pile anchor are shown in Fig. T8.28. Computed stresses of structural components are summarized in Table T8.18. Only the tie-rod force between the sheet pile wall and battered pile anchor was beyond the allowable tensile force. Consequently, it was recommended to replace the existing tie-rods with higher strength tie-rods.

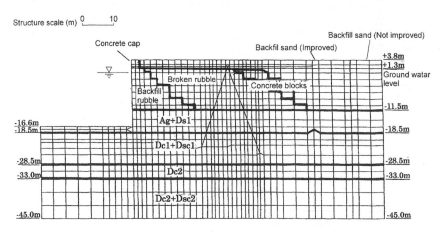

Fig. T8.24. An example of soil layers of sheet pile type quay wall.

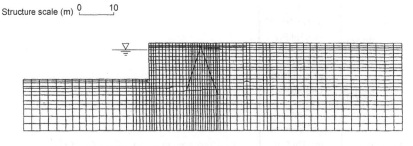

Fig. T8.25. An example of FEM mesh for sheet pile case.

Table T8.17. Parameters for structural materials.

	Shear modulus (kPa)	Poisson's ratio	Density (t/m³)	Cross sectional area per unit breadth (m²/m)	Geometrical moment of inertia per unit breadth (m⁴/m)
Sheet pile	8.1×10^7	0.30	7.85	8.14×10^{-1}	7.31×10^{-2}
Coupled pile	8.1×10^7	0.30	7.85	1.55×10^{-2}	1.89×10^{-3}
Vertical pile	8.1×10^7	0.30	7.85	2.66×10^{-2}	3.18×10^{-3}
Tie rod between sheet pile and coupled pile	6.2×10^7	0.30	7.85	9.65×10^{-4}	0.0
Tie rod between sheet pile and vertical pile	6.2×10^7	0.30	7.85	1.75×10^{-3}	0.0

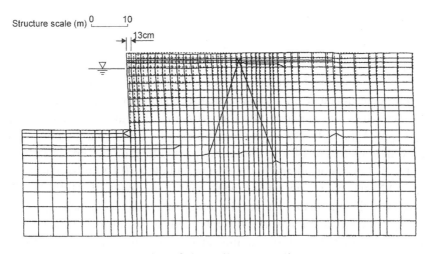

Fig. T8.26. Computed deformation of sheet pile quay wall.

T8.4   PILE-SUPPORTED WHARF (GRADE B)

T8.4.1   Performance requirements and design conditions

A pile-supported wharf with a water depth of 12 m was proposed for construction. The performance grade of this new wharf was designated as Grade B over the life span of 50 years (see Section 3.1 in the Main Text). A seismic hazard and response spectrum analysis estimated that the maximum deck level accelerations are 0.30g and 0.57g for the 50% (L1) and 10% (L2) probability of exceedance, respectively. Damage criteria were determined for the pile-deck structure, as shown in Table T8.19. Damage criteria for Degree II were not defined because the performance of a Grade B structure is specified only by damage degrees I and III (see Section 3.1 in the Main Text).

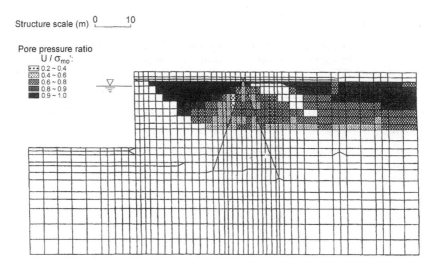

Structure scale (m) 0   10

Pore pressure ratio
U / σ_{mo}':
◻ 0.2 ~ 0.4
▦ 0.4 ~ 0.6
▩ 0.6 ~ 0.8
▨ 0.8 ~ 0.9
■ 0.9 ~ 1.0

Fig. T8.27. Distribution of excess pore water pressures after shaking.

The proposed cross section and plan of the pile-supported wharf are shown in Fig. T8.29. Geotechnical parameters, including the coefficient of subgrade reaction, were determined from a geotechnical investigation and are given in Table T8.20. The wharf was supported by four rows of 1.2 m diameter steel pipe piles. Piles in rows 1 through 3 have a wall thickness of 12 mm, and the piles in row 4 have a wall thickness of 14 mm. Structural parameters for these piles are given in Table T8.21. Loads considered in the design include a 10 kN/m^2 surcharge on the deck, 30 kN/m^2 deadweight of the deck, and crane loads of 2400 kN per unit framework of the pile-deck system. The total load on the deck, including the deck deadweight for a unit framework with a breadth of 5.0 m, is 6600 kN. The unit framework considered for design is indicated by hatching in Fig. T8.29(b).

The Grade B example analyses presented in this section include a simplified analysis used for the preliminary design and a simplified dynamic analysis based on the pushover method to satisfy Grade B design requirements (see Section 5.1 in the Main Text).

T8.4.2 Simplified analysis

The simplified analysis method is being used to estimate the elastic performance of the structure. For this simplified analysis, the natural period of the pile-deck system is computed by

$$T_s = 2\pi \sqrt{\frac{W_g}{g \sum K_{Hi}}} \tag{T8.6}$$

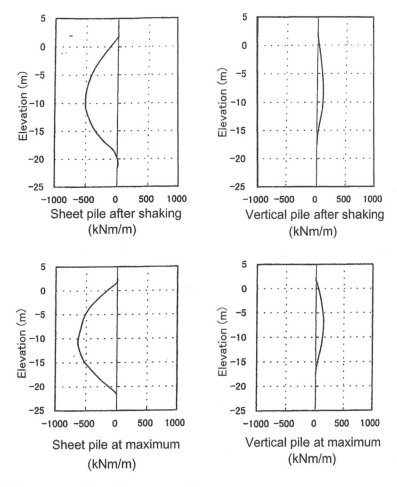

Fig. T8.28. Computed bending moment of structures.

where

T_s: natural period of a pile-deck system (s)
W_g: surcharge and deadweight of deck (kN)
g: acceleration of gravity ($=9.8$ m/s^2)
K_{Hi}: equivalent spring stiffness for lateral reaction by i[th] pile (kN/m)

The equivalent spring stiffness in Eqn. (T8.6) was computed based on an equivalent fixity pile by

$$K_{Hi} = \frac{12EI_i}{\left(l_i + \dfrac{1}{\beta_i}\right)^3} \qquad \text{(T8.7)}$$

where

E: Young's modulus of elasticity (kN/m^2)
I_i: moment of inertia of i-th pile (m^4)

Table T8.18. Computed stresses of structures.

Structure		Unit	Maximum value
Sheet pile		Bending moment	6170
Section modulus	1.22×10^{-1} (m³/m)	(kNm/m)	
Allowable stress	181 (N/mm²)	Fiber stress (N/mm²)	50.6
Vertical pile		Bending moment	1320
Section modulus	6.36×10^{-3} (m³/m)	(kNm/m)	
Allowable stress	279 (N/mm²)	Fiber stress (N/mm²)	208
Coupled pile		Bending moment	350
Sheet pile	3.78×10^{-3} (m³/m)	(kNm/m)	
Allowable stress	206 (N/mm²)	Fiber stress (N/mm²)	93.3
Tie rod between sheet pile and coupled pile		Axial force	1686
Allowable tensile force	1100 (kN)	(kN)	
Tie rod between sheet pile and vertical pile		Axial force	1186
Allowable tensile force	1247 (kN)	(kN)	

Table T8.19. Damage criteria for pile-supported wharf (Grade B).*

Extent of damage		Degree I	Degree II	Degree III	Degree IV
Residual displace-ments	Differential settlement between deck and land behind	Less than 0.1~0.3 m	N/A	N/A	N/A
	Residual tilting towards the sea	Less than 2~3°	N/A	N/A	N/A
Peak response	Piles**	Essentially elastic response with minor or no residual deformation	N/A	Ductile response near collapse (double plastic hinges are allowed to occur at only one pile)	Beyond the state of Degree III

*Modified from the criteria shown in Table 4.3 in the Main Text, due to the specific performance requirements of the pile-supported wharf.
**Bending failure should precede shear failure in structural components.

l_i: unsupported length of i-th pile between the deck and slope surface (m)
$1/\beta_i$: length of equivalent fixity pile below the slope surface (m) (see Eqn. (T7.25) in Technical Commentary 7)

For obtaining the unsupported lengths of piles l_i, a virtual slope surface was considered in design as indicated by the broken line in Fig. T8.29(a)

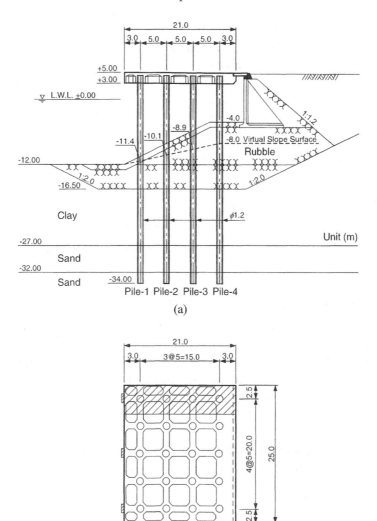

Fig. T8.29. An example of pile-supported wharf (Grade B).
    (a) Cross section.
    (b) Plan.

(see Section T7.3 in Technical Commentary 7). From the geotechnical and structural parameters in Tables T8.20 and T8.21, the parameters needed in Eqns. (T8.6) and (T8.7) were calculated, and are shown in Table T8.22. In particular, β_i for pile rows 1 through 3 was computed using Eqn. (T7.25) in Technical Commentary 7 as

$$\beta_i = \sqrt[4]{\frac{k_{h\text{-sub}}D_p}{4EI_i}} = \sqrt[4]{\frac{29000 \times 1.2}{4 \times 2.06 \times 10^8 \times 0.00723}} = 0.28 \ (\text{m}^{-1}) \tag{T8.8}$$

Table T8.20. Major geotechnical parameters for pile-supported wharf (Grade B).

Soil layers	Density (t/m^3)	Coefficient of subgrade reaction $k_{h\text{-sub}}$ (kN/m^3)	Internal friction angle ϕ (°) or unconfined compressive strength q_u (kN/m^2)
Rubble	1.9	29,000	$\phi = 30°$
Soil layer 1 (Clay)	1.6	29,000	$q_u = 60$ kN/m^2
Soil layer 2 (Sand)	2.0	117,000	$\phi = 35°$
Soil layer 3 (Sand)	2.0	290,000	$\phi = 35°$

Table T8.21. Major pile parameters.

Type of parameter	Pile parameters	
	Piles 1 through 3	Pile 4
Diameter (m)	1.2	1.2
Thickness (m)*	0.011	0.013
Cross sectional area (m^2)	0.0410	0.0484
Moment of inertia (m^4)	0.00723	0.00850
Elastic section modulus (m^3)	0.0120	0.0142
Yield stress (kN/m^2)**	315,000	315,000
Young's modulus (kN/m^2)	2.06×10^8	2.06×10^8

*Cross-section area and moment of inertia are computed by allowing loss of cross section in 1 mm thickness due to corrosion.
**Steel used was SKK490 in JIS-A-5525.

Table T8.22. Equivalent spring coefficients and relevant parameters of piles.

Parameters	Pile 1	Pile 2	Pile 3	Pile 4
l_i (m)	14.4	13.1	11.9	11.0
$1/\beta_i$ (m)	3.6	3.6	3.6	3.8
$l_i + 1/\beta_i$ (m)	18.0	16.7	15.5	14.8
EI_i (kNm^2)	1,490,000	1,490,000	1,490,000	1,750,000
K_{Hi} (kN/m)	3,070	3,840	4,800	6,480

Thus the natural period of the pile-deck system was computed for pile rows 1 through 4 by Eqn. (T8.6) as

$$T_s = 2\pi \sqrt{\frac{6600}{9.8 \times (3070 + 3840 + 4800 + 6480)}} = 1.2 \text{ s} \tag{T8.9}$$

The approximate elastic limit (i.e. threshold level) of this pile-deck system was evaluated based on Eqns. (T7.26) and (T7.27) in Technical Commentary 7 as follows.

The plastic hinge moment without axial load M_{p0} is given by

$$M_{p0} = Z_p \sigma_y \tag{T8.10}$$

where

$Z_p = (4/3)\{r_p^3 - (r_p - t_p)^3\}$: sectional modulus at plastic hinge (m^3)

r_p: pile radius (m)
t_p: wall thickness of the pile (m)
σ_y: yield stress (kN/m^2)

The computed parameters shown in Tables T8.21 and T8.22 resulted in the plastic hinge moment for pile rows 1 through 3 being

$$M_{p0} = \tfrac{4}{3}\{0.60^3 - (0.60 - 0.011)^3\} \times 315000 = 4900 \ (kNm) \tag{T8.11a}$$

and for pile row 4 as

$$M_{p0} = \tfrac{4}{3}\{0.60^3 - (0.60 - 0.013)^3\} \times 315000 = 5770 \ (kNm) \tag{T8.11b}$$

The plastic hinge moment with axial load N is computed by (e.g. Yokota, 1999)

$$M_p = M_{p0} \cos\left(\frac{N}{N_{y0}} \frac{\pi}{2}\right) \tag{T8.12}$$

where

N: axial load (kN)
N_{y0}: yield axial load without bending moment (kN)

The axial loads due to the surcharge, deadweight, and crane, as well as the calculated plastic moments are tabulated in Table T8.23.

The limit lateral load P_u, when plastic hinges form at pile top and in the embedded portion of the piles was computed using Eqns. (T7.27) in Technical Commentary 7 as

$$P_u = \sum \frac{2M_{pi}}{l_i + \dfrac{1}{\beta_i}} = \frac{2 \times 3530}{18.0} + \frac{2 \times 4860}{16.7} + \frac{2 \times 4860}{15.5} + \frac{2 \times 5730}{14.8} \tag{T8.13}$$

$$= 2380 \ (kN)$$

and the approximate elastic limit P_y was computed using Eqn. (T7.26) in Technical Commentary 7 as

$$P_y = 0.82 P_u = 0.82 \times 2380 = 1950 \ (kN) \tag{T8.14}$$

Table T8.23. Plastic hinge moment of piles.

Parameters	Pile 1	Pile 2	Pile 3	Pile 4
M_{p0i} (kNm)	4900	4900	4900	5770
N_i (kN)	6300	1000	1000	1100
N_{y0i} (kN)	12900	12900	12900	15200
M_{pi} (kNm)	3530	4860	4860	5730

Using the peak deck response of 0.30g for the L1 earthquake motion, as determined from the seismic hazard and site response analysis, the seismic load during L1 earthquake motion was estimated as

$$F = 0.30 \times 6600 = 1980 \ (kN) \tag{T8.15}$$

It can be noted that F is nearly equal to P_y in this preliminary study, indicating that the proposed design pile-deck system may satisfy the required performance criteria for Level 1 earthquake shown in Table T8.19.

T8.4.3   Simplified dynamic analysis

As mentioned earlier, the seismic hazard and site response analyses resulted in peak deck responses of 0.30g and 0.57g for the L1 and L2 earthquake motions, respectively. The subsoil, including the slope under the deck, was considered a non-liquefiable granular material.

A pushover analysis was performed using the geotechnical and structural parameters in Tables T8.20 and T8.21. In this analysis, bi-linear (elasto-plastic) springs were used to represent the subgrade and axial reactions of the subsoil. The deck was idealized as a non-linear beam having a tri-linear moment-curvature relationship. The piles were idealized as non-linear beams having a bi-linear moment-curvature relationship, as shown in Fig. T8.30, and specified by the following equations:

$$M = \begin{cases} EZ_e r\phi & (\phi < \phi_p) \\ M_{p0} \cos\left(\dfrac{N}{N_{y0}} \dfrac{\pi}{2} \right) & (\phi \geq \phi_p) \end{cases} \tag{T8.16}$$

where

E: Young's modulus of elasticity (kN/m^2)
Z_e: elastic sectional modulus (m^3)
r: pile radius (m)

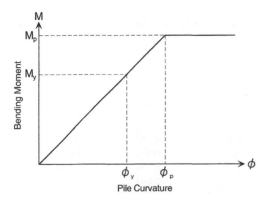

Fig. T8.30. Moment-curvature relationship of piles for pushover analysis.

ϕ: pile curvature (m^{-1})

ϕ_p: pile curvature at the onset of a plastic hinge (m^{-1})

M_{p0}: plastic hinge moment without axial load (kNm) (Eqn. (T8.10))

N: axial load (kN)

N_{y0}: yield axial load without bending moment (kN)

The yield bending moment (M_y), shown in Fig. T8.30, is given by

$$M_y = \left(\sigma_y - \frac{N}{A} \right) Z_e \tag{T8.17}$$

where A: pile cross-sectional area (m^2).

The computer program N-Pier (Yokota et al., 1999) was used for the analysis. This program has been made available (downloadable) free of charge from the relevant web site (http://www.phri.go.jp), although the current version is limited for use on personal computers with a Japanese fonts/characters.

The pushover analysis resulted in the load-displacement curve shown in Fig. T8.31. The P_y value was determined as the 'break-point' in the curve and the P_u value was determined as the point at which a pile develops a double plastic hinge (at the pile cap and in the embedded portion of the pile). It was determined from the curve that points P_y and P_u in this figure were the upper limits for damage Degrees I and III, as defined in Table T8.19.

The seismic loads due to the peak deck responses for the L1 and L2 earthquake motions were compared to the limiting loads for Damage Degrees I and III in Table T8.24. In this comparison, the equivalent elastic limit was chosen at the load level P_y and a displacement ductility factor μ was defined using this equivalent elastic limit as $\mu = d_y/d_u$. Equation (T7.43) in Technical Commentary 7 was used to compute the seismic loads for a ductile response during a L2 earthquake motion. The results tabulated in Table T8.24 indicate that the proposed design satisfies the required performance criterion for peak responses during both the L1 and L2 earthquake motions.

The pushover analysis resulted in tilting of less than $1°$, less than the 2–3% required for the Degree I performance. Differential settlement between the deck and back land was inferred from the horizontal displacement of the deck. Since the peak horizontal displacement of the deck during L1 earthquake motion was estimated as 0.2 m, the differential settlement was considered to be less than this value. This is a crude estimate, and a separate analysis should be performed to evaluate the settlement at the crest of the slope if this is a primary concern for this specific structure. Instead of performing a separate analysis, a relevant earthquake case history was investigated, and it was concluded that the proposed design structure satisfies the criteria with respect to the differential settlement shown in Table T8.19.

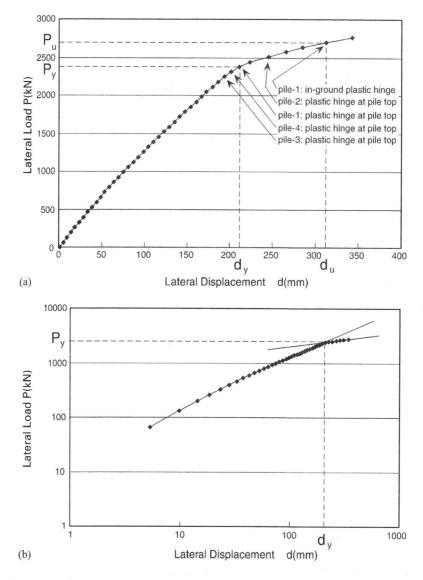

Fig. T8.31. Results of pushover analysis.
(a) Plotting on regular scale.
(b) Plotting on log-log scale.

T8.5   PILE-SUPPORTED WHARF (GRADE S)

T8.5.1   Performance requirements and design conditions

An existing bulkhead with a water depth of -7.0 m was proposed for upgrading by constructing a new pile-supported wharf at a water depth of -14.6 m in front

Table T8.24. Seismic loads and limit loads based on the pushover analysis and the criteria.

Earthquake level	Level 1	Level 2
PGA at deck level	0.30g	0.57g
Ductility factor μ	1.0	1.5
Seismic loads F (kN)	1,980	2,660
Limit loads P_y and P_u (kN)	2,380	2,700
Limit displacements d_y and d_u (m)	$d_y = 0.21$	$d_u = 0.31$
Tilt (°)	< 1	< 1

Table T8.25. Damage criteria for pile-supported wharf (Grade-S).

Extent of damage			Degree I
Residual displacements	Deck	Differential settlement between deck and land behind	Less than 0.3 m
Peak response stresses	Piles	Pile top	Elastic*
	Retaining structure	Sheet pile wall	Elastic*
		Tie rod	Elastic*
		Anchor	Elastic*

*Less than yield stress.

of the existing sheet pile wall. The performance grade of this new wharf was designated as Grade S over the life span of 50 years (see Section 3.1 in the Main Text). Based on seismic hazard analysis, the L2 earthquake motion with a probability of exceedance of 10% was 0.39g at the bedrock. Damage criteria were determined for pile-deck structure and sheet pile retaining structure, as shown in Table T8.25. In particular, 'elastic' in this table was defined as the state where the extreme fibre stress is less than the yield stress.

A proposed cross-section of the pile-supported wharf is shown in Fig. T8.32. Geotechnical investigations shown in Fig. T8.33 revealed a liquefiable layer 2.3 m thick below the elevation of −17.3 m. The effect of this liquefiable layer on the performance of the pile-supported wharf was one of the primary issues investigated during the seismic performance evaluation. Initially, the proposed cross section shown in Fig. T8.32 had partially improved portion of the liquefiable layer by a cement deep mixing stabilization method (CDM). Rubble materials were used above the CDM improved area to form a dike in front of the sheet pile retaining wall. The combined resistance by the sheet pile wall and the CDM portion was the main issue studied in the seismic performance evaluation.

T8.5.2   Dynamic analysis

Since the required performance of this pile supported wharf was Grade S, dynamic analysis was performed for seismic performance evaluation for L2 earthquake

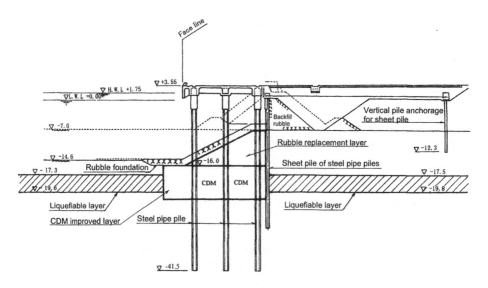

Fig. T8.32. An example of vertical pile supported wharf.

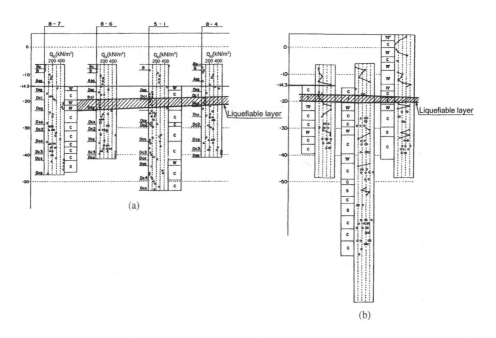

(a)

(b)

Fig. T8.33. Boring logs.
(a) Parallel to the face line of the wharf.
(b) Transverse to the face line of the wharf.

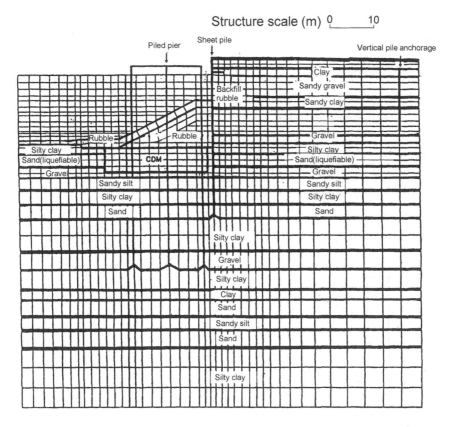

Fig. T8.34. An example of soil layers of vertical pile supported wharf.

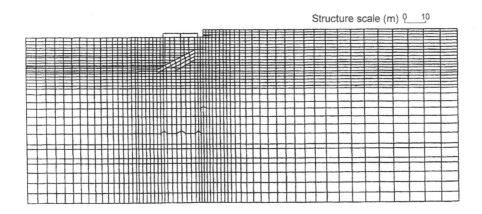

Fig. T8.35. An example of FEM mesh for vertical pile supported wharf.

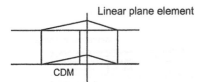

Linear plane element

CDM

Fig. T8.36. Modelling of piles in CDM improved layers.

Table T8.26. Parameters for CDM.

Unconfined compressive strength q_u (kPa)	Young's modulus E (kPa)	Initial shear modulus G_0 (kPa)	Bulk modulus K_0 (kPa)	Internal friction angle ϕ_f (°)	Cohesion c (kPa)
1320	6.6×10^5	2.5×10^5	6.5×10^5	0	662

Soil   Sheet pile   Soil

Fig. T8.37. Modelling of sheet pile.

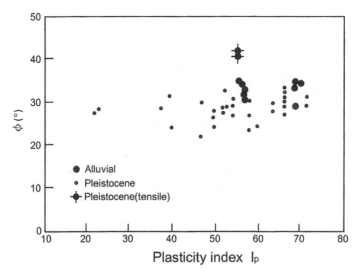

Fig. T8.38. Friction angle and plasticity index for normally consolidated clay.

Table T8.27. Parameters for structures.

	Vertical pile	Sheet pile of steel pipe piles	Vertical pile anchorage	Tie rod	Pier slab
Outside diameter (m)	1.500	1.100	0.800	0.065	
Thickness (m)	0.019	0.016	0.009		
Cross sectional area (m^2)	8.840E − 2	5.449E − 2	2.236E − 2	3.320E − 3	3.30*
Geometrical moment of inertia (m^4)	2.420E − 2	8.00E − 3	1.75E − 3	(Assumed to be 0)	0.958*
Section modulus (m^3)	3.232E − 2	1.460E − 2	4.37E − 2		
Shear modulus (kPa)	7.9E + 7	7.9E + 7	7.9E + 7	7.9E + 7	1.27E +7
Poisson's ratio	0.3	0.3	0.3	0.3	0.20
Density (t/m^3)	7.85	7.85	7.85	(Assumed to be 0)	2.87**
Effective shear area ratio	1/2	1/2	1/2	(Assumed to be 0)	1/1.2

*Per 6 m breadth.
** Including the weight of pavement.

Deformation at the top of pier		
	Residual	Maximum
Horizontal	51	66
Vertical	4	4
		(cm)

Deformation at the top of sheet pile		
	Residual	Maximum
Horizontal	55	76
Vertical	11	13
		(cm)

Structure scale (m) 0 ___ 10
Displacement (m) → 0.50

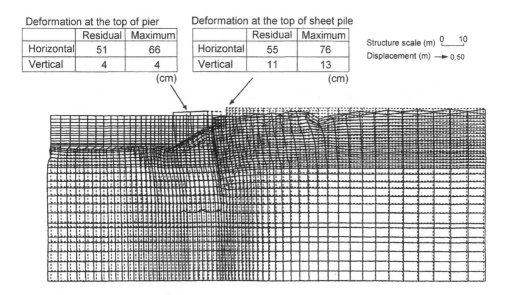

Fig. T8.39. Computed deformation of vertical pile supported wharf.

motion (see Section 5.1 in the Main Text). The method used and the evaluation of input parameters were similar to those for the gravity and sheet pile quay wall analyses discussed in Sections T8.1.5 and T8.3.

The cross section of the pile-supported wharf was idealized into finite elements shown in Figs. T8.34 and T8.35. Modelling of the piles in the CDM improved area is shown in Fig. T8.36. Piles and CDM elements were not connected directly but

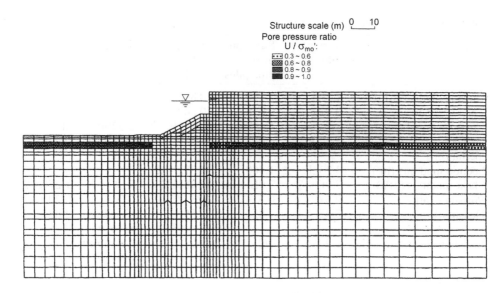

Fig. T8.40. Computed maximum excess pore water pressure ratio of vertical pile supported wharf.

Table T8.28. Computed maximum stresses of structures and yield point.

	Vertical pile of pier									Sheet pile	Vertical pile anchorage	Tie rod
	Bending moment related stress			Axial force related stress			Total maximum stress					
	Sea side	Middle	Land side	Sea side	Middle	Land side	Sea side	Middle	Land side			
Yield stress	—	—	—	—	—	—	315	315	315	315	315	440
Computed stress	284	204	222	−22	23	−4	306	228	227	123	332	229

(N/mm²)

connected through linear plane elements to allow free response of CDM portions between the piles. Parameters for CDM elements based on the unconfined compression strength q_u are summarized in Table T8.26. Modelling of sheet pile at the bulkhead is illustrated in Fig. T8.37. Joint elements are used to simulate the friction at the soil-structure interface. The angle of internal friction for clay layers was determined based on the existing laboratory data shown in Fig. T8.38 (after Tsuchida, 1990). Parameters for structural components are summarized in Table T8.27.

The computed deformation and computed maximum excess pore water pressure ratio are shown in Figs. T8.39 and T8.40. Computed deformation resulted in the differential settlement of 0.5 m between the deck and the land behind. This differential settlement was considered not acceptable based on the damage criteria set forth in Table T8.25. Liquefiable layers were completely liquefied as shown in Fig. T8.40, and it was speculated that liquefaction remediation measures for wider portions of the liquefiable layer could be effective in reducing deformation. Computed maximum stresses and yield stresses of structural components are summarized in Table T8.28. The performance of all the structural components satisfies the damage criteria except for that of the vertical pile anchor. Consequently, liquefaction remediation measures for wider portions of the liquefiable layer were recommended for the next cycle of seismic performance evaluation. The sequence of seismic performance evaluation with a modified cross section was repeated until the damage criteria were satisfied with optimum area of soil improvement.

List of Symbols

A	cross sectional area of a pile
$A_{CPS}(f)$	amplitude of each frequency component (f) of the CPS
A_e	effective shear area (= $0.8A_{gross}$)
A_{gross}	uncracked section area of a pile
A_h	total area of transverse reinforcement, parallel to direction of applied shear cut by an inclined shear crack
A_{hs}	area of stabilized hysteresis loop
A_{sc}	total area of dowel bars in the connection
A_{sp}	spiral or hoop cross section area
A_t	total required area of the headed bond bars
a, b	empirical constants
a'	maximum acceleration of a linear SDOF system
a_{design}	design acceleration
a_{eq}	equivalent acceleration at the layer depth
a_{max}	peak acceleration
$a_{max(d)}$	peak acceleration within the backfill at the elevation of the dredge-line
$a_{max(surface)}$	peak acceleration at ground surface
$a_{max(rock)}$	peak acceleration at bedrock
a_t	threshold acceleration
$a(t)$	time history of acceleration
a_y	yield acceleration of elasto-plastic SDOF system
C_B	correction factor of measured SPT blow count for borehole diameter
C_E	correction factor of measured SPT blow count for hammer energy ratio ER_m
C_F	correction factor for N_{65}, based on the fines content
C_k	R_{max} correction factor, for earthquake motion irregularity
C_M	scaling factor for earthquake moment magnitude
C_N	correction factor of measured SPT blow count for overburden stress
C_q	correction factor for CPT tip resistance
C_R	correction factor of measured SPT blow count for rod length

C_S	correction factor of measured SPT blow count for samplers with or without liners
CC_{max}	maximum value of Cross Correlation function
$CO_{xy}(f)$	coherence function of signals $x(t)$ and $y(t)$
c	cohesion of soil
c'	effective stress cohesion
c_{ex}	depth from extreme compression fiber to neutral axis at flexural strength (see Fig. T5.13)
c_o	concrete cover to center of hoop or spiral bar
c_u	undrained shear strength of soil
c_v	consolidation coefficient
$\boldsymbol{D}$	dead load vector
D	equivalent damping ratio of soil
D	damping ratio of critical of SDOF system
D'	diameter of the connection core, measured to the centerline of the spiral confinement
D^*	reference diameter of a pile
D_{anc}	distance of anchor from tie-rod level of active soil wedge
D_{emb}	depth of wall embedment
D_f	depth of the effective point of rotation of the sheet pile wall below the mudline
D_p	pile (outer) diameter, or equivalent pile width, or gross depth (in the case of a rectangular pile with Spiral confinement in TC5)
D'_p	inner pile diameter
$D(\gamma)$	strain dependent damping ratio of soil
$[\bar{D}(\gamma)]_{lab}$	laboratory measured damping ratio of soil scaled to D_0
D_0	small strain damping ratio of soil
$D1$	thickness of soil deposit below a gravity wall
d	displacement at the top of the wall
d'	maximum displacement of a linear SDOF system
$\bar{d}$	average sliding displacement
d_b	diameter of dowel reinforcement
d_f	depth to fixity of a pile
d_l	leg expansion of a crane
d_m	maximum displacement of elasto-plastic SDOF system.
d_p	nominal diameter of the prestressing strand
d_r	residual displacement of SDOF system ($= d_m - d_y$)
d_u	ultimate displacement of a elasto-plastic SDOF system
d_y	yield displacement of an elasto-plastic SDOF system
$\boldsymbol{E}$	earthquake load vector
E	Young's modulus
E_d	dowel modulus of elasticity
E_n	seismic energy

E_s	equivalent subgrade elastic modulus ($= k_{\text{h-sub}}D_p$)
E_x, E_y	earthquake loads (E) in the principal directions x and y
ER	energy efficiency for SPT procedure
ER_m	energy efficiency of a particular SPT procedure
e	void ratio
e_{ex}	eccentricity
F	seismic load
F'	maximum force of a linear SDOF system
F_c	fines content
F_L	factor of safety against liquefaction
F_m	maximum load on inelastic structural response
F_p	axial prestress force in pile
F_s	factor of safety during an earthquake
F_{so}	factor of safety against overturning
F_{ss}	factor of safety against sliding
F_T	a set of discrete variables describing the fault type
F_y	elastic limit of elasto-plastic SDOF system
f	frequency
f_a	average compression stress at the joint center caused by the pile axial compression force N
f'_c	ultimate (unconfined) compressive strength of concrete at 28 days
f'_{cc}	confined strength of concrete approximated by $1.5f'_c$
f'_{ce}	concrete compression strength with earthquake increase factor
f'_{cm}	concrete compression strength with capacity protection
f_{pu}	ultimate strength of prestressing strand
f_{pue}	prestress strand ultimate strength with earthquake increase factor
f_{pum}	prestress strand ultimate strength with capacity protection
f_r	resonant frequency
f_s	$= 0.0015E_d$
f_y	nominal yield strength of reinforcement
f^o_{yc}	dowel overstrength bar capacity ($= 1.4f_y$)
f_{ye}	reinforcement yield strength with earthquake increase factor
f_{yh}	yield stress of confining steel, transverse spiral or hoop reinforcement
f_{ym}	reinforcement yield strength with capacity protection
f_1, f_2	half-power frequencies
G	equivalent (secant) shear modulus of soil
$G(\gamma)$	strain dependent shear modulus
$\bar{G}(\gamma), G/G_0$	normalised equivalent shear modulus
G_0 (or G_{max})	small strain shear modulus
g	acceleration of gravity ($= 9.8$ m/s^2)
H	height of wall
H_c	hinge height of a crane leg

H_{simple}	a span of a simple beam (i.e. sheet pile wall span between tie-rod and mud line levels)
H_{sub}	thickness of backfill between the mudline and the backfill water table
H_{sur}	thickness of backfill soil above the water table
H_t	height of a tsunami at the shore
H_w	water depth
H_0	height of a tsunami at an offshore site
h_d	depth of the pile cap
h_0	water depth at the offshore site
I	moment of inertia of piles
I_c	soil behaviour type index
I_{eff}	effective moment of inertia
I_i	moment of inertia of i-th pile
I_0 or I_e	maximum intensity of earthquake
I_P	plasticity index
j	an empirical power index to define small strain damping of soil
K	bulk modulus of soil
K_a	static active earth pressure coefficient
K_{ae}	dynamic active earth pressure coefficient
K_c	correction factor for obtaining $(q_{c1N})_{cs}$
K_{Hi}	equivalent spring stiffness for lateral reaction by i-th pile
K_L	total stiffness of the tributary length of wharf
K_p	static passive earth pressure coefficient
K_{pe}	dynamic passive earth pressure coefficient
K_s	correction factor for obtaining $(N_1)_{60cs}$
K_u	rebound bulk modulus.
K_0	earth pressure coefficient at rest
ΔK_{ae}	dynamic component of active earth pressure coefficient
ΔK_{pe}	dynamic component of passive earth pressure coefficient
k	coefficient of permeability
k_c	factor dependent on the curvature ductility within the plastic hinge region, given by Fig. T5.12
$k_{c\text{-type}}$	parameter for specifying subgrade reaction for C-type subsoil
k_d	deck stiffness for force-displacement response of a wharf
k_{DC}	stiffness of wharf for damage control limit
k_e	equivalent seismic coefficient
k_h	horizontal seismic coefficient
k'_h	equivalent horizontal seismic coefficient for submerged condition
$k_{h\text{-sub}}$	coefficient of lateral subgrade reaction
k_I	initial stiffness of inelastic structural response
k_{pt}	additional spring to be added at the deck/pile interface to represent the inelastic stiffness of the top plastic hinge

k_S	stiffness of wharf for serviceability limit
$k_{s\text{-type}}$	parameter for specifying subgrade reaction for S-type subsoil
k_{sub}	initial modulus of subgrade reaction
k_t	threshold seismic coefficient
k_u	unloading stiffness of inelastic structural response
k_v	vertical seismic coefficient
k_v'	equivalent vertical seismic coefficient for submerged condition
L	design live load vector
L	a free pile length/distance from the pile/deck intersection to the point of contraflexure in the pile
L_{ext}	extent of soil improvement
L_L	longitudinal length of wharf segment
L_{max}	seismic stress ratio
L_p	plastic hinge length of piles
L_{span}	rail span for cranes
L_T	transverse length of wharf segment
L_w	travel path of the waveform
L_x	linear spectrum of a signal $x(t)$
l	unsupported length of a pile between the deck and virtual ground surface
l_a	actual embedment length provided
l_e	embedment length of pile in deck
l_i	unsupported length of i-th pile between the deck and slope surface
l_{m1}	depth of contraflexure point
l_3	depth to fixity for pile analysis
M	bending moment
M	Richter magnitude
M^o	overstrength moment (i.e. maximum possible moment) of the plastic hinge
M_{design}	design bending moment for sheet pile wall
M_{FES}	maximum bending moment under free earth support condition
M_J	Japan Meteorological Agency (JMA) magnitude of earthquake
M_L	local magnitude
$M_{mudline}$	driving moment above the mudline level
M_N	nominal flexural strength
M_p	pile bending moment at plastic hinge state
M_{p0}	plastic hinge moment without axial load
M_s	surface wave magnitude of earthquake
M_{simple}	bending moment of sheet pile based on a simple beam analysis
$M_{tie\text{-}rod}$	moment balance about the tie-rod
M_u	ultimate bending moment of a pile
M_w	moment magnitude of earthquake
M_y	yield bending moment

m	an empirical power index to define small strain shear modulus of soil
m_B	long-wave body wave magnitude
m_b	short-wave body wave magnitude
N	SPT penetration resistance (blow count)
N	axial load
N_c	number of cycles
N_{eq}	equivalent number of uniform cycles
N_{ER}	SPT reference penetration resistance (for a particular energy efficiency)
N_i	axial load for the i-th pile
N_l	number of load cycles to cause liquefaction
N_m	measured SPT blow count
N_u	axial compression load on pile, including seismic load
$N_w(f)$	number of wavelengths of a sinusoidal signal (f) while propagating
N_{y0}	yield axial load without bending moment
N_{65}	effective SPT N-value (= Japanese SPT N-value with average SPT energy efficiency ratio of 72% corrected to the effective overburden pressure of 65 kPa)
$(N_1)_{60}$	SPT N-value normalized to the effective overburden pressure of 1 kgf/m^2 (= 98kPa) with SPT energy efficiency ratio of 60%
$(N_1)_{60cs}$	equivalent value of $(N_1)_{60}$ for clean sands
N_{65}^*	equivalent N_{65} value, for $I_P < 10$ or $5 < F_c < 15\%$
N_{65}^{**}	equivalent N_{65} value, for $I_P > 20$ or $F_c > 15\%$
ΔN	increment of N, for calculating N_{65}^{**}
n	mean annual number of events that exceeds magnitude M
n	an empirical power index to define small strain shear modulus of soil
n_{ext}	normalized soil improvement factor ($= L_{ext}/(H + D_{emb})$)
n_i	number of piles in row i in the tributary width
P	subgrade reaction per unit length of a pile at the depth ξ
P_a	static active earth thrust
P_{ae}	dynamic active earth thrust
$(P_{ae})_{mudline}$	dynamic active earth thrust above the mudline level
$P_{foundation}$	equivalent reaction force exerted by the foundation (i.e. mudline support)
P_p	static passive earth thrust
P_{pe}	dynamic passive earth thrust
$P_R(a_{max}, T_L)$	probability of the acceleration a_{max} being exceeded in the exposure time T_L
P_u	the limit lateral load when plastic hinges form at top and embedded portion of all piles
P_{wd}	hydrodynamic force

P_y	approximate elastic limit
ΔP_{ae}	dynamic component of active earth thrust
ΔP_{pe}	dynamic component of passive earth thrust
p	($= P/D_p$) Subgrade reaction per unit area of a pile at the depth x
p'	mean effective stress
p_a	atmospheric pressure
p_{ae}	dynamic active earth pressure
p_{dw}	dynamic fluid pressure
p_{pe}	dynamic passive earth pressure
p_t	nominal principal tension stress in the joint region
p_u	ultimate resistance of soil
p_{wd}	hydrodynamic pressure
Q	an empirical parameter to define small strain damping of soil
q	deviatoric stress
q/σ_r'	deviatoric/radial effective stress ratio
q_c	CPT tip penetration resistance
q_{c1}	reference value of CPT tip resistance
q_{c1N}	normalised and corrected CPT tip resistance
$(q_{c1N})_{cs}$	equivalent value of q_{c1N} for clean sands
q_{sur}	uniformly distributed surcharge
R	a measure of distance
R_{max}	in-situ liquefaction resistance
r	live load reduction factor
r_d	reduction factor for estimating the variation of cyclic shear stress (or acceleration) with depth
r_{EC}	a reduction factor for specifying the effective seismic coefficient as a fraction of a_{design}/g
r_p	pile radius
r_u	pore water pressure ratio
r_{12}	distance between two point, during wave propagation
S	an empirical dimensionless parameter to define small strain shear modulus of soil
S_A	acceleration response spectra
$S_A(T, D)$	acceleration response spectrum
$S_D(T, D)$	displacement response spectrum
S_T	a set of discrete variables describing the site foundation conditions
$S_V(T, D)$	velocity response spectrum
$[S/R]_{xy}$	signal to noise ratio of signal $x(t)$ and $y(t)$
s	spacing of hoops or spiral along the pile axis.
T	natural period of SDOF system
T_{design}	design tie-rod force per meter
T_e	ultimate anchor resistance
T_{FES}	tie-rod force under free earth support condition

T_L	exposure time, life span of structure
T_R	return period
T_s	Natural period of a pile-deck system
T_{simple}	tie-rod force based on a simple beam analysis
T_w	travel time of the waveform
t	time
t_G	degradation parameter
t_p	thickness of steel shell pile
t_{xy}	time delay between signals $x(t)$ and $y(t)$
U_c	coefficient of uniformity
ΔU	excess pore water pressure due to cyclic shearing; subscripts a and p denote pressure within the active and passive soil wedges
u	excess pore water pressure
u_d	pore water pressure in a drained layer
u/σ'_v	excess pore water pressure ratio
V_c	shear strength from concrete
V_i	pile shear force at the limit state deflection Δ_{ls}
V_N	shear strength of pile
V_p	shear strength from axial load
V_{ph}	Fourier component phase velocity
$V_{ph}(f)$	phase velocity of the frequency component (f) of a signal
V_R	Rayleigh wave velocity
V_S	shear wave velocity
V_{shell}	contribution of the shell to the shear strength
V_{S1}	shear wave velocity normalised to a reference effective stress
V_{sk}	shear key force
V_t	shear strength from transverse reinforcement
$V_{\Delta T}$	total segment lateral force at displacement Δ_T for pure translational response
v_{max}	peak velocity
v_j	nominal shear stress in the joint region corresponding to the pile plastic moment capacity
W	width of a gravity wall
W_d	energy damped in the stress-strain loop of soil
W_g	weight of a gravity wall per unit width
W_g	surcharge and deadweight of deck
W_s	energy stored in the stress-strain loop of soil
$x(t)$	a signal for wave propagation analysis
x_r	center of rigidity
$y(t)$	a signal for wave propagation analysis
y_g	ground motion parameter or response spectrum ordinate
Z_e	elastic sectional modulus
Z_p	sectional modulus at plastic hinge

z	elevation below the tie-rod level
$z_{P_{pe}}$	elevation of passive earth thrust below the tie-rod level
$z_{P_{wd}}$	elevation of hydrodynamic force below the tie-rod level
subscript i	i^{th} pile or i^{th} layer
subscript x	horizontal component
subscript y	horizontal component normal to x-direction
α	a factor for specifying the effective seismic coefficient as a fraction of a_{max}/g
α_{ae}	seismic active angle of failure
α_{ax}	angle between line joining the centers of flexural compression in the deck/pile and in-ground hinges, and the pile axis (see Fig. T5.14)
α_{pe}	seismic passive angles of failure
α_t	angle of critical crack to the pile axis for transverse shear (see Fig. T5.13)
α_y	ratio of existing construction exposure time over new construction
β	a parameter for a beam analysis in Winkler foundation $\left(= \sqrt[4]{\dfrac{k_{h\text{-sub}} D_p}{4EI}} \right)$
$1/\beta$	length of equivalent fixity pile below the slope surface
γ	shear strain amplitude
γ_b	buoyant unit weight of soil $(= \gamma_{sat} - \gamma_w)$
γ_{cyc}	cyclic strain amplitude
γ_d	unit weight of dry soil
γ_e	equivalent unit weight of partially submerged soil wedge
$\gamma_{e\text{-sat}}$	equivalent total unit weight for submerged condition
γ_l	linear threshold shear strain
γ_{sat}	saturated total unit weight of soil
γ_v	volumetric threshold shear strain
γ_w	unit weight of seawater
γ_{wet}	unit weight of soil above the water table
Δ	displacement of piles
Δ_{DC}	displacement of a wharf at damage control limit
Δ_g	initial gap between the deck segments of pile-supported wharves and piers
Δ_{ls}	limit state deflection
Δ_{max}	upper bound displacement of deck
Δ_r	residual displacement
Δ_S	displacement of a wharf at serviceability limit
Δ_t	displacement of deck under pure translational response
δ	friction angle at the wall-soil interface (for sign convention, refer to Section T7.1.1 and Fig. T7.1)

δ_G, d_D	cyclic degradation indexes
δ^*	factored wall interface friction angle for passive earth pressure computation
ε	strain in piles
ε_c	strain at 50% peak shear resistance
ε_{cu}	ultimate compressive concrete strain
ε_{max}	strain limit for embedded portion of steel piles
ε_s^p	irrecoverable (or plastic) distortional strain
ε_{sm}	strain at peak stress of confining reinforcement
ε_v	volumetric strain
ε_{vd}	inelastic volumetric strain
ε_v^p	irrecoverable (or plastic) volumetric strain
ε_σ	a random error term with zero mean and σ standard deviation
ζ	elevation beneath the fluid level in front of the sheet pile wall
η	equivalent viscous damping of inelastic response as a percentage of critical damping
Θ_p	inelastic rotation at plastic hinge
λ	wave length
$\lambda(f)$	wavelength of a sinusoidal signal (f) while propagating
λ_R	wave length of Rayleigh waves
μ	ductility factor
μ_b	friction angle at bottom of wall and/or at toe of a crane
μ_ϕ	curvature ductility ($= \phi / \phi_y$)
μ_Δ	displacement ductility factor ($= d_m/d_y$)
ξ	depth from the top of the pile
ξ_{P_a}	elevation of static active earth thrust from the toe of the sheet pile wall
$\xi_{P_{ae}}$	elevation of dynamic active earth thrust from the toe of the sheet pile wall
$\xi_{\Delta P_{ae}}$	elevation of dynamic component of active earth thrust from the toe of the sheet pile
$\xi_{P_{pe}}$	elevation of dynamic passive earth thrust from the toe of the sheet pile wall
$\xi_{P_{wd}}$	elevation of hydrodynamic force from the toe of the sheet pile wall
ρ	bulk soil density
ρ	Rowe's flexibility number for sheet pile design
ρ_l	longitudinal reinforcement ratio, including prestressing steel
ρ_s	effective volume ratio of confining steel
σ	vertical stress on shear plane
σ_a	axial stress in soil specimen
σ_{allow}	allowable stress in piles
σ_c'	effective confining stress; current consolidation pressure in soil
σ_m'	mean effective stress in subsoil
σ_{m0}'	initial mean effective vertical stress in subsoil

σ'_p	maximum consolidation pressure that the soil sample has been subjected to
σ_r	radial stress in soil specimen
σ'_v	vertical effective stress in subsoil
σ'_{v0}	initial effective vertical stress in subsoil
σ_{v0}	in situ total vertical stress in subsoil
σ_y	yield stress in piles
σ'_0	initial effective stress in soil
τ	shear stress
τ_{av}	average shear stress
τ_{cc}	waveform travel time (with respect to CC_{max})
τ_{dmax}	maximum shear stress, from site response analysis
τ_l	cyclic shear stress amplitude
τ_{max}	maximum shear stress
τ/σ'_{v0}	cyclic shear stress ratio
$(\tau_l/\sigma'_c)_{N_1=20}$	cyclic resistance for liquefaction potential assessment or cyclic stress ratio at 20 cycles of loading
$(\tau_l/\sigma'_c)_{NC}$	liquefaction strength for normally consolidated state
$(\tau_l/\sigma'_c)_{OC}$	liquefaction strength for overconsolidated state
Φ	phase of a sinusoidal function
$\Phi_{CPS}(f)$	phase of each frequency component of the CPS
Φ_p	1.0 for assessment, and 0.85 for new pile design
$\Delta\phi(f)$	phase delay of a sinusoidal signal
ϕ	internal friction angle of soil
ϕ	pile curvature
ϕ'	effective stress friction angle of soil
$\phi*$	factored internal friction angle for passive earth pressure computation
ϕ_{LS}	limit state pile curvature for Level 2
ϕ_p	plastic limit state pile curvature ($= (\phi_S$ or $\phi_{LS}) - \phi_y)$
ϕ_S	limit state pile curvature for Level 1
ϕ_u	ultimate pile curvature
ϕ_y	yield pile curvature
ψ	seismic inertia angle
ω	similarity number (i.e. a non-dimensional Rowe's flexibility number) (see Eqn. (T8.4))
$\oplus$	load combination symbol for vector summation

Abbreviations:

APS_{xx}	auto power spectrum function of signal $x(t)$
BE	bender elements
CC_{xy}	cross correlation function of signals $x(t)$ and $y(t)$
CH	cross hole test

CPS_{xy}	cross power spectrum function of signals $x(t)$ and $y(t)$
CPT	cone penetration test
CQC	Complete Quadratic Combination rule
CRR	cyclic resistance ratio (liquefaction resistance)
CSR	cyclic stress ratio (seismic load)
CSS	cyclic simple shear test
CTS	cyclic torsional shear test
CTX	cyclic triaxial test
DFT	direct Fourier transform
DH	down hole test
EAI	effective anchor index for sheet pile design
EPI	embedment participation index for sheet pile design
E1	earthquake motion equivalent to the design seismic coefficient
E2	earthquake motion about 1.5 to 2.0 times larger than E1
FDM	finite difference method
FEM	finite element method
IFT	inverse Fourier transform
LDT	local displacement transducers
L1	level 1 earthquake motion
L2	level 2 earthquake motion
MDOF	multi degree of freedom
MSF	magnitude scaling factor
OCR	overconsolidation ratio ($= \sigma_p'/\sigma_c'$)
PGA	peak ground acceleration
PGA_H	peak ground horizontal acceleration
PGA_V	peak ground vertical acceleration
PGV	peak ground velocity
PGV_H	peak ground horizontal velocity
PGV_V	peak ground vertical velocity
RC	resonant column test
SASW	spectral analysis of surface waves
SC	seismic cone technique
SDOF	single degree of freedom
SH	shear wave horizontal components
SPT	standard penetration test
SPT N-value	standard penetration test blow counts
SSI	soil structure interaction
1D/2D/3D	one, two, and three dimensions

References

Abrahamson, N.A. & Silva, W.J. 1997. Empirical response spectral attenuation relations for shallow crustal earthquakes, *Seismological Research Letters* 68(1).

ACI Committee 116 2000. *Cement and Concrete Terminology,* 73 p.

Alarcon, A., Chameau, J.L. & Leonards, G.A. 1986. A new apparatus for investigating the stress-strain characteristics of sands, *Geotechnical Testing Journal,* ASTM 9(4): 204–212.

Amano, R., Azuma, H. & Ishii, Y. 1956. Aseismic design of walls in Japan, *Proc. 1st World Conference on Earthquake Engineering,* San Francisco, pp. 32-1–32-17.

Ambraseys, N.N. & Menu, J.M. 1988. Earthquake-induced ground displacements, *Earthquake Engineering and Structural Dynamics* 16: 985–1006.

American Petroleum Institute (API) 1993. Recommended Practice for Planning, Designing and Constructing Fixed Offshore Platforms – Working Stress Design, *API Recommended Practice 2A-WSD (RP 2A-WSD),* 20th edition, 191 p.

Anderson, J.G. 1991. Guerrero accelerograph array: seismological and geotechnical lessons, *Geotechnical News* 9(1): 34–37.

Andrus, R.D. & Stokoe, K.H. 1997. Liquefaction resistance based on shear wave velocity, *Proc. NCEER Workshop on Evaluation of Liquefaction Resistance of Soils,* Technical Report NCEER-97-0022, National Center for Earthquake Engineering Research, pp. 89–128.

Annaki, M. & Lee, K.L. 1977. Equivalent uniform cycle concept of soil dynamics, *Journal of Geotechnical Engineering Division,* ASCE 103(GT6): 549–564.

Arango, I. 1996. Magnitude scaling factors for soil liquefaction evaluations, *Journal of Geotechnical Engineering,* ASCE 122(GT11): 929–936.

Arulanandan, K. & Scott, R.F. (eds.) 1993. VELACS – *Verification of Numerical Procedures for the Analysis of Soil Liquefaction Problems,* Balkema, 1801 p.

Aubry, D., Chouvet, D., Modaressi, A. & Modaressi, H. 1985. *GEFDYN_5*: Logiciel d'analyse du comportement statique et dynamique des sols par éléments finis avec prise en compte du couplage sol-eau-air, Rapport Scientifique Ecole Centrale de Paris (in French).

Baldi, G., Ghionna, V.N., Jamiolkowski, M. & Lo Presti, D.C.F. 1988. Parametri di progetto per il calcolo di cedimenti di fondazioni in sabbia da prove penetrometriche, *Proc. I Conv. Naz. G.N.C.S.I.G., CNR 'Deformazioni dei terreni ed interazione terreno-struttura in condizioni di esercizio,'* – SGE, Padova, Vol. 1, pp. 3–18 (in Italian).

Bardet, J.P. 1990. *LINOS – A Non-Linear Finite Element Program for Geomechanics and Geotechnical Engineering,* Civil Engineering Department, University of Southern California.

Bartlett, S.F. & Youd, T.L. 1995. Empirical prediction of liquefaction-induced lateral spread, *Journal of Geotechnical Engineering,* ASCE 121(4): 316–329.

B.C. Hydro 1992. Seismic withstand of timber piles, *Detailed Test Procedures and Test Results*, Vol. 2: Appendices, B. C. Hydro, Hydroelectric Engineering Division, Geotechnical Department.

Bea, R.G. 1997. Background for the proposed International Standards Organization reliability based seismic design guidelines for offshore platforms, *Proc. Earthquake Criteria Workshop – Recent Developments in Seismic Hazard and Risk Assessments for Port, Harbor, and Offshore Structures*, Port and Harbour Research Institute, Japan and Univ. of California, Berkeley, USA, pp. 40–67.

Benuska, L. (ed.) 1990. Loma Prieta Earthquake Reconnaissance Report. *Earthquake Spectra*, Supplement to Vol. 6, EERI, 448 p.

Bishop, A.W. 1955. The use of the slip circle in the stability analysis of slopes, *Geotechnique* 5(1): 7–17.

Blackford, M.E. 1998. International responses to Pacific Tsunami Warnings and Watches, *Proc. UJNR*, 30th Joint Meeting of U.S.–Japan Panel on Wind and Seismic Effects, Gaithersburg, NIST.

Blázquez, R. 1995. Liquefaction modelling in engineering practice, *Proc. 10th European Conference on Earthquake Engineering*, Wien, Vol. 1, Balkema, Rotterdam, pp. 501–513.

Bolton, M.D. & Wilson, J.M.R. 1989. An experimental and theoretical comparison between static and dynamic torsional soil tests, *Géotechnique* 39(4): 585–599.

Boore, D., Joyner, W. & Fumal, T. 1997. Equations for estimating horizontal response spectra and peak acceleration from Western North-America earthquakes: A summary of recent work, *Seismological Research Letters,* 68(1).

Budek, A.M., Benzoni, G. & Priestley, M.J.N. 1997. Experimental investigation of ductility of in-ground hinges in solid and hollow prestressed piles, *Div. of Structural Engineering Report SSRP 97-17*, University of California, San Diego, 108 p.

Buslov, V.M., Rowghani, M. & Weismair, M. 1996. Evaluating earthquake damage to concrete wharves, *Concrete International,* pp. 50–54.

Buslov, V.M. 1996. *Surveys in Port of Eilat*, VBC report.

Byrne, P.M., Jitno, H., Anderson, D.L. & Haile, J. 1994. A procedure for predicting seismic displacements of earth dams, *Proc. 13th International Conference on Soil Mechanics and Foundation Engineering*, New Delhi, India, pp. 1047–1052.

California Division of Mines and Geology 1997. *Guidelines for Evaluating and Mitigating Seismic Hazards in California*, Special Publication, 117 p.

California State Lands Commission, USA 2000. *California Marine Oil Terminal Standard* (under development as of May 2000).

CEN (European Committee for Standardization) 1994. *Eurocode 8: Design Provisions for Earthquake Resistance of Structures. Part 1-1: General Rules – Seismic Actions and General Requirements for Structures* (ENV-1998-1-1); *Part 5: Foundations, Retaining Structures and Geotechnical Aspects* (ENV 1998-5).

CEN (European Committee for Standardization) 1997. *Eurocode 7: Geotechnical design. Part 3: Geotechnical design assisted by field tests.* (ENV 1997-3).

Chang, Y.L. 1937. Lateral pile loading tests, *Transaction of ASCE* 102: 273–276.

Comartin, C.D. (ed.) 1995. Guam Earthquake of August 8, 1993 Reconnaissance Report, *Earthquake Spectra*, Supplement B to Vol. 11, EERI, 175 p.

Cornell, C.A. 1968. Engineering seismic risk analysis, *Bull. Seism. Soc. Am.* 58: 1583–1606.

Cornell, C.A. & Winterstein, S.R. 1988. Temporal and magnitude dependence in earthquake recurrence models, *Bull. Seism. Soc. Am.* 78: 1522–1537.

Coulomb, C.A. 1776. Essai sur une application des regles des maximis et minimis quelques problemes de statique relatifs a l'architecture, *Memoires de l'Academie Royale pres Divers Savants*, Vol. 7.

De Alba, P., Seed, H.B. & Chan, C.K. 1976. Sand liquefaction in large-scale simple shear tests, *Journal of Geotechnical Engineering Division*, ASCE 102(GT9): 909–927.

Dickenson, S.E. & Yang, D.S. 1998. Seismically-induced deformations of caisson retaining walls in improved soils, *Geotechnical Earthquake Engineering and Soil Dynamics III*, Geotechnical Special Publication Vol. No. 75, ASCE, pp. 1071–1082.

Dobry, R., Yokel, F.Y., Ladd, R.S. & Powell, D.J. 1979. Prediction of pore pressure build-up and liquefaction of sands during earthquakes by the cyclic strain method, *Research Report, National Bureau of Standards*, U.S. Department of Commerce, Washington D.C.

Dobry, R., Powell, D.J., Yokel, F.Y. & Ladd, R.S. 1980. Liquefaction potential of saturated sand: the stiffness method, *Proc. 7th WCEE*, Istanbul, Vol. 3, pp. 25–32.

Dobry, R. 1991. Soil properties and earthquake ground response, *Proc. 10th European Conference on Soil Mechanics and Foundation Engineering*, Florence, Vol. 4, pp. 1171–1187.

d'Onofrio, A., Silvestri, F. & Vinale, F. 1999a. Strain rate dependent behaviour of a natural stiff clay, *Soils and Foundations* 39(2): 69–82.

d'Onofrio, A., Silvestri, F. & Vinale, F. 1999b. A new torsional shear device, *Geotechnical Testing Journal*, ASTM 22(2): 101–111.

d'Onofrio, A. & Silvestri, F. 2001. Influence of micro-structure on small-strain stiffness and damping of fine grained soils and effects on local site response, *Proc. 4th International Conference on Recent Advances in Geotechnical Earthquake Engineering and Soil Dynamics*, San Diego.

Drnevich, V.P., Hardin, B.O. & Shippy, D.J. 1978. Modulus and damping of soils by resonant column method, *Dynamic Geotechnical Testing, ASTM STP 654*, American Society for Testing and Materials, pp. 91–125.

Dyvik, R. & Madshus, C. 1985. Lab measurement of G_{max} using bender elements, *Advances in the Art of Testing Soils Under Cyclic Conditions*, ASCE, New York, pp. 186–196.

EAU 1996. *Recommendations of the Committee for Waterfront Structures, Harbours and Waterways*, 7th English edition (English Translation of the 9th German edition), Ernst & Sohn, Berlin, 599 p.

Ebeling, R.M. & Morrison, E.E. 1992. *The Seismic Design of Waterfront Retaining Structures*, U.S. Army Corps of Engineers, Technical Report ITL-92-11/NCEL TR-939, 329 p.

Egan, J.A., Hayden, R.F., Scheibel, L.L., Otus, M. & Serventi, G.M. 1992. Seismic repair at Seventh Street Marine Terminal, *Grouting, Soil Improvement and Geosynthetics*, Geotechnical Special Publication No. 30, ASCE, pp. 867–878.

Elmrabet, T., Levret, A., Ramdani, M. & Tadili, B. 1991. Historical seismicity in Morocco: Methodological aspects and cases of multidisciplinary evaluation, *Seismicity, Seismotectonic and Seismic Risk of the Ibero-Maghrebian Region*. Monografía Num. 8 Institute Geographic National. Spain.

EPRI 1993. *Guidelines for Determining Design Basis Ground Motions*, Report No. TR-102293. Electric Power Research Institute, Palo Alto (California).

Esteva, L. 1970. Seismic risk and seismic design decisions. In R.J. Hansen (ed.), *Seismic Design of Nuclear Power plants*, MIT Press, Cambridge, Massachusetts.

Falconer, T.J. & Park, R. 1983. Ductility of prestressed concrete piles under seismic loading, *PCI Journal* 28(5): 112–114.

Ferritto, J.M. 1997a. Design criteria for earthquake hazard mitigation of navy piers and wharves, *Technical Report* TR-2069-SHR, Naval Facilities Engineering Service Center, Port Hueneme, 180 p.

Ferritto, J.M. 1997b. Seismic design criteria for soil liquefaction, *Technical Report* TR-2077-SHR, Naval Facilities Engineering Service Center, Port Hueneme, 58 p.

Ferritto, J.M., Dickenson, S.E., Priestley, M.J.N., Werner, S.D. & Taylor, C.E. 1999. Seismic criteria for California marine oil terminals, *Technical Report* TR-2103-SHR, Naval Facilities Engineering Service Center, Port Hueneme.

Finn, W.D.L., Lee, K.W. & Martin, G.R. 1977. An effective stress model for liquefaction, *Journal of Geotechnical Engineering Division*, ASCE 103(GT6): 517–533.

Finn, W.D.L., Yogendrakumar, M., Yoshida, N. & Yoshida, H. 1986. *TARA-3*: A program to compute the response of 2-D embankments and soil-structure interaction to seismic loadings, University of British Columbia, Canada.

Finn, W.D.L. 1988. Dynamic analysis in geotechnical engineering, *Earthquake Engineering and Soil Dynamics II: Recent Advances in Ground-Motion Evaluation*, Geotechnical Special Publication No. 20, ASCE, pp. 523–591.

Finn, W.D.L. 1990. Analysis of deformations, stability and pore water pressures in port structures, *Proc. POLA Seismic Workshop on Seismic Engineering*, Port of Los Angeles, pp. 369–391.

Finn, W.D.L. 1991. Assessment of liquefaction potential and post-liquefaction behaviour of earth structures: Developments 1981–1991 (State-of-the-art paper), *Proc. 2nd International Conference on Recent Advances in Geotechnical Earthquake Engineering and Soil Dynamics*, St. Louis, Vol. 3, pp. 1833–1850.

Franklin, A.G. & Chang, F.K. 1977. Earthquake resistance of earth and rockfill dams, Report 5: Permanent displacements of earth dams by Newmark analysis, *US Army Corps of Engineers, Waterways Experiment Station*, Miscellaneous Paper 2-71-17.

Fukushima, Y. & Tanaka, T. 1990. A new attenuation relation for peak horizontal acceleration of strong earthquake ground motion in Japan, *Bull. Seism. Soc. Am.* 80: 757–783.

Gazetas, G., Dakoulas, P. & Dennehy, K. 1990. Empirical seismic design method for waterfront anchored sheetpile walls, *Proc. ASCE Specialty Conference on Design and Performance of Earth Retaining Structures*, ASCE Geotechnical Special Publication No. 25, pp. 232–250.

Goto, S., Tatsuoka, F., Shibuya, S., Kim, Y.S. & Sato, T. 1991. A simple gauge for local small strain measurements in the laboratory, *Soils and Foundations* 31(1): 169–180.

Grünthal, G. (ed.) 1998. *European Macroseismic Scale 1998. Cahier du Centre Européen de Géodynamique et de Séismologie*. Luxembourg. Vol. 15, 99 p.

GSHAP 1999. *Global Seismic Hazard Assessment Program*, http://seismo.ethz.ch/GSHAP.

Gulkan, P. & Sozen, M. 1974. Inelastic response of reinforced concrete structures to earthquake motion, *ACI Journal*, pp. 604–610.

Gutenberg, B. & Richter, C.F. 1944. Frequency of earthquakes in California, *Bull. Seism. Soc. Am.* 34: 1985–1988.

Hall, J.F. (ed.) 1995. *Northridge Earthquake of January 17, 1994* Reconnaissance Report – Vol. 1, *Earthquake Spectra*, Supplement C to Vol. 11.

Hardin, B.O. & Black, W.L. 1968. Vibration modulus of normally consolidated clay, *Journal of Soil Mechanics and Foundations Division*, ASCE 94(SM2): 353–369.

Hardin, B.O. & Black, W.L. 1969. Closure to Vibration modulus of normally consolidated clay, *Journal of Soil Mechanics and Foundations Division*, ASCE 95(SM6): 1531–1537.

Hardin, B.O. & Drnevich, V.P. 1972. Shear modulus and damping in soils: design equations and curves, *Journal of Soil Mechanics and Foundations Division*, ASCE 98(SM7): 667–692.

Hardin, B.O. 1978. The nature of stress-strain behaviour for soils – State of the Art, *Proc. Earthquake Engineering and Soil Dynamics*, ASCE, Pasadena (California), Vol. 1, pp. 3–89.

Heaton, T.H., Tajima, F. & Mori, A.W. 1982. Estimating ground motions using recorded accelerograms, Report by Dames & Moore to Exxon Production Res. Co.; Houston.

Hoar, R.J. 1982. *Field Measurements of Seismic Wave Velocity and Attenuation for Dynamic Analysis*, Ph.D. Diss., The University of Texas at Austin.

Hynes-Griffin, M.E. & Franklin, A.G. 1984. Rationalizing the seismic coefficient method, *US Army Corps of Engineers, Waterways Experiment Station*, Miscellaneous paper GL-84-13, 21 p.

Iai, S. & Koizumi, K. 1986. Estimation of earthquake induced excess pore pressure in gravel drains, *Proc. 7th Japanese Earthquake Engineering Symposium*, pp. 679–684.

Iai, S., Koizumi, K. & Kurata, E. 1987. Basic consideration for designing the area of the ground compaction as a remedial measure against liquefaction, *Technical Note of Port and Harbour Research Institute*, No. 590, pp. 1–66 (in Japanese).

Iai, S., Koizumi, K. & Kurata, E. 1991. Ground compaction area as a remedial measure against liquefaction, *Tsuchi-to-Kiso, Japanese Society of Soil Mechanics and Foundation Engineering* 39(2): 35–40 (in Japanese).

Iai, S., Matsunaga, Y. & Kameoka, T. 1992. Strain space plasticity model for cyclic mobility, *Soils and Foundations* 32(2): 1–15.

Iai, S. & Finn, W.D.L. 1993. Evaluation of seismic sheet pile wall design, *Proc. 4th Canadian Conference on Marine Geotechnical Engineering*, Vol. 1: St. Jone's, Newfoundland, pp. 293–310.

Iai, S. & Kameoka, T. 1993. Finite element analysis of earthquake induced damage to anchored sheet pile quay walls, *Soils and Foundations* 33(1): 71–91.

Iai, S., Matsunaga, Y., Morita, T., Miyata, M., Sakurai, H., Oishi, H., Ogura, H., Ando, Y., Tanaka, Y. & Kato, M. 1994. Effects of remedial measures against liquefaction at 1993 Kushiro-Oki earthquake, *Proc. 5th US-Japan Workshop on Earthquake Resistant Design of Lifeline Facilities and Countermeasures against Soil Liquefaction*, NCEER-94-0026, National Center for Earthquake Engineering Research, pp. 135–152.

Iai, S., Sugano, T., Ichii, K., Morita, T., Inagaki, H. & Inatomi, T. 1996. Performance of caisson type quay walls, *The 1995 Hyogoken-Nanbu Earthquake, -Investigation into Damage to Civil Engineering Structures*, Japan Society of Civil Engineers, pp. 181–207.

Iai, S. & Ichii, K. 1997. Excess pore water pressures behind quay walls, *Observation and Modeling in Numerical Analysis and Model Tests in Dynamic Soil-Structure Interaction Problems*, Geotechnical Special Publication No. 64, ASCE, pp. 11–25.

Iai, S. & Tsuchida, H. 1997. Damage to ports and airports, *Reconnaissance Report on the Northridge, California, Earthquake of January 17, 1994*, Japan Society of Civil Engineers, pp. 157–168 (in Japanese).

Iai, S. 1998a. Rigid and flexible retaining walls during Kobe earthquake, *Proc. 4th International Conference on Case Histories in Geotechnical Engineering*, St. Louis, pp. 108–127.

Iai, S. 1998b. Seismic analysis and performance of retaining structures, *Geotechnical Earthquake Engineering and Soil Dynamics III*, Geotechnical Special Publication No. 75, ASCE, pp. 1020–1044.

Iai, S. & Ichii, K. 1998. Performance based design for port structures, *Proc. UJNR 30th Joint Meeting of United States-Japan Panel on Wind and Seismic Effects*, Gaithersburg, NIST (3–5), pp. 1–13.

Iai, S., Ichii, K., Liu, H. & Morita, T. 1998. Effective stress analyses of port structures, *Soils and Foundations,* Special Issue on Geotechnical Aspects of the January 17, 1995 Hyogoken-Nambu Earthquake, No. 2, pp. 97–114.

Iai, S., Ichii, K., Sato, Y. & Liu, H. 1999. Residual displacement of gravity quaywalls – parameter study through effective stress analysis, *Proc. 7th U.S.-Japan Workshop on Earthquake Resistant Design of Lifeline Facilities and Countermeasures against Soil Liquefaction,* Seattle, MCEER-99-0019, pp. 549–563.

Ichii, K., Iai, S., & Morita, T. 1997. Effective stress analyses on the performance of caisson type quay walls during 1995 Hyogokne-Nanbu earthquake, *Report of Port and Harbour Research Institute* 36(2): 41–86 (in Japanese).

Ichii, K., Sato, Yu., Sato, Yo., Hoshino, Y. & Iai, S. 1999. Site characteristics of strong-motion earthquake stations in ports and harbours in Japan (Part VI), *Technical Note of Port and Harbour Research Institute,* No. 935, 696 p.

Idriss, I.M., Dobry, R. & Singh, R.D. 1978. Nonlinear behaviour of soft clays during cyclic loading, *Journal of Geotechnical Engineering Division,* ASCE 104(12): 353–369.

Idriss, I.M. 1985. Evaluating seismic risk in engineering practice, *Proc. 11th International Conference on Soil Mechanics and Foundation Engineering,* San Francisco, Vol. 1, Balkema, pp. 255–320.

Idriss, I.M. 1991. Earthquake ground motions at soft soil sites, *Proc. 2nd International Conference on Recent Advances in Geotechnical Earthquake Engineering and Soil Dynamics,* St. Louis, Vol. III, pp. 2265–2272.

Imai, T. 1977. P and S wave velocities of the ground in Japan, *Proc. 9th International Conference on Soil Mechanics and Foundation Engineering,* Tokyo, Vol. 2, pp. 257–260.

Inagaki, H., Iai, S., Sugano, T., Yamazaki, H. & Inatomi, T. 1996. Performance of caisson type quay walls at Kobe port, *Soils and Foundations,* Special Issue on Geotechnical Aspects of the January 17 1995 Hyogoken-Nambu Earthquake, pp. 119–136.

Inatomi, T., Uwabe, T., Iai, S., Kazama, M., Yamazaki, H., Matsunaga, Y., Sekiguchi, S., Mizuno, Y. & Fujimoto, Y. 1994. Damage to port facilities by the 1993 Hokkaido-Nansei-Oki earthquake, *Technical Note of the Port and Harbour Research Institute,* No. 791, 449 p. (in Japanese).

Isenhower, W.M. 1979. *Torsional simple shear/Resonant column properties of San Francisco Bay mud,* Ph.D. Thesis, The University of Texas at Austin.

Ishihara, K. 1976. *Foundation of Soil Dynamics,* Kajima Publication, pp. 258–260 (in Japanese).

Ishihara, K. & Takatsu, H. 1979. Effects of overconsolidation and K_0 consolidation on the liquefaction characteristics of sand, *Soils and Foundations* 19(4): 59–68.

Ishihara, K. & Towhata, I. 1980. One-dimensional soil response analysis during earthquakes based on effective stress method, *Journal of the Faculty of Engineering,* University of Tokyo, Vol. 35, No. 4.

Ishihara, K., Yasuda, S. & Yokota, K. 1981. Cyclic strength of undisturbed mine tailings, *Proc. International Conference on Recent Advances in Geotechnical Earthquake Engineering and Soil Dynamics,* St. Louis, Vol. 1, pp. 53–58.

Ishihara, K. 1982. Evaluation of soil properties for use in earthquake response analysis, *Geomechanical Modelling in Engineering Practice,* Balkema, Rotterdam, pp. 241–276.

Ishihara, K. & Towhata, I. 1982. Dynamic response analysis of level ground based on the effective stress method, *Soil Mechanics – Transient and cyclic loads,* John Wiley, pp. 133–172.

Ishihara, K. 1985. Stability of natural deposits during earthquakes, *Proc. 11th International Conference on Soil Mechanics and Foundation Engineering*, San Francisco, Vol. 1, Balkema, pp. 321–376.

Ishikawa, Y. & Kameda, H. 1991. Probability-based determination of specific scenario earthquakes, *Proc. 4th International Conference on Seismic Zonation*, Vol. II, pp. 3–10.

ISSMGE (TC4) 1999. *Manual for Zonation on Seismic Geotechnical Hazards (Revised Version)*, Technical Committee for Earthquake Geotechnical Engineering (TC4), International Society of Soil Mechanics and Geotechnical Engineering, Japanese Geotechnical Society, 209 p.

ITASCA 1995. FLAC – *Fast Lagrangian Analysis of Continua, User's Manual*, Itasca Consulting Group, Inc., Minneapolis, Minnesota.

Japan Cargo Handling Mechanization Association 1996. *Maintenance Manual for Container Cranes*, 158 p. (in Japanese).

Japan Road Association (ed.) 1996. *Specification for Highway Bridge*, Part V. Earthquake Resistant Design, 228 p. (in Japanese).

Japan Society of Civil Engineers 1990. *Preliminary Report of JSCE Reconnaissance Team on July 16, 1990 Luzon Earthquake*, 166 p.

Japanese Geotechnical Society 1989. Effective stress analysis of ground and soil structures, *Proc. Symposium on Seismic Performance of Ground and Soil Structures*, pp. 50–136 (in Japanese).

Japanese Geotechnical Society 1991. Comparative liquefaction analyses, *Proc. Symposium on Remediation of Soil Liquefaction*, pp. 77–198 (in Japanese).

Japanese Society of Soil Mechanics and Foundation Engineering 1982. *Handbook of Soil Mechanics and foundation Engineering*, pp. 1000–1002 (in Japanese).

Japanese Society of Soil Mechanics and Foundation Engineering 1988. *Soft Ground Improvement Method – From Investigation, Design to Construction*, pp. 59–83, pp. 228–232 (in Japanese).

Japanese Society of Soil Mechanics and Foundation Engineering 1991. *Symposium on Liquefaction Remediation of Ground*, pp. 15–40, pp. 215–290 (in Japanese).

Jibson, R.W. 1993. Predicting earthquake-induced landslide displacements using Newmark's sliding block analysis, *Transportation Research Record* No. 1411, Earthquake-Induced Ground Failure Hazards, Transportation Research Board, National Academy Press, Washington, D.C., pp. 9–17.

Joen, P.H. & Park, R. 1990. Flexural strength and ductility analysis of spiral reinforced concrete piles, *PCI Journal* 35(4): 42–61.

Jones, N.P., Thorvaldsdóttir, S., Liu, A., Narayan, P. & Warthen, T. 1995. Evaluation of a loss estimation procedure based on data from the Loma Prieta Earthquake, *Earthquake Spectra* 11(1): 37–61.

Johnson, R.K., Riffenburgh, R., Hodali, R., Moriwaki, Y. & Tan, P. 1998. Analysis and design of a container terminal wharf at the Port of Long Beach, *PORTS '98*, ASCE, pp. 436–444.

Kameda, H. & Ishikawa, Y. 1988. Extended sesmic risk analysis by hazard-consistent magnitude and distance, *Journal of Structural Mechanics and Earthquake Engineering*, Japan Society of Civil Engineers (JSCE), No. 392/I-9: 395–402 (in Japanese).

Kameoka, T. & Iai, S. 1993. Numerical analyses on the effects of initial stress conditions on seismic performance of sheet pile quay walls, *Technical Note of Port and Harbour Research Institute*, No. 751, 29 p. (in Japanese).

Kim, D.S. 1991. *Deformation Characteristics of Soils at Small to Intermediate Strains from Cyclic Tests*, Ph.D. Thesis, The University of Texas at Austin.

Kinoshita, S. 1998. Kyoshin Net (K-NET), *Seismological Research Letters* 69(4): 309–332.

Kishida, T. 1983. On the soil improvement work on second plant in Nagoya port, Nisshin Seifun K.K., (Liquefaction Remediation Measure for Existing Wharf by Gravel Compaction Method), *Proc. 29th National Port and Harbor Construction Conference,* Japanese Association of Ports and Harbors, pp. 82–94 (in Japanese).

Kitajima, S. & Uwabe, T. 1979. Analysis of seismic damage to anchored sheet pile bulk-heads, *Report of Port and Harbour Research Institute* 18(1): 67–127 (in Japanese).

Kowalski, M.J. & Priestley, M.J.N. 1998. Shear strength of ductile bridge columns, *Proc. 5th Caltrans Seismic Design Workshop*, Sacramento.

Kramer, S.L. 1996. *Geotechnical Earthquake Engineering*, Prentice Hall, 653 p.

Kubo, K. 1962. Experimental study on the lateral subgrade reaction of piles – Part 2, *Monthly Report of Transportation Technical Research Institute* 11(12): 533–559 (in Japanese).

Liao, S.S.C. & Whitman, R.V. 1986. Overburden correction factors for SPT in sand, *Journal of Geotechnical Engineering*, ASCE 112(3): 373–377.

Luke, B.A. & Stokoe, K.H. 1998. Application of the SASW Method Underwater, *Journal of Geotechnical and Geoenvironmental Engineering*, ASCE 124(6): 523–531.

Lysmer, J., Udaka, T., Tsai, C.F. & Seed, H.B. 1975. FLUSH: a computer program for approximate 3-D analysis of soil-structure interaction problems, *Report EERC 75-30*, Earthquake Engineering Research Center, University of California, Berkeley, 83 p.

Makdisi, F.I. & Seed, H.B. 1978. Simplified procedure for estimating dam and embank-ment earthquake-induced deformations, *Journal of Geotechnical Engineering Division*, ASCE 104(7): 849–867.

Mancuso, C., Silvestri, F. & Vinale, F. 1995. Italian experiences on dynamic in situ and in laboratory measurements of soil properties, *Proc. 3rd Turkish Earthquake Engineering Conference*, Istanbul, Instanbul Technical University and Chamber of Civil Engineers.

Mancuso, C., Silvestri, F. & Vinale, F. 1997. Soil properties relevant to seismic micro-zonation, *Proc. 1st Japanese-Turkish Conference on Earthquake Engineering*, Istanbul, Istanbul Technical University and Chamber of Civil Engineers.

Mander, J.B., Priestley, M.J.N. & Park, R. 1988. Theoretical stress-strain model for con-fined concrete, *Journal of Structural Engineering*, ASCE 114(8): 1805–1826.

Manja, D. 1999. Personal communication.

Martin, G.R., Finn, W.D.L. & Seed, H.B. 1975. Fundamentals of liquefaction under cyclic loading, *Journal of Geotechnical Engineering Division*, ASCE 101(GT5): 423–438.

Martin, G.R., Finn, W.D.L. & Seed, H.B. 1978. Effects of system compliance on lique-faction tests, *Journal of Geotechnical Engineering Division*, ASCE 104(GT4): 463–479.

Matsuoka, H. 1985. Liquefaction resistance of overconsolidated sand and simple lique-faction prevention method, *Proc. 43rd National Conference of Japan Society of Civil Engineers*, Vol. III, pp. 15–16 (in Japanese).

Matsuzawa, H., Ishibashi, I. & Kawamura, M. 1985. Dynamic soil and water pressures of submerged soils, *Journal of Geotechnical Engineering*, ASCE 111(10): 1161–1176.

Mayne, P.W. & Rix, G. 1993. G_{max}–q_c relationship for clays, *Geotechnical Testing Journal*, ASTM 16(1): 54–60.

McCullough, N.J. & Dickenson, S.E. 1998. Estimation of seismically induced lateral deformations for anchored sheetpile bulkheads, *Geotechnical Earthquake Engineering and Soil Dynamics III*, Geotechnical Special Publication No. 75, ASCE 1095–1106.

McCullough, N.J. 1998. *The Seismic Vulnerability of Sheet Pile Walls*, Master's Thesis Dissertation, Dept. of Civil, Construction and Environmental Engineering, Oregon State University.

McGuire, R.K. 1995. Probabilistic seismic hazard analysis and design earthquakes: closing the loop, *Bull. Seism. Soc. Am.* 85: 1275–1284.

Medvedev, S.V. & Sponheuer, V. 1969. Scale of seismic intensity, *Proc. 4th World Conference on Earthquake Engineering*, Santiago, Chile, pp. 143–153.

Mejia, L.H. & Yeung, M.R. 1995. Liquefaction of coralline soils during the 1993 Guam earthquake, *Earthquake-Induced Movements and Seismic Remediation of Existing Foundations and Abutments*, Geotechnical Special Publication No. 55, ASCE, pp. 33–48.

Memos, C.D. & Protonotarios, J.N. 1992. Patras breakwater failure due to seismic loading, *Proc. 23rd International Conference on Coastal Engineering*, Venice, Ch. 255, pp. 3343–3356.

Meyerhof, G.G. 1957. Discussion, *Proc. 4th International Conference on Soil Mechanics and Foundation Engineering*, Vol. 3, p. 110

Ministry of Transport, Japan, First District Port and Harbour Construction Bureau 1984. *Report on Damage to Akita Port and Restoration – Nihonkai-Chubu Earthquake* (in Japanese).

Ministry of Transport, Japan (ed.) 1989. *Technical Standards for Port and Harbour Facilities and Commentaries*, Japan Port and Harbour Association (in Japanese); English excerpt (1991) by the Overseas Coastal Area Development Institute of Japan, 438 p.

Ministry of Transport, Japan (ed) 1999. *Design Standard for Port and Harbour Facilities and Commentaries*, Japan Port and Harbour Association, 1181 p. (in Japanese). English edition (2001) by the Overseas Coastal Area Development Institute of Japan.

Mitchell, J.K. 1981. State of the art on soil improvement, *Proc. 10th ICSMFE*, Vol. 4, 1981, pp. 510–520.

Mitchell, J.K., Baxter, C.D.P. & Munson, T.C. 1995. Performance of improved ground during earthquakes, *Soil Improvement for Earthquake Hazard Mitigation*, Geotechnical Special Publication No. 49, ASCE, pp. 1–36.

Mizuno, Y., Suematsu, N., & Okuyama, K. 1987. Design method of sand compaction pile for sandy soils containing fines, *Tsuchi-to-Kiso, Japanese Society of Soil Mechanics and Foundation Engineering* 35(5): 21–26 (in Japanese).

Modoni, G., Flora, A., Mancuso, C., Viggiani, C. and Tatsuoka, F. 2000. Pulse wave transmission tests on gravels, *Geotechnical Testing Journal*, ASTM (accepted for publication).

Mok, Y.-J., Sanchez-Salinero, I., Stokoe, K.H. & Roesset, J.M. 1988. In situ damping measurements by cross-hole seismic method, *Proc. Earthquake Engineering and Soil Dynamics II: Recent advances in ground motion evaluation*, ASCE Geotechnical Special Publication, Vol. 20, pp. 305–320.

Mononobe, N. 1924. Considerations on vertical earthquake motion and relevant vibration problems, *Journal of Japan Society of Civil Engineers* 10(5): 1063–1094 (in Japanese).

Mori, K., Seed, H.B. & Chan, C.K. 1978. Influence of sample disturbance on sand response to cyclic loading, *Journal of the Geotechnical Engineering Division*, ASCE 104(GT3): 323–339.

Morita, T., Iai, S., Liu, H., Ichii, K. & Sato, Y. 1997. Simplified method to determine parameter of FLIP, *Technical Note of Port and Harbour Research Institute*, No. 869, pp. 36 (in Japanese).

Mulilis, J.P., Chan, C.K. & Seed, H.B. 1975. The effects of method of sample preparation on the cyclic stress-strain behaviour of sands, *Report No. EERC 75-18*, Earthquake Engineering Research Center, University of California, Berkeley.

Muraleetharan, K.K., Mish, C., Yogachandran, K. & Arulanandan, K. 1988. *DYSAC2: Dynamic Soil Analysis Code for 2-dimensional Problems*, Department of Civil Engineering, University of California, Davis.

Nasu, M. 1984. Liquefaction remedial measures for Tokaido bullet train embankment, *Foundation Engineering* 12(7): 64–68 (in Japanese)

Nagao, T., Koizumi, T., Kisaka, T., Terauchi, K., Hosokawa, K., Kadowaki, Y., & Uno, K. 1995. Evaluation of stability of caisson type quaywalls based on sliding block analysis, *Technical Note of Port and Harbour Research Institute*, No. 813, pp. 302–336 (in Japanese).

Neelakantan, G., Budhu, M. & Richards, R. Jr. 1992. Balanced seismic design of anchored retaining walls. *Journal of Geotechnical Engineering*, ASCE 118(6): 873–888.

Newmark, N.M. 1965. Effects of earthquakes on dams and embankments, 5th Rankine lecture, *Geotechnique* 15(2): 139–160.

New Zealand Standards 1992–1997. NZS 4203 (1992) *General Structural Design Loadings for Buildings;* NZS 3101 Part 1 (1995) *The Design of Concrete Structures;* NZS 3403 Part 1 (1997) *Steel Structures Standard;* NZS 3403 Part 2 (1997) *Commentary to the Steel Structures Standard;* Transit New Zealand (TNZ) *Bridge Design Manual.*

Nishizawa, S., Hashimoto, M., Sakata, Y. & Sonoi, K. 1998. Investigation and analysis of a landing pier of steel pipe piles damaged by the 1995 Hyogoken-Nambu earthquake, *Soils and Foundations*, Special Issue on Geotechnical Aspects of the January 17 1995 Hyogoken-Nambu Earthquake, No. 2, pp. 133–145.

Noda, S., Uwabe, T. & Chiba. T. 1975. Relation between seismic coefficient and ground acceleration for gravity quay wall, *Report of Port and Harbour Research Institute* 14(4): 67–111 (in Japanese).

Noda, S., Iida, T., Kida, H. & Saimura, Y. 1988. Shaking table tests on a countermeasure for liquefaction by drainage piles to quaywalls, *Proc. 23rd National Conference on Soil Mechanics and Foundation Engineering*, pp. 1015–1016 (in Japanese).

Noda, S., Kida, H. & Iida, T. 1990. Experimental study on attachment of drain function to steel members as liquefaction remediation measures, *Proc. 8th Japan Earthquake Engineering Symposium*, pp. 885–890 (in Japanese).

Ohmachi, T. 1999. Formulation of Level 2 earthquake motions for civil engineering structures, *Proc. 7th U.S.-Japan Workshop on Earthquake Resistant Design of Lifeline Facilities and Countermeasures against Soil Liquefaction*, Technical Report MCEER-99-0019, pp. 497–506.

Ohta, Y. & Goto, N. 1978. Empirical shear wave velocity equations in terms of characteristic soil indexes, *Earthquake Engineering and Structural Dynamics* 6: 167–187.

Okabe, N. 1924. General theory on earth pressure and seismic stability of retaining wall and dam, *Journal of Japan Society of Civil Engineers* 10(6): 1277–1323.

Okamoto, S 1973. *Introduction to Earthquake Engineering*, University of Tokyo Press.

Olivares, L. 1996. *Caratterizzazione dell'argilla di Bisaccia in condizioni monotone, cicliche e dinamiche e riflessi sul comportamento del colle a seguito del terremoto del 1980*, Doctoral Thesis, Università degli Studi di Napoli "Federico II" (in Italian).

Omori, H. 1988. Design/construction example for soft soil, Tank Foundation (Soil Strengthening Measure by Pore Water Pressure Dissipation Method), *Foundation Engineering* 16(12): 122–129 (in Japanese).

O'Rourke, T.D., Meyersohn, W.D., Shiba, Y. & Chaudhuri, D. 1994. Evaluation of pile response to liquefaction-induced lateral spreading, *Proc. 5th US-Japan Workshop on Earthquake Resistant Design of Lifeline Facilities and Countermeasures against*

Liquefaction, Technical Report NCEER-94-0026, National Center for Earthquake Engineering Research, pp. 457–479.

Ozutsumi, O., Yuu, K., Kiyama, M. & Iai, S. 2000. Residual deformation analysis of sheetpile quay wall and backfill ground at Showa-Ohashi site by simplified method, *Proc. 12th World Conference on Earthquake Engineering*, Auckland.

Pam, H.J., Park, R. & Priestley, M.J.N. 1988. Seismic performance of prestressed concrete piles and pile/pile cap connections, *Research Report No. 88-3*, Dept. of Civil Engineering, University of Canterbury, 319 p.

Pam, H.J. & Park, R. 1990a. Flexural strength and ductility analysis of spiral reinforced prestressed concrete piles, *PCI Journal* 35(4): 64–83.

Pam, H.J. & Park, R. 1990b. Simulated seismic load test on prestressed concrete piles and pile-pile cap connections, *PCI Journal* 35(6): 42–61.

Park, R., Priestley, M.J.N. & Walpole, W.R. 1983. The seismic performance of steel encased reinforced concrete piles, *Bulletin, NZNSEE* 16(2): 124–140.

Pitilakis, K. & Moutsakis, A. 1989. Seismic analysis and behaviour of gravity retaining walls – the case of Kalamata harbour quaywall, *Soils and Foundations* 29(1): 1–17.

Port and Harbour Research Institute, Japan (ed.) 1997. *Handbook on Liquefaction Remediation of Reclaimed Land*, (translation by US Army Corps of Engineers, Waterways Experiment Station), Balkema, 312 p.

Prakash, S., Wu, Y. & Rafnsson, E.A. 1995. On seismic design displacements of rigid retaining walls, *Proc. 3rd International Conference on Recent Advances in Geotechnical Earthquake Engineering and Soil Dynamics*, St. Louis, pp. 1183–1192.

Prevost, J.H. 1981. *DYNAFLOW: A nonlinear transient finite element analysis program*, Princeton University.

Priestley, M.J.N., Seible, F. & Chai, Y.H. 1992. Design guidelines for assessment retrofit and repair of bridges for seismic performance, *University of California*, San Diego, SSRP-92/01, 266 p.

Priestley, M.J.N., Seible, F. & Calvi, G.M. 1996. *Seismic Design and Retrofit of Bridges*, John Wiley and Sons, New York, 686 p.

Priestley, M.J.N. 1997. Comments on the ATC32 seismic design recommendations for new bridges, *Proc. 4th Caltrans Seismic Design Workshop*, Sacramento.

Priestley, M.J.N. & Seible, F. 1997. *Bridge Foundations under Seismic Loads – Capacity and Design Issues,* Sequad Consulting Engineers, Solana Beach, Report 97/10.

Priestley, M.J.N. 1999. *POLA Berth 400. Dynamic Analyses of Linked Wharf Segments*, Seqad Consulting Engineers, Solana Beach, Report 99/01.

Puertos del Estado, Madrid (Spain) 2000. *ROM 0.6, Acciones y Efectos Sísmicos en las Obras Marítimas y Portuarias*. (English translation will be available).

Pyke R. 1979. Nonlinear soil models for irregular cyclic loadings, *Journal of Geotechnical Engineering Division*, ASCE 105(GT6): 715–726.

Rampello, S., Silvestri, F. & Viggiani, G. 1994. The dependence of small strain stiffness on stress state and history for fine-grained soils: the example of Vallericca clay, *Proc. 1st International Symposium on Pre-failure Deformations of Geomaterials*, Sapporo, Vol. 1, Balkema, Rotterdam, pp. 273–279.

Reese, L.C., Cox, W.R. & Koop, F.D. 1974. Analysis of laterally loaded piles in sand, *Proc. 6th Annual Offshore Technology Conference*, Vol. 2, Paper No. 2080, Houston, Texas, pp. 473–483.

Reese, L.C. & Wang, S. 1994. *Documentation of Computer Program LPILE*, Ensoft, Inc., Vers 4.0, Austin, TX, 366 p.

Richards, R. Jr. & Elms, D. 1979. Seismic behavior of gravity retaining walls, *Journal of Geotechnical Engineering Division*, ASCE 105(GT4): 449–464.

Richart, F.E., Hall, J.R. & Woods, R.D. 1970. *Vibrations of Soils and Foundations*, Prentice-Hall Inc., Englewood Cliffs (New Jersey).

Richter, C.F. 1958. *Elementary Seismology*, W.H. Feeman and Company, San Francisco.

Rix, G. & Stokoe, K.H. 1991. Correlation of initial tangent modulus and cone penetration resistance, *Proc. International Symposium on Calibration Chamber Testing*, Elsevier Publishing, New York, pp. 351–362.

Robertson, P.K. & Wride, C.E. 1997. Cyclic liquefaction and its evaluation based on the SPT and CPT, *Proc. NCEER Workshop on Evaluation of Liquefaction Resistance of Soils*, Technical Report NCEER-97-0022, National Center for Earthquake Engineering Research, pp. 41–88.

Robertson, P.K., Campanella, R.G., Gillespie, D. & Rice, A. 1985. Seismic CPT to measure in-situ shear wave velocity, *Measurement and Use of Shear Wave Velocity for Evaluating Dynamic Soil Properties*, ASCE, New York, pp. 18–34.

Robertson, P.K., Woeler, D.J. & Finn, W.D.L. 1992. Seismic cone penetration test for evaluating liquefaction potential under cyclic loading, *Canadian Geotechnical Journal* 29(4): 686–695.

Roth, W., Fong, H. & de Rubertis, C. 1992. Batter piles and the seismic performance of pile supported wharves, *PORTS '92*, ASCE, pp. 336–349.

Rowe, P.W. 1952. Anchored sheet pile walls, *Proc. ICE*, Vol. 1, Part 1, pp. 27–70.

Rowe, P.W. 1955. A theoretical and experimental analysis of sheet pile walls, *Proc. ICE*, Vol. 4, Part 1.

Sabetta, F. & Pugliese, A. 1996. Estimation of response spectra and simulation of non-stationary earthquake ground motions, *Bull. Seism. Soc. Am.* 86: 353–362.

Sagaseta, C., Cuellar, V. & Pastor, M. 1991. Modelling stress-strain-time behaviour of natural soils – cyclic loading, *Proc. 10th European Conference on Soil Mechanics and Foundation Engineering*, Florence (Italy), Vol. 3, Balkema, Rotterdam, pp. 981–999.

Sasaki, Y. & Taniguchi, E. 1982. Shaking table tests on gravel drains to prevent liquefaction of sand deposits, *Soils and Foundations* 22(4): 1–14.

Sato, S., Mori, H. & Okuyama, K. 1988. Consideration on improvement effect of sandy soil by the vibro rod method, *Proc. 43rd National Conference of JSCE*, III, pp. 106–107 (in Japanese).

Sato, Yu., Ichii, K., Hoshino, Y., Sato, Yo., Iai, S. & Nagao, T. 1999. Annual report on strong-motion earthquake records in Japanese ports (1998), Technical Note of Port and Harbour Research Institute, No. 942, 230 p.

Schnabel, P.B., Lysmer, J. & Seed, H.B. 1972. SHAKE: a Computer Program for Earthquake Response Analysis of Horizontally Layered Sites, *Report No. EERC 72-12*, Earthquake Engineering Research Center, University of California, Berkeley.

Scott, R.F. 1997. Crane response in 1995 Hyogoken Nanbu earthquake, *Soils and Foundations* 37(2): 81–87.

SEAOC 1995. *Performance Based Seismic Engineering of Buildings*, Structural Engineers Association of California, Sacramento, California.

Seed, H.B. & Whitman, R.V. 1970. Design of earth retaining structures for dynamic loads, *ASCE Specialty Conference on Lateral Stresses in the Ground and Design of Earth Retaining Structures*, Ithaca, pp. 103–147.

Seed, H.B. & Idriss, I.M. 1971. Simplified procedure for evaluating liquefaction potential, *Journal of Soil Mechanics and Foundations Division*, ASCE 97(SM9): 1249–1273.

Seed, H.B., Idriss, I.M., Makdisi, F. & Banerjee, N. 1975. Representation of irregular stress time histories by equivalent uniform stress series in liquefaction analyses, *Report*

No. EERC 75-29, Earthquake Engineering Research Center, University of California, Berkeley.

Seed, H.B. & Booker, J.R. 1977. Stabilization of potentially liquefiable sand deposits using gravel drains, *Journal of Geotechnical Engineering Division*, ASCE 103(GT7): 757–768.

Seed, H.B. 1979. Soil liquefaction and cyclic mobility evaluation for level ground during earthquakes, *Journal of Geotechnical Engineering*, ASCE 105(GT2): 201–255.

Seed, H.B. & Idriss, I.M. 1982. Ground motions and soil liquefaction during earthquakes, *Earthquake Engineering Research Institute Monograph Series*, Oakland, California, 134 p.

Seed, H.B., Idriss, I.M. & Arango, I. 1983. Evaluation of liquefaction potential using field performance data, *Journal of Geotechnical Engineering*, ASCE 109(3): 458–482.

Seed, H.B., Tokimatsu, K., Harder, L.F. & Chung, R.M. 1985. The influence of SPT procedures in soil liquefaction resistance evaluations, *Journal of Geotechnical Engineering*, ASCE 111(12): 1425–1445.

Sekita, Y., Tatehata, H., Anami, T. & Kadowaki, T. 1999. Outline and performance of new tsunami forecast service in the Japan Meteorological Agency, *Proc. 31st Joint Meeting of U.S.-Japan Panel on Wind and Seismic Effects*, UJNR, Tsukuba.

Shibuya, S., Mitachi, T., Fukuda, F. & Degoshi, T. 1995. Strain rate effects on shear modulus and damping of normally consolidated clay, *Geotechnical Testing Journal*, ASTM 18(3): 365–375.

Shimizu, T. & Takano, Y. 1979. Design and construction of LNG tank foundation on loose sand – Indonesia, *Tsuchi-to-Kiso, Japanese Society of Soil Mechanics and Foundation Engineering* 27(1): 21–26 (in Japanese).

Simonelli, A.L. & Mancuso, C. 1999. Prove dinamiche in sito per la misura delle proprietà meccaniche dei terreni (in italian), *Advanced Professional Training Course on Problemi di ingegneria geotecnica nelle aree sismiche,* CISM (International Centre for Mechanical Sciences), Udine, Italy.

Skempton, A.W. 1986. Standard penetration test procedures and the effects in sands of overburden pressure, relative density, particle size, aging and overconsolidation, *Geotechnique* 36(3): 425–447.

Sritharan, S. & Priestley, M.J.N. 1998. Seismic testing of a full scale pile-deck connection utilizing headed reinforcement, *Div. of Str. Eng.*, TR98/14, University of California, San Diego, 34 p.

Steedman, R.S. 1984. *Modelling the behaviour of retaining walls in earthquakes*, PhD Thesis, Cambridge University.

Steedman, R.S. & Zeng, X. 1990. The seismic response of waterfront retaining walls, *Design and Performance of Earth Retaining Structures*, Geotechnical Special Publication No. 25, ASCE, pp. 872–886.

Steedman, R.S. & Zeng, X. 1996. Rotation of large gravity walls on rigid foundations under seismic loading, *Proc. ASCE Annual Convention*, Washington.

Steedman, R.S. 1998. Seismic design of retaining walls, *Geotechnical Engineering*, Proc. Institution of Civil Engineers, Vol. 131, pp. 12–22.

Stark, T.D. & Olson, S.M. 1995. Liquefaction resistance using CPT and filed case histories, *Journal of Geotechnical Engineering*, ASCE 121(GT12): 856–869.

Stokoe, K.H., Wright, S.G., Bay, J.A. & Roesset, J.M. 1994. Characterisation of geotechnical sites by SASW method, *Geophysical characterisation of sites*, R.D. Woods (ed.), Balkema, Rotterdam, pp. 15–25.

Sugano, T., Kitamura, T., Morita, T. & Yui, Y. 1998. Study on the behavior of steel plate cellular bulkheads during earthquake, *Proc. 10th Japan Earthquake Engineering Symposium*, Vol. 2, pp. 1867–1872.

Sugano, T. & Iai, S. 1999. Damage to port facilities, *The 1999 Kocaeli Earthquake, Turkey – Investigation into Damage to Civil Engineering Structures*, Earthquake Engineering Committee, Japan Society of Civil Engineers, pp. 6-1–6-14.

Sugano, T., Kaneko, H. & Yamamoto, S. 1999. Damage to port facilities, *The 1999 Ji-Ji Earthquake, Taiwan – Investigation into Damage to Civil Engineering Structures*, Earthquake Engineering Committee, Japan Society of Civil Engineers, pp. 5-1– 5-7.

Suzuki, Y., Tokito, K., Suzuki, Y. & Babasaki, J. 1989. Applied examples of antiliquefaction foundation ground improvement method by solidification method, *Foundation Engineering* 17(9): 87–95 (in Japanese).

Suzuki, Y., Tokimatsu, K., Koyamada, K., Taya, Y., & Kubota, Y. 1995. Filed correlation of soil liuqefaction based on CPT data, *Proc. International Symposium on Cone Penetration Testing, CPT'95*, Linkoping, Sweden, Vol. 2, pp. 583–588.

Takahashi, K., Kikuchi, Y. & Asaki, Y. 1993. Analysis of flexural behavior of anchored sheet pile walls, *Technical Note of Port and Harbour Research Institute*, No. 756, 32 p. (in Japanese).

Takeda, T., Sozen, M.A. & Nielsen, N.N. 1970. Reinforced concrete response to simulated earthquakes, *Journal of Structural Division*, ASCE 96(ST12): 2557–2573.

Talaganov, K.V. 1986. Determination of liquefaction potential of level site by cyclic strain, *Report IZIIS 86-129*, Institute of Earthquake Engineering and Engineering Seismology, Univ. Kiril and Metodij, Skopje, Yugoslavia.

Tanaka, S. & Inatomi, T. 1996. Seismic response analysis of gantry cranes on container quay walls due to the Great Hanshin earthquake, *Proc. 11th World Conference on Earthquake Engineering*, Acapulco, Paper No. 1250.

Tanaka, S. & Sasaki, C. 1989. Sandy ground improvement against liquefaction at Noshiro thermal power station, *Tsuchi-to-Kiso, Japanese Society of Soil Mechanics and Foundation Engineering* 37(3): 86–90 (in Japanese).

Tatsuoka, F., Jardine, R.J., Lo Presti, D., Di Benedetto, H. & Kodaka, T. 1997. Characterising the pre-failure deformation properties of geomaterials, *Proc. 14th ICSMFE*, Hamburg, Theme Lecture, Vol. 4, pp. 2129–2164.

Tchebotarioff, G.P. 1949. *Large Scale Earth Pressure Tests with Model Flexible Bulkheads*, Princeton Univ.

Terzaghi, K. 1955. Evaluation of coefficients of subgrade reaction, *Géotechnique* 5: 297–326.

Tokimatsu, K., Kuwayama, S. & Tamura, S. 1991. Liquefaction potential evaluation based on Rayleigh wave investigation and its comparison with field behaviour, *Proc. 2nd International Conference on Recent Advances in Geotechnical Earthquake Engineering and Soil Dynamics*, St. Louis (Missouri), Vol. 1, pp. 357–364.

Towhata, I. & Ishihara, K. 1985. Modelling soil behaviour under principal stress axes rotation, *Proc. 5th International Conference on Numerical Methods in Geomechanics*, Nagoya, Balkema, Rotterdam, pp. 523–530.

Towhata, I. & Islam, S. 1987. Prediction of lateral movement of anchored bulkheads induced by seismic liquefaction, *Soils and Foundations* 27(4): 137–147.

Toyota, Y., Nakazato, S., & Nakajima, Y. 1988. Consideration on behavior of existing steel pipe pile wall wharf during installation of vibro rod method, *Proc. 43rd National Conference of JSCE*, III, pp. 756–757 (in Japanese).

Tsinker, G.P. 1997. *Handbook of Port and Harbor Engineering, Geotechnical and Structural Aspects*, Chapman & Hall, 1054 p.

Tsuchida, H., Noda, S., Inatomi, T., Uwabe, T., Iai, S., Ohneda, H. & Toyama, S. 1985. Damage to port structures by the 1983 Nipponkai-Chubu earthquake, *Technical Note of Port and Habour Research Institute*, No. 511, 447 p. (in Japanese).

Tsuchida, H., de La Fuente, R.H., Noda, S. & Valenzuela, M.G. 1986. Damage to port facilities in Port of Valparaiso and Port of San Antonio by the March 3, 1985 earthquake in middle of Chile, *4as Jordanadas Chileanas de Sismologia e Intenieria Antisismica & International Seiminar on the Chilean March 3, 1985 earthquake.*

Tsuchida, H. 1990. Japanese experience with seismic design and response of port and harbor structures, *Proc. POLA Seismic Workshop on Seismic Engineering*, Port of Los Angeles, pp. 138–164.

Tsuchida, T. 1990. Study on deformation of undrained strength of clayey ground by means of triaxial tests, *Technical Note of Port and Harbour Research Institute*, No. 688, 199 p. (in Japanese).

U.S. Navy 1982. *DM7.2, Foundation and Earth Structures*, US Dept. of the Navy, Naval Facilities Engineering Command, pp. 7.2-209.

U.S. Navy 1987. *Naval Facilities Engineering Command Military Handbook* MIL-HDBK-1002/3 'Structural Engineering Steel Structures'.

Uwabe, T. 1983. Estimation of earthquake damage deformation and cost of quaywalls based on earthquake damage records, *Technical Note of Port and Harbour Research Institute*, No. 473, 197 p. (in Japanese).

Vazifdar, F.R. & Kaldveer, P. 1995. The great Guam earthquake of 1993 port disaster and recovery, *PORTS '95*, ASCE, pp. 84–94.

Viggiani, G. & Atkinson, J.H. 1995. Interpretation of bender element tests, *Géotechnique* 45(1): 149–154.

Vucetic, M. 1992. Soil Properties and Seismic Response, *Proceedings of the X World Conference on Earthquake Engineering*, Madrid, Balkema, Vol. 3, pp. 1199–1204.

Vucetic, M. 1994. Cyclic threshold shear strains in soils, *Journal of Geotechnical Engineering Division*, ASCE 120(12): 2208–2228.

Vucetic, M. & Dobry, R. 1991. Effects of the soil plasticity on cyclic response, *Journal of Geotechnical Engineering Division*, ASCE 117(1): 89–107.

Werner, S.D. 1986. A case study of the behavior of seaport facilities during the 1978 Miyagi-ken Oki earthquake, *Proc. Lifeline Seismic Risk Analysis-Case Studies*, ASCE, pp. 113–130.

Werner, S.D. (ed.) 1998. *Seismic Guidelines for Ports*, Technical Council on Lifeline Earthquake Engineering, Monograph No. 12, ASCE.

Westergaard, H.M. 1933. Water pressure on dams during earthquakes, *Transactions of ASCE* 98: 418–472.

Whitman, R.V. & Liao, S. 1984. Seismic design of gravity retaining walls, *Proc. 8th World Conference on Earthquake Engineering*, San Francisco, Vol. III, pp. 533–540.

Whitman, R.V. & Liao, S. 1985. Seismic design of retaining walls, *US Army Corps of Engineers*, *Waterways Experiment Station*, Miscellaneous Paper GL-85-1.

Whitman, R.V. & Christian, J.T. 1990. Seismic response of retaining structures, *Proc. POLA Seismic Workshop on Seismic Engineering*, Port of Los Angeles, pp. 427–452.

Wilson, E.L., Der Kiureghian, A. & Bayp, E.P. 1981. A replacement of the SRSS methods in seismic analysis, *Earthquake Engineering and Structural Dynamics*, Vol. 9, pp. 187–194.

Wittkop, R. 1994. Application of VELACS philosophy to Port of Los Angeles Pier 400 Project, in *VELACS – Verification of Numerical Procedures for the Analysis of Soil Liquefaction Problems*, K. Arulanadan and R.F. Scott (eds), Balkema, pp. 1647–1655.

Wood, H.O. & Neumann, F. 1931. Modified Mercalli intensity scale of 1931, *Bull. Seism. Soc. Am.* 21: 277–283.

Wu, S.C., Cornell, C.A. & Winterstein, S.R. 1995. A hybrid recurrence model and its implication on seismic hazard results, *Bull. Seism. Soc. Am.* 85: 1–16.

Wyllie, L.A., Abrahamson, N., Bolt, B., Castro, G., Durkin, M.E., Escalante, L., Gates, H.J., Luft, R., McCormick, D., Olson, R.S., Smith, P.D. & Vallenas, J. 1986. The Chile Earthquake of March 3, 1985, *Earthquake Spectra*, 2(2): EERI, 513.

Wyss, M. 1979. Estimating maximum expectable magnitude of earthquakes from fault dimensions, *Geology* 7: 336–340.

Yamanouchi, K., Kumagai, T., Suzuki, Y., Miyasaki, Y. & Ishii, M. 1990. Liquefaction remedial measures using grid-shaped continuous underground wall, *Proc. 25th National Conference on Japanese Society of Soil Mechanics and Foundation Engineering*, pp. 1037–1038 (in Japanese).

Yamazaki, H., Zen, K. & Kagaya, H. 1992. Liquefaction characteristics of overconsolidated sand, *Proc. 27th National Conference on Soil Mechanics and Foundation Engineering*, pp. 845–846 (in Japanese).

Yamazaki, H., Zen, K. & Koike, F. 1998. Study of the liquefaction prediction based on the grain size distribution and the SPT N-value, *Technical Note of Port and Harbour Research Institute*, No. 914 (in Japanese).

Yang, D.S. 1999. *Deformation Based Seismic Design Models for Waterfront Structures*, Ph.D. dissertation, Department of Civil, Construction and Environmental Engineering, Oregon State University, Corvallis, 260 p.

Yokota, H., Kawabata, N., Akutagawa, H., Kurosaki, K., Tsushima, T., Harada, N. & Yato, A. 1999. Verification of seismic performance of an open piled pier by an elasto-plastic method and a simplified method, *Technical Note of Port and Harbour Research Institute*, No. 943, 40 p. (in Japanese).

Yoshimi, Y. & Tokimatsu, K. 1991. Ductility criterion for evaluating remedial measures to increase liquefaction resistance of sands, *Soils and Foundations* 31(1): 162–168.

Youd, T.L. & Idriss, I.M. (eds.) 1997. *NCEER Workshop on Evaluation of Liquefaction Resistance of Soils*, Technical Report NCEER-97-0022, National Center for Earthquake Engineering Research, pp. 1–40.

Zen, K., Yamazaki H. & Umehara, Y. 1987. Experimental study on shear modulus and damping ratio of natural deposits for seismic response analysis, *Report of Port and Harbour Research Institute* 26(1): 41–113 (in Japanese).

Zen, K. 1990. Development of premix method as a measure to construct a liquefaction-free reclaimed land, *Tsuchi-to-Kiso*, *Japanese Society of Soil Mechanics and Foundation Engineering* 38(6): 27–32 (in Japanese).

Zen, K., Yamazaki, H. & Sato, Y. 1990. Strength and deformation characteristics of cement treated sands used for premixing method, *Report of Port and Harbour Research Institute* 29(2): 85–118 (in Japanese).

Zienkiewicz, O.C. & Bettess, P. 1982. Soils and other saturated media under transient, dynamic conditions: General formulation and the validity of various simplifying assumptions, *Soil Mechanics – Transient and Cyclic Loads*, John Wiley & Sons, pp. 1–16.

Zienkiewicz, O.C., Chan, A.H.C., Pastor, M., Schrefler, B.A. & Shiomi, T. 1999. *Computational Geomechanics with Special Reference to Earthquake Engineering*, John Wiley & Sons, 383 p.

Zmuda, R., Weismair, M. & Caspe, M. 1995. Base isolating a wharf using sliding friction isolators at the Port of Los Angeles, *PORTS '95*, ASCE, pp. 1263–1274.

Zwamborn, J.A. & Phelp, D. 1995. When must breakwaters be rehabilitated/repaired? *PORTS '95*, ASCE, pp. 1183–1194.

Index

T - #0314 - 101024 - C0 - 254/178/26 [28] - CB - 9789026518188 - Gloss Lamination